T5-ARO-535

Actinide Speciation
in High Ionic Strength Media

Experimental and Modeling Approaches
to Predicting Actinide Speciation and Migration
in the Subsurface

Actinide Speciation in High Ionic Strength Media

Experimental and Modeling Approaches
to Predicting Actinide Speciation and Migration
in the Subsurface

Edited by

Donald T. Reed

Chemical Technology Division
Argonne National Laboratory
Argonne, Illinois

Sue B. Clark

Department of Chemistry
Washington State University
Pullman, Washington

and

Linfeng Rao

Chemical Sciences Division
Lawrence Berkeley National Laboratory
Berkeley, California

KLUWER ACADEMIC / PLENUM PUBLISHERS
NEW YORK, BOSTON, DORDRECHT, LONDON, MOSCOW

Library of Congress Cataloging-in-Publication Data

Actinide speciation in high ionic strength media : experimental and
   modeling approaches to predicting actinide speciation and migration
   in the subsurface / edited by Donald T. Reed, Sue B. Clark, and
   Linfeng Rao.
        p.   cm.
    Includes bibliographical references and index.
    ISBN 0-306-46185-4
    1. Actinide elements--Environmental aspects. 2. Alpha-bearing
wastes--Environmental aspects.   I. Reed, Donald Timothy, 1956-   .
II. Clark, Sue B., 1961-   . III. Rao, Linfeng.
TD196.R3A267 1999
621.48'38--dc21   99-36676
                                                               CIP

Proceedings of an American Chemical Society Symposium on Experimental and Modeling Studies of
Actinide Speciation in Non-Ideal Systems, held August 26–28, 1996, in Orlando, Florida

ISBN: 0-306-46185-4

© 1999 Kluwer Academic / Plenum Publishers
233 Spring Street, New York, N.Y. 10013

10 9 8 7 6 5 4 3 2 1

A C.I.P. record for this book is available from the Library of Congress.

Printed in the United States of America

PREFACE

The management and disposal of radioactive wastes are key international issues requiring a sound, fundamental scientific basis to insure public and environmental protection. Large quantities of existing nuclear waste must be treated to encapsulate the radioactivity in a form suitable for disposal. The treatment of this waste, due to its extreme diversity, presents tremendous engineering and scientific challenges. Geologic isolation of transuranic waste is the approach currently proposed by all nuclear countries for its final disposal. To be successful in this endeavor, it is necessary to understand the behavior of plutonium and the other actinides in relevant environmental media.

Conceptual models for stored high level waste and waste repository systems present many scientific difficulties due to their complexity and non-ideality. For example, much of the high level nuclear waste in the US is stored as alkaline concentrated electrolyte materials, where the chemistry of the actinides under such conditions is not well understood. This lack of understanding limits the successful separation and treatment of these wastes. Also, countries such as the US and Germany plan to dispose of actinide-bearing wastes in geologic salt deposits. In this case, understanding the speciation and transport properties of actinides in brines is critical for confidence in repository performance and risk assessment activities. Many deep groundwaters underlying existing contaminated sites are also high in ionic strength. Until recently, the scientific basis for describing actinide chemistry in such systems was extremely limited.

Establishing the nature of actinide species prevalent in the subsurface and, perhaps more importantly, the key factors that define actinide speciation are important aspects of almost all nuclear waste related activities. The speciation of transuranics in the environment — that is, their oxidation states, interactions with organic and inorganic ligands, and their degree of aggregation — defines their mobility and bioavailability in the environment. This information is needed to accurately predict radionuclide migration, to define and develop waste remediation strategies, to develop better *in situ* barriers to radionuclide transport, to perform risk-based evaluations for remediation, and to identify/develop needed reprocessing technology.

This book is a current compilation of manuscripts describing modeling and experimental approaches for actinide chemistry in high ionic strength media. The first section of this book is a series of invited papers that address general aspects of actinide chemistry in high level nuclear waste and in environmental systems where high ionic strength is important.

The two subsequent sections of papers address the special problems of actinide behavior in concentrated electrolytes and/or alkaline conditions. In the second section, papers on actinide solubility and organic complexation in high ionic strength media are presented. Redox effects will, in large part, define the overall solubility, and hence mobility, of the actinides. Effects linked to the potential presence of microbes and colloids are presented in the third section. The presence of colloidal species very quickly lead to the issues of actinide transport and mobility in the subsurface, and these are overriding concerns addressed by these authors.

The contributions for this book were developed from presentations made at the 211th National American Chemical Society Meeting, August 1996, in a symposium entitled "Experimental and Modeling Studies of Actinide Speciation in Non-Ideal Systems". These papers pertain to general aspects of actinide chemistry in high ionic strength media and have undergone peer review. The research described at the American Chemical Society Meeting and reported herein is intended to help provide a scientific basis for predicting actinide speciation in the U. S. Department of Energy's Waste Isolation Pilot Plant, the proposed nuclear repository for transuranic waste in Southeastern New Mexico. This symposium was organized by Sue Clark of Washington State University, Barbara Stout of University of Cincinnati, Linfeng Rao of Lawrence Berkeley National Laboratory, and was sponsored by the Division of Nuclear Science and Technology.

Argonne, Illinois   Donald T. Reed

Pullman, Washington   Sue B. Clark

Berkeley, California   Linfeng Rao

ACKNOWLEDGMENTS: The editors would like to thank all the authors who made contributions to this book. Additionally, we would like to thank those who assisted in the peer-review process. Lastly, we owe special thanks to the U.S. Department of Energy, and in particular, the WIPP Project Office, which supported much of the research reported herein.

CONTENTS

ACTINIDE CHEMISTRY IN HIGH IONIC STRENGTH MEDIA: GENERAL ASPECTS

ACTINIDE COMPLEXATION AND SOLUBILITY

ACTINIDE COLLOIDAL AND MICROBIOLOGICAL INTERACTIONS

I. Actinide Chemistry in High Ionic Strength Media: General Aspects

NEAR FIELD AND FAR FIELD
INTERACTIONS AND DATA NEEDS
FOR GEOLOGIC DISPOSAL OF NUCLEAR WASTE

Gregory R. Choppin

Department of Chemistry
Florida State University
Tallahassee, Florida 32306-3006

ABSTRACT

The behavior of radionuclides released in the near field of a repository will be influenced by the geochemical media as modified by radiolysis, elevated temperatures, and the materials used in the encapsulation, backfill, etc. By contrast, the behavior in the far field should resemble a more normal environmental situation. The difference in the nature of the data base required for satisfactory modeling of radionuclide migration in these two situations is reviewed. The value of data from natural analogue sites to validate the far field modeling and from the Oklo site to validate the near field modeling is discussed.

INTRODUCTION

The research and development programs on the disposal of nuclear wastes being pursued by a number of nations vary widely in nature and scope yet possess many common features. Unprocessed spent fuel elements and/or high level wastes from processing are to be immobilized in some "insoluble" matrix and placed in an underground geological repository[1]. Disposal sites in different types of geologic media are being studied in the various national programs. In this paper, the focus is on the types of interactions which radionuclides may undergo those geological media. The differences expected between the situations in the near field (the vicinity of the burial site affected by thermal and radiation effects from the wastes) and in the far field are reviewed with emphasis on the types of data required for valid modeling of the long term behavior of released radionuclides.

In the U.S., the waste would consist of the unreprocessed irradiated fuel elements from civilian nuclear reactors which would be encased in an "inert" barrier material. In the case of defense-related wastes, the high-level radioactivity from reprocessing would be converted to compounds that would be fixed in a solid waste form such as a borosilicate glass, a ceramic, or a synthetic "rock". This waste form material must have a low solubility, be relatively

Actinide Speciation in High Ionic Strength Media, edited by Reed *et al.*
Kluwer Academic / Plenum Publishers, New York, 1999

chemically inert to the environment of the storage site, be stable to heat and radiation, and have good heat conduction and mechanical and structural stability. The waste form would be enclosed by a metal jacket or canister designed to last at least a few thousand years. or, hopefully, much longer, after which groundwater might contact the waste form and possibly start a slow leaching and dissolution process[2].

As an additional barrier to release of the radioactive species to the surface environment, the canister would be surrounded with an overpack (i.e., backfill) of clay or some other material with good ion exchange or sorptive properties[3]. This material would retain the various radioactive cations as they are leached from the glass by the groundwater. This backfill would serve not only as a chemical barrier to the migration of radionuclides into the groundwater but also to conduct heat from the canister, and to hinder the flow of water around the canister, thereby slowing its rate of dissolution. It could also provide an elastic support for the canister to prevent cracking of the waste form in case of rock movements in rigid formations such as granite.

NEAR FIELD

In the vicinity of the burial site of the high level wastes, the temperature will be elevated due to the heating from the released radioactive decay energy. Further, at the depths (300-1000 m) planned for most proposed repositories, the estimated temperatures may vary (depending on the predisposal storage time of the nuclear waste, the spacing of waste packages, etc.) from 60 to 300°C while the water pressure would range from 100 kPa (1 atm) to 10 MPa. Such pressures and temperatures represent conditions in which chemical behavior has not been studied extensively[4]. Geochemically, under such conditions, mineral-water reactions are slow, and most systems are metastable as kinetics dominate equilibrium factors. Common processes include precipitation of amorphous solids, recrystallization of metastable minerals and of carbonaceous materials into graphite, dehydration of some clays and zeolites, changes in ground water conditions, etc. Laboratory studies are required to provide data for behavior under such temperature-pressure conditions. Moreover, these experiments should be conducted in a manner that, if possible, allows confidence in extrapolation of the data from a few years of laboratory observations to the 10^4 to 10^6 years required for the repository to function safely.

For most planned repositories the water is reducing and of low ionic strength. The effects of elevated temperatures and of radiolysis also must be included in modeling the behavior of the radionuclides in the site. By contrast, in the Waste Isolation Pilot Plant (WIPP) site where actinide defense wastes are to be emplaced without vitrification, the temperature will not be elevated by radioactive decay energy nor will radiolysis be a significant factor as the level of radiation will be much lower. This repository is in bedded salt so any water present has a high ionic strength; in addition, chemicals used in nuclear separation systems will be included in the wastes and their possible effect on the long term release of the radionuclides must be evaluated. Table 1 compares the composition of the ground water in the YMP and WIPP sites and in a typical deep crystalline rock of the general nature being considered in other countries.

To summarize, the data needed to assess effects in the near field on vitrified wastes are the following:

- corrosion rate of container;
- effect of such corrosion on pH, E_H of water;
- glass dissolution and backfill retention rate;
- effect of released silica, etc. on sorption properties and colloid formation;

Table 1. Ground Water Composition (mg/L)

Species	YMP[10]	Deep Crystalline Rock[11]	WIPP*[12]	WIPP‡[12]
Ca^{2+}	1-20	25-50	600	1.16×10^3
Mg^{2+}	0.05-2	5-20	3.5×10^4	1.8×10^3
Na^+	45-95	10-100	4.2×10^4	5.5×10^4
K^+	1-5	1-5	3.0×10^4	1.2×10^3
$Fe^{2+,3+}$	0.01-.05	1-20	---	---
$HCO_3^-+CO_3^{2-}$	120-170	60-400	860	50
Cl^-	5.5-7.7	5-50	1.9×10^5	8.8×10^4
SO_4^{2-}	18-28	1-15	4.2×10^4	7.2×10^4
pH	6.9-7.7	7.2-8.5	7.3	7.1
Ionic Strength	low(<0.01m)	low(<0.01m)	7.78m	2.97m

* Salado Brine A; ‡ H-17 Brine.

- water movement;
- effect of higher temperature and radiation on all of above.

The rate and mechanisms of dissolution of borosilicate glass as a function of water flow, silica concentration, temperature, etc. are an area that is being studied intensively to evaluate the release rate of the radioactivity from the glass, the chemical form of the dissolution products, and what waste form may best resist release[5]. Ceramics and "SYNROCK"[6] also are being investigated for use as waste forms and require similar release rate studies to allow assessment of their value.

After the water passes the canister and leaches radionuclides from the waste form, it would flow through the backfill barrier of clay. The cationic species such as ^{90}Sr, ^{239}Pu, ^{235}U, ^{237}Np, etc. can be retained by ion exchange with Na^+ in the clay barrier. However, the extent of this ion exchange depends on such factors as temperature, concentration, and speciation (e.g., the oxidation state, the degree of hydrolysis and complexation, etc.), sorption on colloids, and other surfaces, etc., of the released radioactivities. These same factors would determine the behavior of the species in the water leaving the barrier and migrating in the geologic medium. The solution chemistry of the radionuclides in neutral and basic waters must be studied to measure hydrolytic behavior as well as complexation by anionic species such as CO_3^{2-}, HCO_3^-, F^-, Cl^-, and SO_4^{2-} which are present in the groundwaters[7]. Additionally, studies are needed on the effects of binding of these species by organic polyelectrolytes such as humic acid and adsorption by colloids such as hydrous silica and by the solid surfaces of the minerals of the rock formation. In many sites, there are natural zeolites and clays such as smectites and mordenite (Na, Ca, K + $NaAlSi_{4.5-5.0}O_{11-12} \cdot 3.2-3.5H_2O$), clinoptilolite (Na, K, Ca + $NaAlSi_{4.2-5.0}O_{10.4-12.0} \cdot 3.5-4.0H_2O$), and heulandite (Na, K, Ca + $NaAlSi_{3.5-4.0}O_{10.5-12} \cdot 3.5-4.0H_2O$) which are known for their high sorptive capability. Their sorptive properties are particularly high for strontium, cesium, and barium whose affinity may be due mainly to ion-exchange reactions.

These natural minerals have sorptive properties comparable to those of materials being considered for use as the backfill barriers placed around the capsules containing the radioactivity. Finally, possible consequences of biological activity, either natural or introduced by man must be evaluated. Figure 1 shows the primary factors involved in determining the probablility of migration of radionuclides in the near field.

The effect of the elevated temperatures resulting from the energy released by the radioactive decay could be a modification in the geology of the area, the diffusion rate of water within the site, the chemical speciation and redox of the radionuclides, the solubility and/or sorption of the species formed by the radionuclides, etc. If the temperature exceeds the boiling point of the water in the formation, the result could be a dry repository with reduced or no migration or even release from the canister of the radionuclides[8]. This could be achieved in the YMP by stowing the canisters more densely. However, it is not certain how much the higher temperature required for a "dry" repository over long times might alter the stability and porosity of the geologic material, which might result in allowing liquid to seep back into the tunnels.

WIPP NEAR FIELD

Since the waste material deposited in the WIPP repository will not be contained in glass or some other inert waste form, but simply placed in steel drums, the previous discussion on canister corrosion and waste form leaching is not pertinent to that site. Moreover, the temperature will not be increased by the radioactive decay as that will be much lower in WIPP. Finally, the geologic material is embedded salt with high ionic strength pore water. The repository will be oxidizing after closure, but as the metallic drums corrode and the cellulose materials oxidize, the site will become anoxic. The drum corrosion will release Fe^{2+} and Ni^{2+} whose interaction with the residual organic ligands (from separation systems) would compete with the complexation of the actinide cations by these ligands. A list of the data needs for

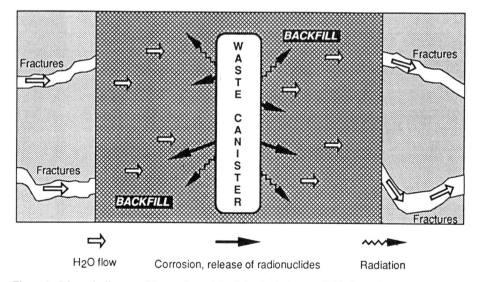

Figure 1. Schematic diagram of factors determining behavior in the near field of a nuclear waste repository.

assessing the near field effects in WIPP would include:

- rate of change from oxidizing to reducing;
- effect of microbial degradation of organic material (e.g., gas generation);
- effect of Mg^{2+} on organic ligands;
- rate of water diffusion;
- solubilizing effect of organic ligands;
- solubilities of actinides under the pH, E_H, ionic strength conditions;
- colloid formation.

FAR FIELD

The far field region is defined as that beyond which the effects induced by the presence of the buried high level radioactive waste are absent. The temperature, E_H, pH, geology, etc. are those normal for the undisturbed site. To estimate the probability of migration of radionuclides through the far field region to the ecosystem, data is required on:

- rate of release from near field;
- flow rate of underground water;
- speciation and solubility of radionuclides;
- diffusion rates of the radionuclides.

The geochemical modeling calculations based on measured equilibrium data are of primary importance in the safety assessments of proposed nuclear waste repositories. The modeling calculations use data from laboratory and field studies to predict the solubilities, and nuclide migration of material which might be released from nuclear repositories over thousands and, even, hundreds of thousands of years. It is not possible to demonstrate rigorously that these models are accurate as they may simplify the natural system, use incorrect data or misrepresent or ignore important processes which occur over long time periods.

NATURAL ANALOGUES

Some validation of the results of the modeling calculations can be obtained from careful comparison of the calculated values with those measured in appropriate geologic sites, known as *natural analogues*. These natural analogue sites are areas in which uranium or thorium ores have been present for geologic time periods; if these sites have not been affected by human activities, the record of geological, long term effects should be well preserved. A number of such sites are being studied around the world. In these natural analogue sites. the mobilization and fixation of uranium involves complexation, redox and retention on minerals via adsorption, and ion exchange. In clay media, the redox potential is strongly buffered if significant amounts of organic substances are present.

In a formation in Brazil in which the water is reducing, most of the thorium, the rare earths, and, to a lesser extent, the uranium, is associated with goethite (FeOOH) particles and transport by organic colloids is much less important[9]. In another region of this formation. the thorium and rare earths are associated with organic (humic) colloids and have a higher mobility. At an analogue site in Scotland, the liquid from the ore passes into a peat bog in which the uranium becomes associated predominantly with humic material. The Th is found on Fe/Al oxyhydroxide colloids and particles. In many clay deposits, the organic material is most significant in maintaining a reducing potential which restricts actinide migration and provides a sorption source of the mobilized fraction, confirming the value of such material as the backfill in the near field[9] An important observation in these sites is the role of the humic type organics which can reduce the species or interact with them to result in increased migration or in increased sorption (decreased migration) on repository surfaces.

With proper regard for the differences in sites, the information from natural analog studies can be of significant value in validating the calculations of probable radionuclide release rates from the far field of nuclear repositories when such releases can be attributed to normal conditions (i.e., no unusual natural events such as earthquakes, volcanic activity, flooding, etc.). However, since these natural analogue sites have normal ambient temperatures and the geological media have not been altered by radiolytic or thermal effects, they cannot be used to validate models for near field effects.

These natural analog sites do not include neptunium, which in oxic waters forms the much more mobile NpO_2^+. However, release of Np from the buried wastes to the oxic far field would be controlled by the very insoluble Np(IV) hydrolysis product in the reducing near field of most repositories.

OKLO

Fortunately, there is an analog site that can be used to measure long term effects in hot, altered repositories. The Oklo natural reactors[9]operated over a period of 100,000 to 500.000 years, fissioning about 12 metric tons of uranium and producing over 1 metric ton of ^{239}Pu. The average energy release in the reactor zone was ca. 50 W m^{-2} which is several times greater than the release estimated for geologic waste repository sites. It is estimated that the liquids in the mineral grain inclusions had temperatures of 450-600° C which is 2 to 3 times that expected in a repository near field. Moreover, there is evidence of significant dissolution and alteration of minerals in Oklo due to radiation and thermal conditions. As much as 10^{12} liters of water passed through the reactor zones and circulated to a distance of 30 m from these. Thus, the conditions were considerably more perturbed from the natural ones and much less favorable for radionuclide retention in the Oklo site than those estimated for the planned waste repositories.

The uranium and rare earth elements show evidence of a small amount of localized redistribution, but the majority of these elements were retained within the reactor zone. By contrast, the fission product rare gases, halogens, molybdenum, alkali and alkaline earth elements migrated greater distances from the reactor zones, apparently while the area was hot. Essentially 100% of the Pu, 85 - 100% of the Nd, 75 - 90% of the Ru and 60 - 85% of the Tc were retained within the reactor zones. The migrating fission products were held within a few tens of meters of these zones. The water leaving the reactor zones had 2×10^{-8} g U per m^3 and 10^{-10} M concentrations of Tc, Ru, and Nd (the Tc and Ru were in the anionic forms, TcO_4^- and RuO_4^-). The migration rate of Tc and Ru seems to have been ~ 10^{-5} m y^{-1} in water which itself had a flow of 5 m y^{-1}. Thermodynamic calculations of the temperature dependent solubilities indicate that the loss of the fission products was diffusion controlled, whereas, retention in the surrounding rocks was due to deposition (which would vary with temperature) from an aqueous solution. The lack of migration of the actinides and the much slower release of Tc agree with the predictions of laboratory studies and support the value of such studies in validating the safety of nuclear repositories.

In summary, the data from Oklo site would seem to provide a reliable base for evaluating laboratory and modeling data for the behavior of radionuclides in the near field environment of HLW geologic repositories.

RELEASE SCENARIOS

In this paper, we have discussed data needs to model the release of radionuclides over long periods of time from an undisturbed geologic repository. Migration to the surface environment or to an aquifer which is likely to be used for purposes related to human activities can be modeled if the proper data in sufficient quantity and quality is available. The value of these modeling estimates of release over time can be assessed to some extent by comparison with data from Oklo (for near field) and natural analogue sites (for far field).

A second scenario for release of the buried radionuclides to the environment has been proposed in which unexpected events disturb the conditions. Some probability can be assigned at specific sites for their disturbance by events such as earthquake and volcanic activity. However, estimating the probability of release of radionuclides to the surface by inadvertent human intrusion (e.g., via drilling of bore holes) can be done, at best, only qualitatively as it must be based on unmeasurable and unpredictable events. While such events may, in fact, be the most likely source of releases, the experimental physical scientist must leave this evaluation to others who are trained in such tasks.

ACKNOWLEDGEMENT

The preparation of this paper was performed as part of a contract from the USDOE-OBES Division of Chemical Sciences.

REFERENCES

1. G. Choppin, J. Rydberg and J. O. Liljenzin, "*Radiochemistry and Nuclear Chemistry*", 2nd ed., 1995, Butterworth-Heinemann Ltd., Oxford, Chapter 21 (and references therein).

2. UNIPUB, "*Safety Principles and Technical Criteria for Underground Disposal of High Level Radioactive Wastes*", IAEA, Vienna, 1989, vol. 99.

3. J. Westsik, L. A. Bray, F. N. Hodges, E. J. Wheelwright, in "*The Scientific Basis for Nuclear Waste Management*", ed. S. Topp, Mater. Res. Soc. Symp. Proc., 1982, vol. 329.

4. D. C. Hoffman, and G. R. Choppin, *J. Chem. Ed.*, 63, 1986, 1059.

5. E. Vernaz and J. L. Dussossoy, "*Basic Mechanisms of Aqueous Corrosion of Waste Glasses*", CEC Contract, Commission of the European Communities, 1990.

6. A. E. Ringwood, "*Safe Disposal of High Level Nuclear Reactor Wastes: A New Strategy*", ANU Press, Canberra, 1978.

7. G. R. Choppin and P. J. Wong, *J. Radioanal. Nucl. Chem.*, 203, 1996, 575.

8. L. D. Ramspott, in "*Proc. of the 2nd Annual High-Level Radioactive Waste Management Conf.*", 1991, Am. Nucl. Soc., La Grange Park, Il., p. 1602.

9. Ref. 1, Chapter 22.

10. W. R. Daniels, et. al., *"Summary Rept. on Geochemistry of Yucca Mtn. And Environs"*, Los Alamos Natl. Lab. Rept. LA-9328-MS, Dec. 1982.

11. J. Rydberg, *"Groundwater Chemistry Of a Nucl. Waste Repository in Granite Bedrock"*, U.C. Radiation Lab Rept. UCRL-53155, Sept. 1981.

12. D. E. Hobart, C. J. Bruton, F. J. Millero, I.-M. Chou, K. M. Trauth, and D. R. Anderson, *"Estimates of the Solubilities of Waste Element Radionuclides in WIPP Brines"*, Sand96-0098, Sandia Nat. Lab., May, 1996.

MODERN SPECIATION TECHNIQUES APPLIED TO ENVIRONMENTAL SYSTEMS

Heino Nitsche,[1] Robert J. Silva,[2] Vinzenz Brendler,[1] Gerhard Geipel,[1] Tobias Reich,[1] Yuri A. Teterin,[3] Michael Thieme,[4] Lutz Baraniak,[1] and Gert Bernhard[1]

[1]Forschungszentrum Rossendorf e. V., Institute of Radiochemistry, Dresden, Germany
[2]San Jose State University, Chemistry Dept., San Jose, CA, U.S.A.
[3]Russian Research Center "Kurchatov Institute", Moscow, Russia
[4]Technische Universität Dresden, Institute of Materials Science, Dresden, Germany

INTRODUCTION

Many nuclear facilities in the U. S., Europe and elsewhere have reached the end of their designed life expectancy. Often contamination extends beyond the main technical parts of the installations slated for controlled decommissioning or is contained by methods not totally reliable for long-term storage. Large cleanup programs are already under way in the U.S. for former nuclear sites (USDOE, 1989; USDOE, 1990). Other areas contaminated by nuclear accidents, the worst case resulting from Chernobyl, will have to undergo remediation actions (Eisenbud, 1987). And finally, regions affected by the very first step in the nuclear cycle, i.e., the mining and processing of uranium ore, are also listed for restoration measures (BfS, 1992).

Forty-five years of uranium mining in the southern parts of the former East Germany have left the population there an extensive number of mines, rock piles and mill tailings (BMWi, 1995; WISMUT, 1994). The WISMUT company mined 220,000 tons of uranium between 1945 and 1990. This company, now run by the federal government of Germany, is still responsible for an area of about 1000-1200 km^2, where there are more than 800 mine tailing piles containing 500 million tons of rock material. The mines contain a total shaft length of 1400 km and reach a depth of up to 1800 m. The mill tailings contain about 100 million m^3 of contaminated residue. The primary source of external contamination is from ground water that has flooded the mine shafts, as well as the seepage waters from mines and mill tailing piles. Uranium, its decay products and arsenic are the main potential hazards to the biosphere (BfS, 1992).

In order to insure efficient and effective remediation actions for a given contaminated site, the course of a restoration strategy under consideration has to be predicted so that it can be used as the basis for comparison with alternative strategies and for a cost benefit evaluation. This action requires extensive modeling, coupling chemical speciation processes with transport and risk assessment (Hodgkinson, 1988). Such predictive methods must be based on realistic

Actinide Speciation in High Ionic Strength Media, edited by Reed *et al.*
Kluwer Academic / Plenum Publishers, New York, 1999

11

models and reliable data sets. The latter incorporate data bases describing the chemistry of the system, i.e., thermodynamic and kinetic data. Therefore, the radiochemist is faced with the following challenging tasks: (a) finding a model that accurately represents the physico-chemical phenomena and parameters governing the participation of radioactive substances in the near-field chemistry of a given site, and (b) determining all the numerical data required by the model.

Finding an appropriate model is an important first step in defining the radioactive source term, and the model must take into account such complex features as the interactions with dissolved organic and biological materials in addition to the mineral matrix. The source term must be defined in its initial state and predictions made on its evolution in time and space. Therefore, particular emphasis must be placed on the time dependencies and kinetics of the processes involved. There are many different physical and chemical processes that can alter the source term of the radioactive contamination (Silva, 1995). These processes can influence the release and the formation of distinct chemical species and ultimately their transport in many ways, e.g., change the pH or Eh, introduce new complexing agents, shift chemical equilibria, enhance or inhibit sorption on to mineral surfaces, produce co-precipitation, induce colloid or aerosol formation, and establish new diffusion gradients. In order to model these processes, we need to know the speciation of all relevant chemical components, i.e., quantitative information about the chemical state and distribution of radionuclide species in solution and on solids. In addition to concentrations, species identification (stoichiometry, charge, oxidation state) and structural information (ionic, polynuclear, colloidal, bond type and characteristics) are required (Silva, 1995). But independent of a particular analytical approach, the researcher should always compare experiment and modeling and use the feedback to improve both.

The determination of numerical data is no easier than the identification of the major interactions with the geomedia and solution components along a transport path. One especially needs thermodynamic data on radionuclide complex stabilities, solubilities, sorption behavior, etc, as well as kinetic data on phase transition rates and reversibilities. The quality and extent of the available data bases are often not satisfactory. So the radiochemist must not only validate existing data but also add to the data bases through experimentation.

Considering the needs outlined above, and the fact that the research involves real-world samplings with all their restrictions and difficulties, one can specify essential features required of modern speciation determining methods. In order to avoid altering of the speciation, the methods should preferably be non-destructive, non-invasive and in-situ. Usually one is not able to correct at a later time the errors that occurred in the very first step, i.e., the process of sampling in the environment. Therefore, careful sampling is extremely important. Contaminant concentration levels are often rather low, from 10^{-5} to 10^{-10} mol/L (Silva, 1995; Nitsche, 1991). Under these conditions, it is rather difficult to achieve high accuracy and reproducibility. And finally, large sample sets may need to be processed which require automation and on-line data processing. The above stated sensitivity demand is clearly a goal which by now is not reached by most of the available techniques. Therefore, we have to refer in the following chapters also to methods with a lesser resolution. But on the other side, the current development in many techniques will sooner or later bring them into the required range of sensitivity. An example is the current establishment of new beam lines for advanced EXAFS and XANES applications, utilizing the newest generation of synchrotrons like at ESRF in Grenoble / France, combined with the development of new more sensitive detectors.

There are numerous methods for the detection and quantification of radionuclides in the environmental, fewer for the characterization of radionuclide speciation and structure, and fewer still that provide both pieces of information. Several new developments and refinements to existing methods have enabled us to obtain structural as well as quantitative information. This paper will concentrate on several methods that we have used extensively during the last few years. It describes sequential extraction, synchrotron based X-ray absorption spectroscopy (XANES and EXAFS), X-ray photoelectron spectroscopy (XPS), and laser-induced spectroscopy (TRLFS and LIPAS). All chapters are in the context of the remediation of former uranium mining and milling sites in the state of Saxony, Germany. Because the main contaminations are caused by uranium and arsenic, this paper focusses mainly on these elements. The exception is the XPS chapter stemming from a close collaboration with scientists investigating the consequences of the Chernobyl nuclear accident.

SEQUENTIAL EXTRACTION

Sequential extraction consists of a multi-step treatment of a rock or soil sample using increasingly aggressive chemical leaching to selectively solubilize various geologic components. At the same time, this technique may involve the transfer of sorbed radionuclides to the solution, whereby their solid host phases may be classified or at best identified. It involves the transfer of radionuclides from a solid phase to the solution phase. This approach assumes that the chemical reactions are highly selective and no redistribution of the dissolved species occurs. This may not, however, always be the case (Shan and Chen, 1993).

A broad variety of sequential extraction procedures are described in the literature (Tessier et al., 1979, Beneš et al., 1981, Förstner, 1983, Plater et al., 1992, Margane and Boenigk, 1993, Shan and Chen, 1993, Fujiyoshi et al., 1994, Rüde and Puchelt, 1994). They can be grouped according to the mineral components of the material, the treatment principles, or the analytical methods. In general, most of the procedures are quite similar and are based on the same sequence: (1) the exchange of weakly bound cations by treating with concentrated salt solutions, mostly at pH 7, (2) the dissolution of carbonates in weakly acidic media, (3) the dissolution of iron and manganese oxides using acidic solutions with reducing and complexing components, such as NH_3OHCl, oxalic acid, ascorbic acid, citrate plus dithionite, (4) the oxidative destruction of organic complexants by hydrogen peroxide plus nitric acid, and (5) the digestion of the remaining residual with concentrated strong acids. Additional steps had been introduced if the presence of humic substances was expected. At each of the above steps, the procedure is divided into extraction, solid/solution phase separation and analysis of the solution phase.

Table 1. Sequential extraction steps.

Step	Extractant	Time	Target
1	1 M NH$_4$Ac[1], pH = 7	1h	water-soluble salts and species bound by ion exchange
2	1 M NaAc, Hac added for pH = 5	5 h	bound carbonates
3	0.1 M NH$_3$OHCl + 0.01 M HNO$_3$, pH = 2; replaced by 0.2 M (NH$_4$)$_2$oxalate + 0.2 M oxalic acid + 0.1 M ascorbic acid, pH = 3.0, at 96°C	12 h 1 h	amorphous/crystalline iron and manganese oxides
4	30% H$_2$O$_2$, pH = 2, 85°C	2 h	organic components (or sulfides)
5	0.25 M EDTA + 1.7 M NH$_3$, 85°C	5 h	barium sulfate
6	Concentrated HNO$_3$ or aqua regia, digestion using a micro wave apparatus	40 min	remaining residue, e.g., crystalline (alumo) silicates

[1] Ac =CH$_3$-COO$^-$

We have performed chemical extraction experiments on several activity-rich sedimentary rocks from the uranium mining area in the upper Elbe river valley (Saxony / Germany). The 80-100 million years old rift valley is built of older cenomanian sandstone separated from the younger partly limonite-rich turonian sandstone by a massive layer of claystone. The experiments, as described by Thieme et al., 1995, were performed in order to obtain information mainly on the binding state of radium, but also on that of uranium, each in the context of the release of typical mineral-forming elements. A sequential six-step treatment, based on Beneš et al., 1981, and Förstner, 1983, was used. The extraction solutions, contact times and the compounds or species targeted are shown in Table 1. In accordance with Beneš et al., 1981, a dilute sulfate solution (5×10^{-4} M) was added in steps (1) through (4) to minimize the unwanted dissolution of $BaSO_4$. In addition, the rock samples were digested directly and analyzed for comparison with the extraction results.

Our experiments with the rock samples, all from the Königstein uranium mine, were divided into the following series (lithologic symbols in parentheses, specifying turonian or cenomanian origin):

A. Series I to III: five activity-rich rock types, samples treated after decompactation and milling. Further, untreated samples from these series were leached with HNO_3 (series I), HNO_3+HCl (series II), and HNO_3+HF (series III).

    (A1)   turonian claystone ($t_1 t$),

    (A2)   cenomanian sandstone ($c_1 s_{qu}$), in-situ leached with sulfuric acid,

    (A3)   near-field sandstone ($c_1 s_{qu}$), not subjected to mining activities,

    (A4)   sandstone ("Wurmsandstein", $c_1 c_{wu}$), veined,

    (A5)   sandstone ("Wechsellagerung", $c_1 c_{wl}$), first layers on the granitic base stratum, with mixed compositions.

B. Series IV: three activity-rich sandstones, initially treated in the laboratory with 0.025 M H_2SO_4 for 9 weeks. This is the procedure used in the mine for uranium leaching. The samples (B1), (B2), and (B3) are of the same type as (A3), (A4), and (A5), respectively.

C. Series V: four turonian unleached sediment samples. Prior to the present experiments, these rocks had been contacted with ^{226}Ra containing solutions to produce radium sorption onto the rocks (Baraniak et al., 1995).

    (C1)   sandstone containing limonite ($t_1 s_{Li}$),

    (C2)   lime marl ("Pläner", $t_1 p$),

    (C3)   porous lime marl ("Pläneräquivalent", $t_1 p_{\ddot{A}}$),

    (C4)   claystone ($t_1 t$).

Aqueous phases from the extraction steps were centrifuged and filtered (0.45 µm pore size). The radium was isolated from the filtrates by sulfate precipitation in the presence of Pb^{2+} and Ba^{2+} in acidic medium. The precipitation had to be modified following steps 3 and 5 to reduce the masking effects due to the presence of complexants. The sulfate precipitate was dissolved in EDTA+NH_3 (Stewart et al., 1988, Chalupnik and Lebecka, 1992) and a second sulfate precipitation (Ba, Ra) was performed. The resulting solids were taken up in a liquid scintillation cocktail (gel forming type) and alpha counted in a Beckman 6000 LS liquid scintillation counter capable of energy analyses. The energy spectra revealed a very good separation of radium. The ingrowth of its dacay products, such as ^{222}Rn or ^{218}Po, after the re-precipitation was accounted for by a special algorithm. The radium determinations were standardized in parallel samples for each batch experiment, using known amounts of ^{226}Ra. The standard deviation obtained from 28 samples was 2.5%. Measurements on blank samples allowed us to set a lower detection limit of 0.01 kBq/kg.

In addition to radium, the extraction behaviors of Mg, Ca, Ba, Fe, Al and U were followed by measuring their concentrations via atomic absorption spectrometry (AAS 4100, Perkin-Elmer), ICP-MS (ELAN 5000, Perkin-Elmer), and X-ray fluorescence analysis (Spectrace Instrument 5000).

The fractions of elements obtained in the respective extraction steps formed patterns that, in most cases, were characteristic of the element and the rock type. The results for the extractions of radium are presented as bar charts in Figures 1 and 2. The following observations were made from the results focussing on radium:

• The sums of concentrations from the extraction fractions for a given component agree with the values obtained for the direct acid leaching of the rocks (HNO_3, HNO_3+HF, HNO_3+HCl), with the exception of the mixture containing HF. According to these data, the amounts of radium released from the different rocks types lie in between 1.5 and 30 kBq/kg. They may be ranked in the following order:

$$c_1 c_{wu} \text{ (A4)} > c_1 c_{wl} \text{ (A5)} > c_1 s_{qu} \text{ (A3)} > t_1 t \text{ (A1)} > c_1 s_{qu} \text{ (A2)}$$

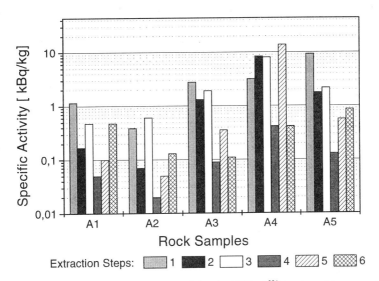

Figure 1. Activity concentrations (mean of three parallel series I-III) of [226]Ra released in sequential extraction treatments on different activity-containing rock types: Fractions from steps one to six.

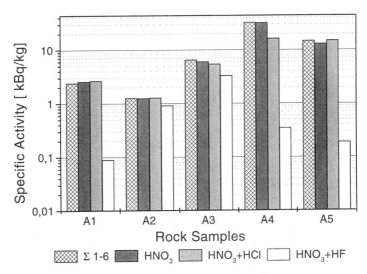

Figure 2. Activity concentrations (mean of three parallel series I-III) of [226]Ra released in sequential extraction treatments on different activity-containing rock types: Sums of the individual fractions and release values of treatments using concentrated acids (HNO_3, HNO_3, and HF, and HCl + HNO_3).

- With most sedimentary rock types, the largest radium release is in the first extraction step, i.e. under comparatively mild extraction conditions. This implies that a considerable fraction of radium is bound rather loosely by ion exchange. According to X-ray diffraction, kaolinite is present as the main matrix mineral. It is a bi-layered silicate that is able to fix radium (Jasmund and Lagaly, 1993), but has a low cation-exchanging capacity (Beneš et al., 1984). The major binding site for radium is presumed to be this clay component. According to the measured extraction patterns, calcium is probably bound together with radium. Unlike the clear preferential release of radium in extraction step (1), the extraction of uranium needed much stronger conditions, as Figure 2 demonstrates.
- A different pattern was seen with the sandstone sample $c_1 c_{wu}$. Here, the radium extraction was maximum in step (5), i.e., radium was bound considerably stronger than in the cases discussed above. Because the pattern is similar to that of barium, we suspect that radium is incorporated in $BaSO_4$.

The results of the series IV are shown in Figures 3 and 4, with the following features seen:

- The aqueous fractions of the 0.025 M H_2SO_4 leach step (Fig. 3) were found to be very low in radium (near the detection limit). This result is a consequence of prior in-situ leaching and in agreement with mining experience.
- The striking similarity between the extraction patterns of radium and barium (Fig. 4) leads one to the conclusion that radium is bound to $BaSO_4$ in all these cases. This includes the leached sandstone sample $c_1 s_{qu}$ (A2) (cf. series I-III), which contains $Ba(Ra)SO_4$ from the sulfuric acid treatment.
- The extraction behavior of uranium was found to be quite different from that of radium. The sulfuric acid leaching in the laboratory treatment led to a large release of uranium in the order of about 1g/kg, which exceeded the releases in the following steps by 1-2 orders of magnitude.

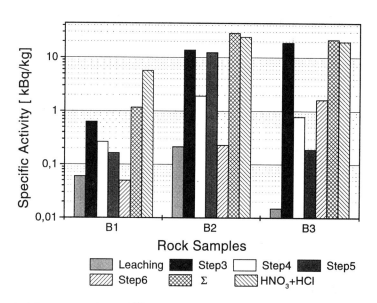

Figure 3. Activity concentrations of ^{226}Ra released in sequential extraction treatments including H_2SO_4 leaching on three different activity-containing rock types (series IV, averages of duplicate measurements): Fractions of the individual steps, their sums and the release values of treatments using concentrated acids ($HNO_3 + HCl$).

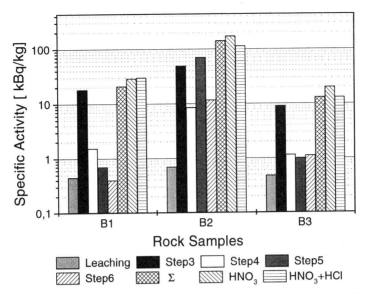

Figure 4. Concentrations of barium released in sequential extraction treatments including H_2SO_4 leaching on three different activity-containing rock types (series IV, averages of duplicate measurements): Fractions of the individual steps, their sums and the release values of treatments using concentrated acids (HNO_3 and HNO_3 + HCl).

As for the radium-loaded turonian rock types (series V), the extraction results gave two groups:

- The first, composed of limonite containing sandstone (t_1s_{Li}) and lime marl (t_1p) was characterized by ease of dissolution of radium in the following order:

  step (1) > step (2) > step (3).

- In contrast, the second group, containing porous lime marl ($t_1p_{\ddot{A}}$) and claystone (t_1t), showed the following sequence in the radium extraction concentrations:

  step (3) > step (4) > step (5) > step (6) > step (2) > step (1).

The porous lime marl is known to have a high sorption coefficient of radium (Baraniak et al., 1995). Obviously, these rock types exhibit a considerably stronger fixation of radium than the ones in the first group. The reason for this difference is likely connected to differences in the mineralogical composition. However, unequivocal conclusions await the results of sorption studies on single minerals and autoradiographic examinations of compact rocks after radium loading.

. In the examples presented here, an adapted sequential extraction procedure was used to estimate the concentrations of radium and accompanying elements that were bound in different sedimentary rocks. From their extraction behavior, general statements could be made regarding the conditions for mobilizing radium, other contaminants and some geomatrix elements. We found that multi-element analysis of the extract of each step provide us with information about the solid phase components and the mechanismns involved in the immobilization of contaminants in a given geologic matierial.

Similar experience is described by Clark et al., 1996. There contaminated soil from a radioactive waste disposal site was analyzed by optimized extraction sequences (Miller et al., 1986, Tessier et al., 1988). Through the determination of the main naturally-occuring elements of the soil, e.g. Fe, Al, Mn, Zr, and the contaminants Cs, U and Pb in all the extracts, they could distinguished between the part of the elements that represents the soil and the part that belongs to the contamination. The majority of radioactive caesium, for example, was exchanged by 1 M NH_4 acetate and the natural caesium was released by digestion with 30% H_2O_2, followed by leaching with 20% HNO_3 and by dissolution of the residue in an aqua regia-

HF mixture. The soil matrix-bound uranium was set free in the extraction step for the crystalline oxides (0.15 M sodium citrate, 0.005 M citric acid, 0.75 g sodium dithionite) and in the final dissolution, whereas uranium contamination was mobilized with pure water and 0.5 M $Ca(NO_3)_2$. Lead contaminations were evident in the treatments with 1 M NH_4 acetate, 1 M acetic acid and 0.4 M $NH_2OH \cdot HCl$ in 25% acetic acid indicating exchangeable and Fe/Mn oxides bound lead. The background part was detected after matrix dissolution. In this way the proposed scheme provided evidence on the soil phases dissolved in each step and the mobilization of the contaminants that can be transported in the aqueous phase under normal geochemical conditions.

In our case, the extraction data were used with a two-fold aim: first, to estimate the radium mobilization in the process of flooding the underground-leached Königstein uranium mine and, second, to optimize the remediation strategy for this site. For the flooding process that consists in the dilution of the sulfuric acid leaching solution in the porous sediments with low-mineralized groundwater accompanied by a decrease of sulfate content from 2000 ppm to 15 ppm, it was concluded, that the radium mobilization amounts only to about 1% of the total radium inventory increasing the radium concentration in the mine water by two orders of magnitude (from ≤ 0.7 to nearly 50 kBq/m^3). Furthermore, the infiltration of a barium salt or $Ba(OH)_2$ solution into the leached sediments counteracts the loss of sulfate concentration in the $Ba(Ra)SO_4$ dissolution and prevents, therefore, the radium mobilization and its migration into the groundwater table to a certain degree.

X-RAY ABSORPTION SPECTROSCOPY

Although a number of traditional spectroscopies have contributed significantly to our knowledge of the speciation of radionuclides and other heavy metals in the environment, some methods are not able to provide information on the local structure of a species or fail to detect a species in a particular form or oxidation state.

In recent years, synchrotron-based x-ray absorption spectroscopy (XAS) has proven quite valuable for studying the speciation of environmental contaminants (SSRL, 1995). In XAS the sample is irradiated by monochromatized synchrotron radiation and the absorption spectrum of a given element is measured as a function of x-ray photon energy. The detailed structure in absorption intensities within 20 eV above and below the absorption edge of a particular atomic shell is referred to as x-ray absorption near-edge structure (XANES). The absorption structure associated with the spectral range which lies 20 - 1000 eV past the absorption edge contains the so called extended x-ray absorption fine structure (EXAFS).

XANES contains information about the electronic structure and local geometry of the absorbing atom. For example, XANES spectroscopy can be used to determine the oxidation state of a particular type of absorbing atoms and symmetry adapted by their surrounding. Features in the EXAFS region are weak modulations of the absorption cross section and can be described by the following formula (Koningsberger and Prins, 1988):

$$\chi(k) = S_0^2 \sum_{i=1}^{n} \frac{N_i S_i(k,R_i) F_i(k,R_i)}{kR_i^2} \exp\left(\frac{-2R_i}{\lambda_i(k,R_i)}\right) \exp(-2\sigma_i^2 k^2)\sin(2kR_i + \phi_i(k,R_i)) \qquad (1)$$

where k is the photoelectron wave number, N_i is the coordination number for atom i, F_i is the effective EXAFS scattering amplitude function, R_i is the absorber-backscatterer distance for atom i, λ_i is the photoelectron mean free path, σ_i is the Debye-Waller factor, and ϕ_i is the scattering EXAFS phase function.

By fitting the theoretical Eq. (1) to the experimental EXAFS spectrum (George and Pickering, 1995), the type and number of near-neighbor atoms and their distance to the central atom can be determined. The coordination number, N, and the inner-atomic distance, R, can be obtained to within $\pm 15-25\%$ and ± 0.02 Å, respectively. Scattering amplitude and phase are functions of the atomic number of the scattering atom and can be accurately calculated by modern scattering computer codes like FEFF (Zabinsky et al., 1995) and GNXAS (Westre et al., 1995). Therefore, the atomic number, Z, of near-neighbor atoms may be identified to within ±5-10.

EXAFS does not depend on long-range order of atoms and can therefore provide structural information on amorphous solids or species in solution. This makes XAS a valuable tool for the study of radionuclides and heavy metals in environmental systems, e.g., sorption onto minerals and soil or species in liquid or solid matrices (Eller et al., 1985, Petit-Maire et al., 1989, Waychunas and Brown, 1990, Combes et al., 1992, Waite et al., 1994, Allen et al., 1995). It should be noted that the sensitivity of XANES and EXAFS spectroscopies that can be routinely obtained corresponds roughly to heavy metal concentrations of 10^{-5} and 10^{-4} mol/L in solutions, respectively. In some environmentally relevant samples these metal concentration will be lower than these limits. However, in the near future the detection limit for heavy-metal contaminants will be further reduced by the development of new detector systems, faster counting electronics, and the increase of photon flux provided by the high brightness of third generation synchrotron sources. For example, a practical uranium detection limit of 5 μmol/L has been achieved with a novel design of a solid-state Ge fluorescence detector system (Allen et al., 1996a, Bucher et al., 1996).

The analysis of arsenic species, which may be present at toxic concentrations in tailing waters of former uranium mines in the Eastern part of Germany, may serve as an illustration for the determination of the oxidation states in aqueous solutions by XANES spectroscopy (Denecke et al., 1997a). Arsenic is associated with uranium ores in the Erzgebirge, Germany. Many years of uranium mining and processing resulted in 10 ppm or more arsenic in tailing pond waters of decommissioned uranium ore processing plants. The knowledge of arsenic speciation in these contaminated waters is essential for assessing the toxicity, bioavailability, migration behavior of arsenic, and for selecting and optimizing water purification methods (Cullen and Reimer, 1989, Thayer, 1995).

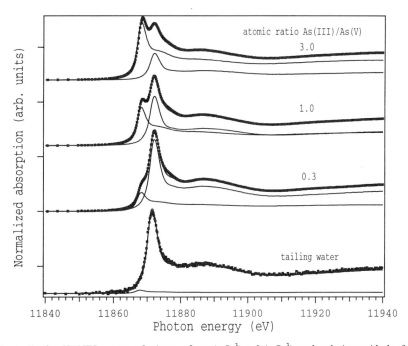

Figure 5. As K-edge XANES spectra of mixtures from AsO_3^{3-} and AsO_4^{3-} stock solutions with the following atomic ratio (from top to bottom): 3.0, 1.0, and 0.3, and of a tailing water. The solid line is the fit to the experimental data (dots). The spectra of the individual As(III) and As(V) components are also shown as solid lines.

Arsenic *K*-edge absorption spectra of metallic As and solutions of 0.02 mol/L arsenite, AsO_3^{3-}, and arsenate, AsO_4^{3-}, were measured in transmission mode with the Si(311) double-crystal monochromator RÖMO II at Hamburger Synchrotronstrahlungslabor, HASYLAB. The normalized spectra given in Fig. 5 show a chemical shift toward higher energies and an increase in white line intensity when the positive effective atomic charge increases by going from As(0) to As(V). If the position of the absorption edge is defined as the inflection point of the rising edge (maximum of first derivative), the *K*-edge positions of As(III) and As(V) relative to As(0) increase by 3.1 and 6.6 eV, respectively. Thus, the determination of the oxidation state distribution of arsenic in aqueous samples is feasible. If more than one arsenic species is present in a solution, one would like to determine their relative concentrations as well. In Fig. 6, the arsenic *K*-edge spectra of three solutions of mixtures of AsO_3^{3-} and AsO_4^{3-} are shown. Arsenite and arsenate stock solutions were mixed to give solutions with atomic ratios As(III)/As(V) of 3, 1, and 0.33, but a constant total As concentration of 0.02 mol/L. By fitting the sum of the spectra of individual As(III) and As(V) components to the experimental data (see Fig. 6), the following atomic ratios As(III)/As(V) were determined: 2.50, 0.90, and 0.31. These values agree to within 16% or better with the true As(III)/As(V) ratios.

A sample taken from the surface of a uranium mill-tailing water basin contained 7.7×10^{-4} mol/L arsenic. As can be seen from Fig. 6, the arsenic *K*-edge spectrum of this sample measured in fluorescence mode resembled that of a As(V) solution. The spectrum has an intense white line and is shifted 6.7 eV higher in energy relative to metallic As. The fit of the spectra resulted in a 4% contribution of an As(III) species. However, such a small amount may be a result of the larger fit error due to a lower signal-to-noise ratio of the fluorescence spectrum as compared to transmission spectra. Therefore, it can be concluded that under the given conditions the arsenic is oxidized to As(V), i.e., the mill-tailing water contained the AsO_4^{3-} species to at least 96%.

This approach has also been applied to determine the selenium speciation following bacterial metabolization (Buchanan et al., 1995) and in soil samples from the Kesterson Reservoir, California (Pickering et al., 1995). The speciation of technetium in cement waste forms and the efficiency of various agents to reduce Tc(VII) have been studied by Tc *K*-edge

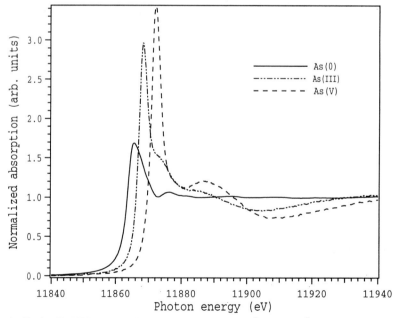

Figure 6. As *K*-edge XANES spectra of metallic arsenic and AsO_3^{3-} and AsO_4^{3-} solutions. The spectra have been normalized to have equal intensity at 11950 eV.

XANES (Allen et al., 1997). XANES spectroscopy was also used to determine the ratio of U(IV)/U(VI) in soils from the United States Department of Energy's former uranium production facility located at Fernald, OH, contaminated with ~10 to 8000 ppm uranium (Morris et al., 1996). The U(IV)/U(VI) oxidation state distribution of soil samples was determined by analysis of uranium L_{III}-edge x-ray absorption spectra. Although some samples contained significant amounts of U(IV), the majority of the uranium was present in the hexavalent oxidation state in the top 0.25-0.3 m of the soil. The spatial distribution of U(IV) and U(VI) within sediments from the Fernald Site and from the Savannah River Site, near Aiken, SC, was studied by micro-XANES spectroscopy (Bertsch et al., 1997).

An example for the application of EXAFS spectroscopy to speciation studies is the measurement of structural parameters for uranium in aqueous complexes with simple carboxylic acids and humic acid. Humic acids are polyfunctional and polyelectrolytic organic macromolecules present in soil, sediment, and water. The speciation of heavy metals in the geosphere can be significantly influenced by the strong complexation of metal ions and humic acids (Choppin and Allard, 1985, Choppin, 1988). Therefore, uranyl humates are important species to be considered in the migration/retardation of uranium in the geosphere. The most important functional groups of humic acids responsible for metal complexation are carboxylic and phenolic OH groups (Choppin and Allard, 1985). Despite the strong interest in uranium complexation with humic acids in recent years, much remains unknown about the structure of the complexes formed. Depending on the coordination geometry taken by the ligands, characteristic interatomic distances between uranium and its neighboring oxygen and carbon atoms can be observed by EXAFS spectroscopy. EXAFS measurements of simple uranyl carboxylates can guide the interpretation of EXAFS determinations of structural parameters of uranyl humate complexes.

Table 2. Sample composition and calculated U(VI) speciation using published complex formation constants.

No.	Ligand (L)		U(VI)	pH	U(VI):L ratio	Reference
	Acid	mol/L	mol/L		(of complex)	
1	acetic	1.10	0.05	3.7	1:3	Ahrland, 1951
2	maleic	0.01	0.001	4.2	1:1	Rajan and Martell, 1967
3	malonic	0.20	0.05	3.9		Rajan and Martell, 1967

Aqueous solutions of uranyl complexes with acetic, maleic, and malonic acids were prepared by dissolving uranyl nitrate hexahydrate and the corresponding acid in bidestilled water. The desired uranium to ligand ratio (U:L) and pH value were adjusted so that only one uranyl species would be present in the solution. This is an essential precondition for the measurement of structural parameters. If significant amounts of more than one uranium species are present in solution, the EXAFS analysis may fail because the XAS spectrum is an average over all species. Thus, optimized conditions were calculated using the software package RAMESES (Leung et al., 1988) and published complexation an dissociation constants for the carboxylic acids (Ahrland, 1951, Rajan and Martell, 1967). The speciation calculations also included the hydrolysis equilibrium of the uranyl ion using Nuclear Energy Agency's Uranium Thermodynamic Database (Grenthe et al., 1991). According to the speciation calculations, more the 95% of the uranium in solution was present in form of the desired uranyl complex with a specific U:L ratio (see Table 2). Uranyl humate was prepared by dissolving 70 mg synthetic humic acid (Pompe et al., 1996) and adding uranyl perchlorate solution. The pH of the solution was adjusted to 3.9 and the ionic strength to 1 mol/L. The uranium solution concentration was 10^{-3} mol/L and corresponded to 7.5% of the proton exchange capacity of the humic acid. Uranium L_{III}-edge XAS measurements were performed using the Si(220) double-crystal monochromator on beamlines 4-1 and 2-3 at the Stanford Synchrotron Radiation Laboratory, SSRL. Samples with 0.05 mol/L uranium were measured in transmission mode and the remaining samples in fluorescence mode. The EXAFS spectra were analyzed according to standard procedures (Koningsberger and Prins, 1988) using the suite of

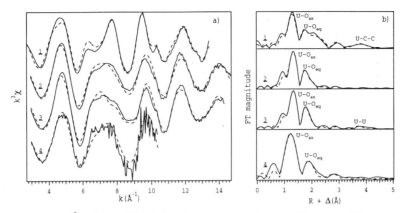

Figure 7. a). U L_3-edge k^3-weighted EXAFS data for aqueous uranyl acetate (1), maleate (2), malonate (3) and humate (4). b). Corresponding Fourier transforms of the data shown in a). The solid lines are the experimental data. The dashed lines are the best theoretical fit of the data.

programs EXAFSPAK (George and Pickering, 1995).Theoretical scattering phases and amplitudes were calculated using the scattering code FEFF6 (Mustre de Leon et al., 1991).

Fig. 7a shows the raw U L_{III}-edge k^3-weighted EXAFS data for the uranyl acetate (1), maleate (2), and malonate (3) solutions. The corresponding Fourier transforms (FT's) are shown in Fig. 7b. The first peak in the FT of samples 1 - 3 at 1.3 Å (uncorrected for scattering phase shift) is due to oxygen atoms, O_{ax}, of the linear uranyl group. The second uranium coordination shell consists of oxygen atoms, O_{eq}, lying in the equatorial plane of the uranyl group. This O_{eq} coordination shell gives rise to a FT peak near 2 Å. This FT peak is broadened and shifted toward a larger distance for sample 1. The appearance of a peak at 3.8 Å in the FT of samples 1 and 3 indicates the presence of a third coordination shell. Sample 2 does not show any peak in this region. The raw k^3-weighted EXAFS spectrum of uranyl humate (4) solution exhibits spectral features similar to the uranyl maleate solution (2). Both FT's show evidence of similar interatomic distances for the first two coordination shells.

Table 3. EXAFS structural parameters of aqueous uranyl complexes[1].

No.	Uranyl complex	U-O_{eq}			U-X	
		R(Å)	N	σ^2(Å^2)	X	R(Å)
1	acetate	2.44	6	0.008	C	4.34
2	maleate	2.37	6	0.010		
3	malonate	2.36	5	0.007	U	3.96
4	humate	2.38	4	0.008		

[1]U-O_{ax} (samples 1-4): N=2, R=1.78±0.02 Å, and σ^2=0.002 Å^2.

Structural parameters of the first two to three coordination shells were obtained by modeling and curve-fitting the raw EXAFS data to Eq. 1. The results are summarized in Table 3. The bond distance between the uranium and the O_{ax} atoms of the uranyl group is 1.78±0.02 Å for samples 1-3. Five to six atoms surround the uranyl group in the equatorial plane. The error in determining the coordination number is ±20%. The interatomic distance U-O_{eq} is 2.44±0.02 Å for uranyl acetate (1) which is the typical bond length for oxygen/metal distance to a COO⁻-group with symmetric bidentate geometry. The assignment of bidentate coordination is supported by the detection of an additional carbon coordination shell at 4.3 Å. Normally, light elements such as carbon and oxygen are not detected in room temperature solutions at distances greater than 3 Å. For spectrum 1 it can be shown that the EXAFS amplitude associated with the FT peak at 3.8 Å is enhanced by multiple scattering (MS) along a linear U-C-C arrangement. A structural model for the uranyl acetate solution (1), where three acetate ions assume a bidentate geometry, is shown in Fig. 8. For monodentate coordination of the COO⁻-group, the U-O$_{eq}$ bond length is much shorter, typically 2.37 Å. Therefore, the COO⁻-groups of maleic and malonic acids appear to act as monodentate ligands to the uranyl ion. The analysis of the nature of the peak centered at 3.8 Å in the FT of sample 3 revealed that it is due to U-U interaction with a bond length of 3.96 Å and not due to U-C-C MS effects. A similar U-U interaction was also observed in uranyl complexes with malic, citric, and tartaric acids. Here, two uranyl groups are bond to oxygen atoms of two acid molecules and are linked by the α-hydroxyl group of the acids (see Allen et al., 1996b, for a structural model).

For uranyl humate solution (4) structural parameters similar to those measured for the uranyl maleate (2) solution were obtained. In the equatorial plane, the uranyl ions are

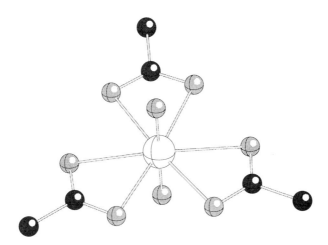

Figure 8. Structural model for uranyl triacetate having the empirical formula UO$_2$(Ac)$_3$⁻.

surrounded by approximately 4-5 oxygen atoms at a distance of 2.38±0.02 Å (Table 3). Due to the possibility for the uranyl ion to interact with one or two of the numerous COO⁻ and/or OH-groups of a humic acid molecule, the measured structural parameters have to be considered as an average value. However, under the conditions chosen for our experiment, most of uranyl ions would be expected to be bound to humic acid molecules, only a small amount of other uranyl species might be present in the solution. Based on a comparison with the structural parameters of the equatorial oxygen shell in model solutions 1-3, we conclude that the majority of the COO⁻-groups of humic acids form monodentate complexes with the uranyl ion. EXAFS and IR spectroscopic studies of several solid uranyl humates showed that the predominant coordination mode of the COO⁻-groups is the same as in humic acid solution (Denecke et al., 1997b). This approach of comparing EXAFS spectra of model compounds with those from environmental samples has also been applied to the determination of lead speciation in contaminated soils Manceau et al., 1996). Zinc K-edge EXAFS measurements were performed to study the complexation mechanism of zinc by natural humic acids over a wide range of zinc concentrations (Sarret et al., 1997).

X-ray absorption spectroscopy has also been effectively applied to study the solution speciation of other radionuclides, e.g., Np and Pu (Allen et al., 1996c, Clark et al., 1996).

X-RAY PHOTOELECTRON SPECTROSCOPY

Although the electronic structure and the physicochemical properties of actinide compounds have been studied over the last decades, many aspects of their structure and the nature of their chemical bonds remain unclear (Teterin et al., 1996a, Teterin and Gagarin,

1996). The development of precise X-ray photoelectron spectrometers (Siegbahn et al., 1967, Siegbahn et al., 1969) and the use of relatively fast methods for the calculation of the electronic structure of actinide compounds, e.g., Xα-methods (Boring et al., 1975, Ellis et al., 1975, Ryzhkov et al., 1991, Ryzhkov et al., 1992), have led to the study the electronic structure of such compounds by X-ray photoelectron spectroscopy (XPS) on a qualitatively new level.

Photoelectron spectroscopy is based upon the photoeffect (Siegbahn et al., 1967, Siegbahn et al., 1969). If a solid, liquid or gaseous sample is irradiated by characteristic X-rays (for example AlK$_{\alpha1,2}$ with an energy of $h\nu$ = 1486.6 eV) electrons with a kinetic energy between 0 - 1480 eV are ejected as a result of the photoelectric effect. The kinetic energy of the photoelectrons, E$_K$, can be measured to an accuracy of ±0.1 eV. By means of the Einstein equation, the electron binding energy E$_B$ in eV (ionization potential) can be determined as:

$$E_B = h\nu - E_K - \phi_{sp}.$$

If the studied sample is a metal, ϕ_{sp} is the electron work function for the spectrometer material. In this case the electron binding energy is usually given relative to the Fermi level and the spectrometer is calibrated to the binding energy of the Au4f$_{7/2}$ electrons of gold. In the case of insulators the electron binding energy is usually given relative to a known reference value, e.g.,

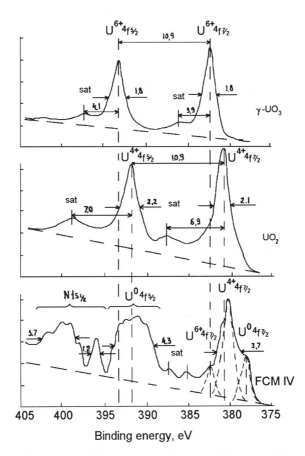

Figure 9. XPS spectra of U4f electrons for UO$_2$ and γ-UO$_3$ and the FCM sample IV.

the binding energy of the C1s electrons of hydrocarbons on the sample surface (Teterin et al., 1996a, Teterin and Gagarin, 1996).

The photoelectron spectrum of a compound gives the dependence of the number of photoelectrons (intensity I in relative units) on the binding energy (eV) of these electrons (Fig. 9). The sensitivity of the method is of the order of 0.1 to 1 percent of the element in the sample. Photoelectron spectroscopy is largely a surface technique which allows the study of sample surface layers of approximately 15-50 Å thickness. In order to study the bulk composition of a sample, layer-to-layer etching of its surface or sample grinding are performed. In our laboratory the photoelectron spectra considered in the present work were obtained at room temperature in a vacuum of 1.3×10^{-7} Pa with an electrostatic spectrometer HP5950A (Hewlett-Packard) using monochromotized $AlK_{\alpha1,2}$ (1486.6 eV) X-ray excitation. A gun of low-energy electrons was used to compensate the electrostatic sample charging. The spectrometer resolution measured as the full-width-at-half-maximum of the $Au4f_{7/2}$-electron line was 0.8 eV. The values of binding energies E_B(eV) were taken relatively to the binding energy of C1s- electrons of hydrocarbonates on the sample surface, which is accepted to be equal to 285.0 eV. If the binding energy of a gold substrate equals $E_B(Au4f_{7/2})$=83.8 eV, than the measured binding energy of the adsorbed layer of hydrocarbons is E_B(C1s)=284.7 eV. The errors in determination of electron binding energy values did not exceed 0.1 eV and that of for relative intensities of lines was less than 10 %. The samples were prepared for XPS measurement from finely dispersed powders milled in an agate mortar. A small amount of the obtained powder was pressed into indium to get a dense layer with a mirror-like surface. Such a sample preparation and spectra registration allows to avoid differential sample charging and to obtain reproducible spectra (Teterin et al., 1996a, Teterin and Gagarin, 1996). Although the sensitivity of the method is not very high when compared to optical methods, the obtained results are interpreted unambiguously with a high degree of reliability.

The results of XPS studies on actinide compounds have shown that not only the electron spectra of the outer valence molecular orbitals (OVMOs) but also the inner valence molecular orbitals (IVMOs) and deeper laying levels of the atoms can contain information on their physicochemical properties (Verbist et al., 1974, Veal et al., 1975, Pireaux et al., 1977a, Pireaux et al., 1977b, Veal et al., 1977a, Veal et al., 1977b, Teterin et al., 1980, Courteix et al., 1981, Teterin et al., 1981a, Teterin et al., 1981b, Teterin et al., 1984a, Teterin et al., 1984b, Teterin et al., 1985, Teterin and Baev, 1986, Teterin et al., 1996a, Teterin and Gagarin, 1996). In addition to the chemical shift of the electron levels and their line intensity, the characteristic fine structure of these spectra contain valuable information. For actinide compounds, almost all electron spectra with a binding energies between 0 - 1250 eV show fine structure, which is used to study compounds by XPS (Teterin and Baev, 1986).

At the present time, the relationships between the parameters of the spectral fine structure and the oxidation state of actinide ions, the near-neighbor environment in compounds, their magnetic properties, and the nature of their chemical bonds have been established (Veal et al., 1975, Teterin et al., 1980, Teterin et al., 1981a, Teterin et al., 1981b, Teterin et al., 1984b, Teterin et al., 1984a, Teterin et al., 1985, Teterin and Baev, 1986, Teterin et al., 1996a, Teterin and Gagarin, 1996). On the basis of the spectral fine structure, the effectiveness of the formation of IVMO's from An 6p and L ns atomic orbitals from neighboring actinide (An) and ligand (L) atoms was shown experimentally for actinide compounds (Teterin et al., 1980, Teterin et al., 1981a, Teterin et al., 1984b, Teterin et al., 1984a, Teterin et al., 1985, Teterin and Baev, 1986, Teterin et al., 1996a, Teterin and Gagarin, 1996). Methods for the determination of oxidation states (Teterin et al., 1980, Teterin et al., 1981b, Teterin and Baev, 1986) and bond lengths (Veal et al., 1975, Teterin and Baev, 1986, Nevedov et al., 1996, Teterin et al., 1996a) in amorphous actinide compounds have been developed.

As an example one can consider the XPS study of fuel-containing mass (FCM), which was formed as a result of the accident at block IV of Chernobyl Nuclear Power Plant (CNPP) (Teterin et al., 1994). The line intensity of inner-shell electrons is proportional to the number of atoms of a given element in the sample. It is known that energy shifts depend upon oxidation state and chemical form. The $U4f_{7/2}$ photoelectron spectra of samples I - VI showed evidence of four different chemical states. In case of sample IV we observed only three chemical states (see Fig. 9). In order to interpret these spectra, the binding energy of each $U4f_{7/2}$ component was compared to experimental binding energies of several uranium oxides and other uranium compounds which contained uranium in various formal oxidation states. Thus, a formal oxidation state was assigned to each chemical state observed for the uranium, i.e., U(VI), U(IV), U(III), and U(0)/UC_n. The symbol U(0)/UC_n stands for the chemical states of metallic uranium or uranium carbide. XPS can be used to determine the relative concentration of elements in FCM samples. For samples I - VI, which were obtained from different parts of CNPP block IV, the following atomic compositions relative to uranium were determined (Teterin et al., 1994):

$$U_1Zr_3Ca_{11}Na_{13}Si_{56}N_4C_{23}O_{172} \tag{I}$$
$$U_1Zr_3Ca_{12}Na_9Si_{58}N_{0.5}C_{13}O_{177} \tag{II}$$
$$U_1Zr_3Ca_{16}Na_{16}Si_{71}N_1C_{18}O_{251} \tag{III}$$
$$U_1Zr_2Ca_{17}Na_{11}Si_{85}N_9C_{22}O_{228} \tag{IV}$$
$$U_1Zr_1Ca_{12}Si_{60}N_4C_{30}O_{236} \tag{V}$$
$$U_1N_9C_{466}O_{41} \tag{VI}$$

It should be noted that the quantitative elemental analysis of samples can be carried out with greater accuracy by other methods (optical and chemical). However, the present work shows that XPS is a fast and simple method for elemental analysis of all chemical elements of the periodic table (except hydrogen). It follows from the spectra of the FCM samples that a variety of oxidation states were observed for every element. In particular, in the spectrum of U4f electrons, which consists of the $U4f_{7/2}$ and $U4f_{5/2}$ doublet due to spin-orbit interaction, several of theses doublets were observed (Fig. 9). They are shifted in energy relative to each other because each FCM sample contains uranium in several chemical states. In Fig. 9 the U4f electron spectra of UO_2 and γ-UO_3 are shown for comparison. The relative contents of the different uranium chemical states in the FCM samples I - VI is shown in Table 4.

On the basis of these results, it was concluded that all elements (except oxygen), present in FCM are in an oxidation states which are lower than that of their stabile oxides. Under the influence of the environment conditions and bacteria, the elements can be oxidized to their stabile oxide forms. Thereby thermal energy can be released. These effects have to be taken into account during the neutron-flow control and the thermosetting process of spent fuel.

Table 4. Relative content of uranium ions U^{n+} (%) and overall content of uranium ΣU in FCM samples I - VI, (mass %).

Sample	U^{6+}	U^{4+}	U^{3+}	U^0, UC_n	ΣU
I	5.6	26.5	51.4	16.5	4.0
II	11.3	48.3	30.0	10.4	4.0
III	10.7	31.1	50.0	8.2	3.0
IV	12.4		65.0	22.6	3.1
V	4.5	38.1	51.6	5.8	3.6
VI	30.8	42.3	21.8	5.1	3.5

Another example for the application of XPS is the study of "new product", which formed at the surface of FCM of block IV (Teterin et al., 1994). At the end of 1990, somewhere on the surface of the hardened melt of FCM in the room below the machine room, a yellow crystalline substance, "new product", was found. Similar "new products" appeared in other places later. Since the "new product" contained uranium, it became obvious that under the influence of the environment (bacteria, temperature, rain, humidity) the FCM was modified, transported, and accumulated in certain places of CNPP block IV. To understand the reason and mechanism of the formation of the "new product", it was necessary to determine in detail the relative concentration of the elements, their oxidation states, the structure of the uranium near-neighbor environment, and the nature of the chemical bonds .

The qualitative and quantitative XPS analysis of several "new product", samples, VII - IX, yielded the following elemental compositions relative to uranium (Teterin et al., 1994):

$$U_1Na_{69}S_9K_{1.5}C_{11}O_{88} \quad (VII)$$
$$U_1Na_{27}S_1C_9O_{47} \quad (VIII)$$
$$U_1Na_{10}K_{1.5}C_6O_{39} \quad (IX)$$

It followed from the analysis of the spectral fine structures of IVMOs and core electron lines that the "new product" contained uranyl compounds:

$$Na_4UO_2(CO_3)_3{\cdot}8Na_2CO_3{\cdot}9Na_2SO_4{\cdot}31NaOH{\cdot}nH_2O \quad (VII)$$
$$Na_4UO_2(CO_3)_3{\cdot}6Na_2CO_3{\cdot}Na_2SO_4{\cdot}9NaOH{\cdot}nH_2O \quad (VIII)$$
$$Na_4UO_2(CO_3)_3{\cdot}3Na_2CO_3{\cdot}nH_2O, \quad (IX)$$

where Na can be replaced partially by K. It should be noted, that the content of the "new product" (samples VII-IX) is suspicious, in particular because of the presence of Na^+ and CO_3^{2-} groups. This can be explained by the fact that large quantities of various compounds including soda ($Na_2CO_3{\cdot}10H_2O$) were dropped from helicopters on the melting block IV of Chernobyl NPP after the accident.

Based on the fine-structure parameters of the photoelectron spectra of inner valence electrons of hexavalent uranium compounds, a method for the determination of bond lengths in axial (Veal et al., 1975, Nevedov et al., 1996, Teterin et al., 1996a, Teterin and Gagarin, 1996, Teterin et al., 1996b) and equatorial (Nevedov et al., 1996, Teterin et al., 1996a, Teterin and Gagarin, 1996, Teterin et al., 1996b) directions of the uranyl group was developed. Also,

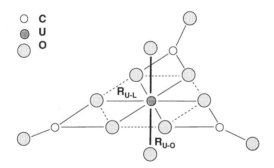

Figure 10. Proposed molecular structure of the "new product" based on the R_{U-O} and R_{U-L} bond distances of 1.74 Å and 2.39-2.60 Å as determined by XPS. The structural model was derived from the known crystal structure of $Na_4UO_2(CO_3)_3$.

the presence of CO_3^{2-}, OH^-, and SO_4^{2-} groups in a compound and their relative amount can be determined from the spectral features of inner and valence electrons. It is possible to compare the spectral features of well-characterized compounds, e.g., $Na_4UO_2(CO_3)_2$, with the corresponding characteristics of our samples (Teterin and Baev, 1986, Teterin et al., 1996a). Based on such a comparison we assumed the same structure as in $Na_4UO_2(CO_3)_2$ (see Fig. 10) for the near-neighbor uranium surrounding in our samples. Such an approach of structure determination of amorphous compounds is very important, because the possibilities of other methods, e.g., x-ray diffraction, are limited.

For the "new product" the near-neighbor uranium environment was determined. The U-O bond length within the uranyl group (R_{U-O}) is 1.74 Å. For the equatorial oxygen atoms, the U-O bond length (R_{U-L}) is 2.39 - 2.60 Å (Fig. 10). These distances are in agreement with the interatomic distances in $Na_4UO_2(CO_3)_2$ Also, the corresponding spectral features of "new product" and $Na_4UO_2(CO_3)_2$ agree. We found that the "new product" contained significantly more uranium (19.3 mass %) than the FCM (about 4 mass %). It should be noted that dissolved uranyl compounds are present in hydrothermal flows under the influence of the environment. This leads to migration, crystallization, and precipitation of uranyl compounds in several places, where they can accumulate to relatively large amounts.

The third example for the application of XPS is the study of the interaction of uranyl ions with surfaces of calcite (Teterin et al., 1995, Teterin et al., 1996b) and diabase (Teterin et al., 1996b) . In particular, an amorphous surface layer consisting of uranyl compounds, where the oxygen atoms of OH^- and CO_3^{2-} groups are located in the equatorial plane of the uranyl group, were found on the minerals. The U-O bond lengths in the axial (R_{U-O}) and equatorial (R_{U-L}) planes are 1.82 and 2.30 Å, respectively. It should be noted that structural parameters of amorphous compounds are difficult to measure with methods other than XPS and EXAFS.

In conclusion, we would like to add that XPS (Veal et al., 1977a, Veal et al., 1977b, Teterin et al., 1984a, Teterin et al., 1984b, Teterin and Baev, 1986, Ryzhkov et al., 1992, Teterin and Gagarin, 1996, and references therein) can be applied effectively to the study of compounds of other actinides like Th, Np, Pu, Am, Cm, and Bk.

LASER-INDUCED SPECTROSCOPY

Laser-induced spectroscopies have recently gained increased interest, especially since certain solid crystals have become widely available for the production of tunable-energy lasers. This allows for laser spectroscopic investigations over wide wavelength ranges without the use of dye lasers with all their associated limitations and drawbacks. Laser-induced spectroscopy can be categorized into at least four different experimental concepts, all of which have been successfully applied to radionuclide speciation studies. They are: Time-Resolved Laser-Induced Fluorescence Spectroscopy (TRLFS) (Beitz et al., 1988, Czerwinski et al., 1994, Eliet et al., 1995, Kato et al., 1994, Klenze et al., 1991, Moulin et al., 1995, Wang et al., 1984, Brendler et al., 1996, Bernhard et al., 1996, Geipel et al., 1996), Laser-Induced PhotoAcoustic Spectroscopy (LIPAS) (Beitz et al., 1988, Kim et al., 1990, Kimura et al., 1992, Klenze and Kim, 1989, Okajima et al., 1991), Thermal Lensing Spectroscopy (TLS) (Berthoud et al., 1983, Enokida et al., 1992, Grenthe et al., 1989, Gutzman et al., 1993, Moulin et al., 1988), and Laser Breakdown Spectroscopy (LBS) (Wachter and Cremers, 1987, Archontaki and Crouch, 1988, Cremers, 1987, Ito et al., 1995).

We will focus on the first two methods with which we have the most experience. Fig. 11 gives a schematic diagrams of the experimental setups which we use in both our TRLFS and LIPAS investigations. A pulsed Nd:YAG laser is used to pump an optical parametric oscillator (OPO) crystal with a frequency doubled beam (525 nm wavelength). An external fourth harmonic generator (FHG) with a fixed output is also available. Most system functions, e.g. wavelength setting, wavelength scanning or energy table mode, are set by a controller which is connected to a personal computer as a user interface. A signal from the advanced Q-switch of the Nd:YAG laser or from the fast photo diode is used to initiate the start of the experiments. The laser output energy can be reduced to useful levels for the fluorescence measurements (0.1 to 3 mJ/pulse) via a telescope and for the LIPAS measurements (1 to 5 mJ/pulse) via a variable attenuator. A more detailed description of the experimental setup can be found in Geipel et al., 1997.

For TRLFS experiments, the sample solution is placed inside a standard quartz cuvette and excited with a laser pulse. For energy correlation, the true laser energy is recorded using an energy meter. A portion of the fluorescent light is transported to a triple-grated spectrograph (150 lines/mm with 275 or 500 mm focal length) via an optical fiber cable, orthogonal to the laser beam, where the light is split into wavelength dependent components. The intensities of the various wavelength components are then measured using an intensified diode array. The 700 photo diodes of the array allow one to cover a wavelength range of 418.4 nm with a resolution of 0.6 nm. For time-resolved fluorescence measurements, time windows are set for collection of data by the diode array. In order to avoid overloading the detection system with

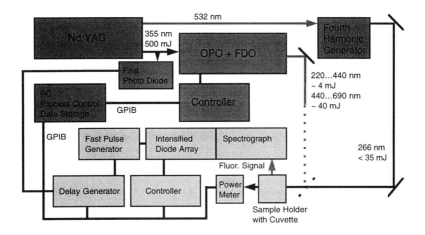

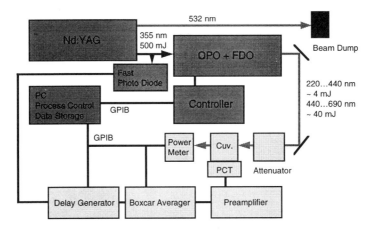

Figure 11. Schematic diagrams of the experimental setup for TRLFS (*upper*) and LIPAS (*lower*) investigations. OPO+FDO: Optical Parametric Oscillator with Frequency Doubling Option, GPIB: General Purpose Interface board, PCT: Piezo Ceramic Transducer.

the very large initial light flash at time zero, a delay in the time of the start pulse is possible over the range 5 ns to 95 ms. Time window widths can be varied from 5 ns to 2 ms and can be stepped along in equal time widths to within 1 ns. Usually, each single spectrum is collected over 100 laser shots to increase the number of recorded counts per wavelength setting and thus reduce the error.

Typical two dimensional spectra from TRLFS measurements are shown in Fig. 12. Each spectrum is made up of signals originating from all the different individual species in that sample. Each curve is the fluorescence spectrum taken during sequential time-window widths of 50 ns. The fluorescence signal exhibits an exponential decrease (decay) with time. Because different species produce different energy spectra and different fluorescence lifetimes, two types of information are obtained from a set of measurements. Consequently, peak deconvolution of nearly identical energy spectra resulting from different solution species is made much easier because of the different lifetimes. One can often further enhance the measurements because the signal is also dependent on the excitation wavelength. The TRLFS spectrum given in Fig. 12 is of a typical mine seepage water. Such waters can be characterized as follows: ionic strength of about 0.1 M; pH from 7.8 to 8.2; main solution components of

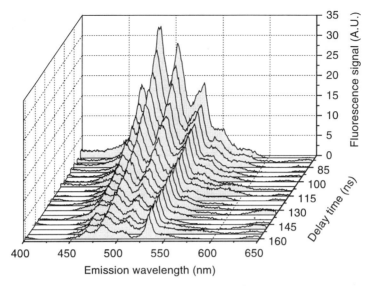

Figure 12. Two-dimensional TRLFS spectrum (266 nm excitation wavelength) of a seepage water containing 1.1×10^{-5} M uranium at pH 8.14 and an ionic strength of 0.1 M.

magnesium, calcium, carbonate and sulfate; high organic content; aerobic conditions; solution in contact with a variety of mineral phases.

Computer modeling with the geochemical speciation code EQ3/6 version 7.2a from Wolery, 1992, using available thermodynamic data from the NEA data base (Grenthe et al., 1992) indicates that, under such conditions, the uranium would be present primarily in the hexavalent oxidation state, i.e. the uranyl ion UO_2^{2+}, and the uranyl speciation would be dominated by a uranyl carbonate complex, as shown in the upper part of Fig. 13. It is known that the uranyl carbonate complexes exhibit only a very weak or no fluorescence (Kato et al., 1994) but, as Fig. 12 clearly indicates, there is a very distinct fluorescence signal. Furthermore, the decay lifetime was about 40 ns, much higher than one would expect for lifetimes for possible organic complexing agents like humic acid (Kumke et al., 1994). Experiments were conducted to attempt to understand the apparent paradox.

An artificial seepage water was prepared which contained only the main components Ca^{2+}, Mg^{2+}, CO_3^{2-} and SO_4^{2-} together with UO_2^{2+}, and the fluorescence spectrum was measured. The resulting spectrum was the same as that of the original uranium-containing seepage water. By varying the concentrations of these five components, a region was defined where the solution species was formed that gave rise to this fluorescence spectrum. It became clear that this species must contain Ca^{2+}, UO_2^{2+} and CO_3^{2-}. A substitution of Mg^{2+} for Ca^{2+} in the solution produced totally different fluorescence spectra. In addition, experiments were performed where both the artificial and real seepage waters containing uranium were placed in contact with both cation and anion resin. Eighty percent of the fluorescent species remained in solution indicating a neutral complex. All samples were passed through 0.45 μm diameter filters before spectroscopic measurements. Some samples were also passed through 1.5 nm diameter filters and they continued to exhibit the same fluorescence spectrum. Therefore, it was concluded that the fluorescing species was not of colloidal dimensions nor did it arise from uranyl sorbed onto calcite or other mineral particles. Moreover, we found that the minerals Liebigite and Zellerite give fluorescence spectra similar to the one obtained from the residue of dried seepage water. Thus, a proposed structure for the newly discovered complex that appeared logical was $[Ca_2(UO_2)(CO_3)_3](aq)$. Fluorescence spectra were taken as a function of calcium, carbonate

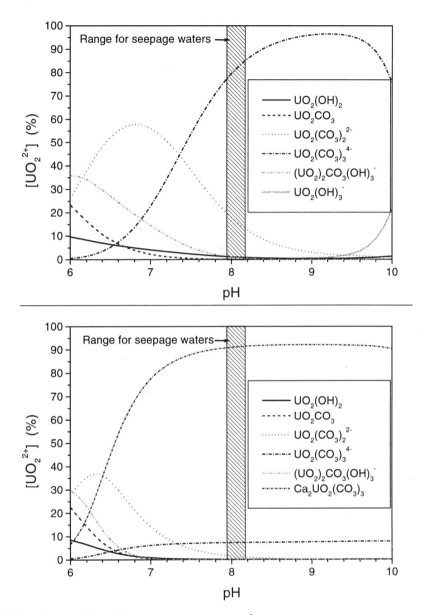

Figure 13. Uranium speciation of a seepage water with 1.1×10^{-5} M uranium at pH 8.14 and an ionic strength of 0.1 M. The calculation was performed with EQ3/6 using the NEA data base without (*lower*) and with (*upper*) the newly found complex.

and uranium concentrations. From these results, the stoichiometry was confirmed, the major emission wavelengths were found to occur at 464, 484, 503, 524 and 555 nm, the fluorescence lifetime was measured as 64 ± 17 ns, and the complex formation constant, log ß, was determined to be 26.8 ± 0.7 at 0.1 M ionic strength. For more details, see Bernhard et al., 1996.

This new species and its formation constant was added to our NEA-based data set and the uranium speciation in our seepage waters was re-calculated using again the EQ3/6 code. The results are given in the lower part of Fig. 13. The speciation differs significantly from the earlier calculations that ignored the new complex. The new complex is indeed the dominant species over a wide pH range.

Unfortunately, not all radionuclide species emit fluorescence light when excited. Under these conditions, the use of laser-induced photoacoustic spectroscopy (LIPAS) has become the absorption spectroscopy method of choice. In this method, the energy transferred to solution ions by excitation of excited states via the pulsed-laser beam, de-excite through non-radiative processes that generate heat between laser beam bursts. This heat produces a sudden expansion in the volume of aqueous solution excited resulting in a pressure pulse that propagates as an acoustic wave. The pressure wave is detected with a piezoelectric transducer. Thus, LIPAS is not restricted to fluorescent systems. Unlike conventional absorption spectroscopy, where the absorption is measured as the difference in intensity of light passing in and out of the sample cell (for dilute samples a difference between two large numbers), photoacoustic spectroscopy measures the amount of energy absorbed by the solute ions in the path of the laser beam directly and, therefore, the acoustic signal is directly proportional to the amount of absorbed light. One obtains the same spectral information as with conventional UV/Vis absorption spectroscopy, but at a factor of 100 to 1000 lower concentration levels. Data acquisition and

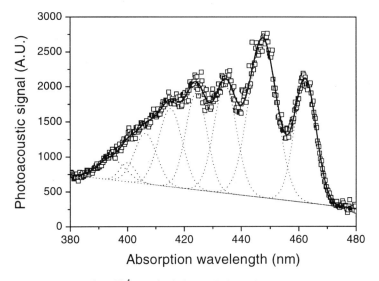

Figure 14. LIPAS spectrum for a 10^{-4} uranyl solution at pH 8.5 with a carbonate content of 2.7×10^{-3} M.

processing for the identification of solution species and the measurement of complex stability constants proceed in the same way as in conventional absorption spectroscopy.

For LIPAS measurements, the laser beam intensity is attenuated before entering the sample cuvette and measured with a power meter after passing through the sample. This signal is later used to normalize data to compensate for variations in the beam intensity from pulse to pulse. The acoustic wave is detected with the ceramic piezoelectric transducer. After preamplification, which includes an impedance change and frequency filtering, the electrical signal is measured in a boxcar averager. The acoustic wave signal looks much like a sine wave that decreases in amplitude with time, however, the peaks tend to be distorted and shift in time depending on the cell geometry. Fortunately, the first positive maximum in the wave form is reasonably stable and a window is set to detect the amplitude of this portion of the wave form only. Because the photoacoustic wave travels at the speed of sound in the aqueous phase and is detected by the transducer between 5 to 30 µs after the laser pulse, depending on the distance from the solution volume excited by the laser beam to the transducer, a delay generator is used to adjust for the time difference. In comparison to conventional dye lasers, the above described set-up with an OPO allows for a much broader wavelength range, covering, e.g., the whole absorption spectrum of the uranyl ion.

The LIPAS spectrum for the $UO_2(CO_3)_3^{4-}$ species is easily detectable at an uranyl concentration of 10^{-4} M (pH 8.5 at 2.7×10^{-3} M HCO_3^-), as shown in Fig. 14. The detection limit for $UO_2(CO_3)_3^{4-}$ was determined to be 5×10^{-6} M. The same spectrum as above can be found in

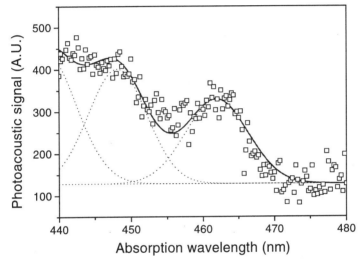

Figure 15. LIPAS spectrum for an original uranium mine tailing water (2.5×10^{-5} M UO^{2+}, pH 8.46).

original tailing waters, see Fig. 15, with a uranyl concentration as low as 2.5×10^{-5} M. The wavelength range below 440 nm is omitted because contributions from organics will hide the uranyl carbonate spectrum. This gave evidence that in such highly carbonized waters at higher pH the species $UO_2(CO_3)_3^{4-}$ is the dominant uranyl carbonate species, with no hydrolysis species being present. Conventional UV/Vis techniques will fail in such cases, because the low uranyl concentrations would require addition of complexing agents to yield higher molar absorption coeffcients, but this inevtably destroys the original speciation. Also TRLFS would encounter problems due to the non-existing fluorescence of the higher uranyl carbonates.

CONCLUSIONS

To reveal environmental speciation patterns, it is essential to combine different analytical methods. They should complement each other and deliver both structural information and quantitative distributions. Recent work has shown that a combination of sequential extraction techniques, synchrotron based X-Ray spectroscopies, X-ray photoelectron spectroscopy and laser-induced spectroscopies is a good approach for this goal. Each of them alone are already powerful speciation tools, in use for several years now (or even decades in the case of sequential extraction), but their development is still in progress. Further improvements are to be expected concerning accuracy, detection limits or elemental range. Some of the directions for enhancement planned by the Forschungszentrum Rossendorf are: (1) the establishment of a synchrotron beamline dedicated to radioactive work, (2) a combination of LIPAS with "One point" excitation spectroscopy, and (3) the step to femto-seconds TRLFS using a pulse laser with a CCD camera (2D) with a gate width < 100 ps).

REFERENCES

Ahrland, S., 1951, *Acta Chem. Scand.* 5:199.

Allen, P.G., Bucher, J.J., Clark, D.L., Edelstein, N.M., Ekberg, S.A., Gohdes, J.W., Hudson, E.A., Kaltsoyannis, N., Lukens, W.W., Neu, M.P., Palmer, P.D., Reich, T., Shuh, D.K., Tait, C.D., and Zwick, B.D., 1995, Multinuclear NMR, Raman, EXAFS, and x-ray diffraction studies of uranyl carbonate complexes in near-neutral aqueous solution. X-ray structure of $[C(NH_2)_3]_6(UO_2)_3(CO_3)_6 \cdot 6.5H_2O$, *Inorg. Chem.* 34:4797.

Allen, P.G., Shuh, D.K., Bucher, J.J., Edelstein, N.M., Reich, T., Denecke, M.A., and Nitsche, H., 1996a, EXAFS determinations of uranium structures: The urayl ion complexed with tartaric, citirc, and malic acids, *Inorg. Chem.* 35:784.

Allen, P.G., Veirs, D.K., Conradson, S.D., Smith, C.A., and Marsh, S.F., 1996b, Characterization of aqueous plutonium(IV) nitrate complexes by extended x-ray absorption fine structure spectroscopy, *Inorg. Chem.* 35:2841.

Allen, P.G., Siemering, G.S., Shuh, D.K., Bucher, J.J., Edelstein, N.M., Langton, C.A., Clark, S.B., Reich, T., and Denecke, M.A., 1997, Technetium speciation in cement waste forms determined by x-ray absorption fine structure spectroscopy, *Radiochim. Acta* 76:77.

Archontaki, H.A., and Crouch, S.R., 1988, Evaluation of an isolated droplet sample introduction system for laser-induced breakdown spectroscopy, *Applied Spectroscopy* 42:741.

Baraniak, L., Thieme, M., Bernhard, G., Nindel, K., and Schreyer, J., 1995, Sorption behaviour of radium on sandy and clayey sediments of the upper saxon elbe river valley, Annual Report Research Center Rossendorf, Inst. Radiochem.

Beitz, J.V., Bowers, D.L., Doxtader, M.M., Maroni, V.A., and Reed, D.T., 1988, Detection and speciation of transuranium elements in synthetic groundwater via pulsed-laser excitation, *Radiochim.Acta* 44/45:87.

Beneš, P., Borovec, Z., and Strejc, P., 1984, Interaction of radium with freshwater sediments and their mineral components, II. Kaolinite and montmorillonite, *J. Radioanal. Nucl. Chem.* 82:275.

Beneš, P., Sedlácek, J., Šebesta, F., Šandrik, R., and John, J., 1981, Method of selective dissolution for characterization of particulate forms of radium and barium in natural and waste waters, *Water Res.* 15:1299.

Bernhard, G., Geipel, G., Brendler, V., and Nitsche, H., 1996, Speciation of uranium in seepage waters from a mine tailing pile studied by time-resolved laser-induced fluorescence spectroscopy (TRLFS), *Radiochim. Acta*, 74:87.

Bertsch, P.M., Hunter, D.B., Nuessle, P.R., and Clark, S.B., 1997, Molecular characterization of contaminants in soils by spatially resolved XRF & XANES spectroscopy, *J. Phys. IV France* 7:817.

Berthoud, T., Mauchien, P., Omenetto, N., and Rossi, G., 1983, Determination of low levels of uranium(VI) in water solutions by means of the laser-induced thermal lensing effect, *Anal.Chim.Acta* 153:265.

BfS, 1992, Radiologische Erfassung, Untersuchung und Bewertung bergbaulicher Altlasten, BfS Schriften 8/92 (ISSN 0937-449), Bundesamt für Strahlenschutz, Salzgitter/Germany

BMWi, 1995, WISMUT - Progress in Decomissioning and Rehabilitation, Doc. No. 370 (ISSN 0342-9288), German Federal Ministry of Economics, Bonn/Germany

Boring, M., Wood, J.H., and Miscowitz, J.M., 1975, Self-consistent field calculation of the electronic structure of the uranyl ion (UO_2^{2+}), *J. Chem. Phys.* 63:638.

Brendler, V., Geipel, G., Bernhard, G., and Nitsche, H., 1996, Complexation in the system $UO_2^{2+}/PO_4^{3-}/OH^-$(aq): Investigations at very low ionic strengths, *Radiochim. Acta*, 74:75.

Buchanan, B.B., Bucher, J.J., Carlson, D.E., Edelstein, N.M., Hudson, E.A., Kaltsoyannis, N., Leighton, T., Lukens, W., Shuh, D.K., Nitsche, H., Reich, T., Roberts, K., Torretto, P., Woicik, J., Yang, W.S., Yee, A., and Yee, B.C., 1995, A XANES and EXAFS investigation of the speciation of selenite following bacterial metabolization, *Inorg. Chem.* 34:1617.

Bucher, J.J., Allen, P.G., Edelstein, N.M., Shuh, D.K., Madden, N.W., Cork, C., Luke, P., Pehl, D., and Malone, D., 1996, A multichannel monolithic Ge detector system for fluorescence x-ray absorption spectroscopy, *Rev. Sci. Instrum.* 67:1.

Chalupnik S., and Lebecka, J.M., 1992, Determination of ^{226}Ra, ^{228}Ra and ^{224}Ra in water and aqueous solutions by liquid scintillation counting, Proc. Liquid Scintillation Spectrometry 1992, Radiocarbon 1992, 397.

Choppin, G.R., 1988, Humics and radionuclide migration, *Radiochim. Acta* 44/45:23.

Choppin, G.R., and Allard, B. (1985). Complexes of actinides with naturally occurring organic compounds. Handbook on the physics and chemistry of the actinides. A. J. Freeman and C. Keller. Amsterdam, Elsevier Science Publisher.

Clark, D.L., Conradson, S.D., Ekkberg, S.A., Hess, N.J., Neu, M.P., Palmer, P.D., Runde, W., and Tait, C.D., 1996a, EXAFS studies of pentavalent neptunium carbonato complexes. Structural elucidation of the principal constituents of neptunium in groundwater environments, *J. Am. Chem. Soc.* 118:2089.

Clark, S.B., Johnson, W.H., Malek, M.A., Serkiz, S.M., and Hinton, T.G., 1996b, A comparison of sequential extraction techniques to estimate geochemical controls on the mobility of fission product, actinide, and heavy metal contaminants in soil. *Radiochim. Acta* 74:173.

Combes, J.M., Chisholm-Brause, C.J., Brown, G.E., Parks, G.A., Conradson, S.D., Eller, P.G., Triay, I.R., Hobart, D.E., and Meijer, A., 1992, EXAFS spectroscopic study of neptunium(V) sorption at the α-FeOOH/water interface, *Environ. Sci. Technol.* 26:376.

Courteix, D., Chayrouse, J., Heintz, L., and Baptist, R., 1981, XPS study of plutonium oxides, *Sol. State Com.* 39:209.

Cremers, D.A., 1987, The analysis of metals at a distance using laser-induced breakdown spectroscopy, *Applied Spectroscopy* 41:572.

Cullen, W.R., and Reimer, K.J., 1989, Arsenic speciation in the environment, *Chem. Rev.* 89:713.

Czerwinski, K.R., Buckau, G., Scherbaum, F., and Kim, J.I., 1994, Complexation of the uranyl ion with aquatic humic acid, *Radiochim.Acta* 65:111.

Denecke, M.A., Friedrich, H., Reich, T., Bernhard, G., Knieß, T., Rettig, D., Zorn, T., and Nitsche, H., 1997, Application of x-ray absorption spectroscopy for in-situ determination of the arsenic oxidation state in solids and solutions, *submitted to Fresenius J. Anal. Chem.*

Denecke, M.A., Pompe, S., Reich, T., Moll, H., Bubner, M., Heise, K.H., Nicolai, R., and Nitsche, H., 1997b, Measurement of the structural parameters for the interaction of uranium(VI) with natural and synthetic humic acids using EXAFS, *Radiochim. Acta (in press)*.

Eisenbud, M., 1987. Environmental Radioactivity from Natural, Industrial, and Military Sources, Academic Press, N.Y.

Eliet, V., Bidoglio, G., Omenetto, N., Parma, L., and Grenthe, I., 1995, Characterisation of hydroxide complexes of uranium(VI) by time-resolved fluorescence spectroscopy, *J.Chem.Soc., Faraday Trans.* 91:2275.

Eller, P.G., Jarvinen, G.D., Purson, J.D., Penneman, R.A., Ryan, R.R., Lytle, F.W., and Greegor, R.B., 1985, Actinide valences in borosilicate glass, *Radiochim. Acta* 39:17.

Ellis, D.E., Rosen, A., and Walch, P.F., 1975, Application of the Dirac-Slater model to molecules, *Int. J. Quant. Chem. Symp.* 9:351.

Enokida, Y., Shiga, M., and Suzuki, A., 1992, Determination of Uranium(VI) over wide concentration ranges in aqueous nitric acid and 30% tri-n-butyl phosphate by laser-induced thermal lensing spectroscopy, *Radiochim.Acta* 57:101.

Förstner, U., 1983, Bindungsformen von Schwermetallen in Sedimenten und Schlämmen: Sorption/Mobilisierung, chemische Extraktion und Bioverfügbarkeit, *Fres. Z. Anal. Chem.* 316:604.

Fujiyoshi, R., Okamoto, T., and Katayama, M., 1994, Behavior of radionuclides in the environment. 2. Application of sequential extraction to Zn(II) sorption studies, *Appl. Rad. Isot.* 45:165.

Geipel, G., Bernhard, G., Brendler, V., and Nitsche, H., 1996, Sorption of uranium(VI) on rock material of a mine tailing pile: solution speciation by fluorescence spectroscopy, *Radiochim. Acta*, 74:235.

Geipel, G., Brachmann, A., Brendler, V., Bernhard, G., and Nitsche, H., 1997, Uranium(VI) sulfate complexation studied by time-resolved laser-induced fluorescence spectroscopy (TRLFS), *Radiochim. Acta*, 75:199.

George, G.N., and Pickering, I.J. (1995). A suite of computer programs for analysis of x-ray absorption spectra, Stanford Synchrotron Radiation Laboratory, Stanford, California 94309, USA.

Grenthe, I., Bidoglio, G., and Omenetto, N., 1989, Use of thermal lensing spectrophotometry (TLS) for the study of mononuclear hydrolysis of uranium(IV), *Inorg.Chem.* 28:71.

Grenthe, I., Fuger, J., Lemire, R.J., Muller, A.B., Nguyen-Trung, C., and Wanner, H., 1991, *Chemical thermodynamics of uranium*, Elsevier Science, Amsterdam.

Gutzman, D.W., and Langford, C.H., 1993, Application of thermal lens spectrometry to kinetic speciation of metal ions in natural water models with colloidal ligands, *Anal.Chim.Acta* 283:773.

Hodgkinson, D.P., Robinson, P.C., Tasker, P.W., Lever, D.A., 1988. A review of modelling requirements for radiological assessment, *Radiochim. Acta,* 44/45:317.

Howatson, J., Grev, D.M., and Morosin, B., 1975, Crystal and molecular structure of uranyl acetate dihydrate, *J. Inorg. Nucl. Chem.* 37:1933.

Hudson, E.A., Allen, P.G., Terminello, L.J., Denecke, M.A., and Reich, T., 1996, Polarized x-ray absorption spectroscopy of the uranyl ion: Comparison of experiment and theory, *Phys. Rev. B* 54:156.

Hudson, E.A., Rehr, J.J., and Bucher, J.J., 1995, Multiple-scattering calculations of the uranium L_3-edge x-ray-absorption near-edge structure, *Phys. Rev. B* 52:13815.

Ito, Y., Ueki, O., and Nakamura, S., 1995, Determination of colloidal iron in water by laser-induced breakdown spectroscopy, *Anal.Chim.Acta* 303:401.

Jasmund, K., and Lagaly, G., 1993, *Tonminerale und Tone,* Darmstadt: Steinkopff.

Kato, Y., Meinrath, G., Kimura, T., and Yoshida, Z., 1994, A study of U(VI) hydrolysis and carbonate complexation by time-resolved laser-induced fluorescence spectroscopy (TRLFS), *Radiochim.Acta* 64:107.

Kim, J.I., Stumpe, R., and Klenze, R., 1990, Laser-induced photoacoustic spectroscopy for the speciation of transuranic elements in natural aquatic systems, *Topics Curr.Chem.* 157:129.

Kimura, T., Serrano, J., Nakayama, S., Takahashi, K., and Takeishi, H., 1992, Speciation of uranium in aqueous solutions and in precipitates by photoacoustic spectroscopy, *Radiochim.Acta* 58/59:173.

Klenze, R., and Kim, J.I., 1989, Speciation of transuranic ions in groundwater by laser-induced photoacoustic spectroscopy (LPAS), *Mat.Res.Soc.Symp.Proc.* 127:985.

Klenze, R., Kim, J.I., and Wimmer, H., 1991, Speciation of aquatic actinide ions by pulsed laser spectroscopy, *Radiochim.Acta* 52/53:97.

Koningsberger, D.C., and Prins, R., 1988, *X-ray absorption, Principles, application, techniques of EXAFS, SEXAFS and XANES,* John Wiley & Sons, New York.

Kumke, M.U., Löhmannröben, H.-G., and Roch,Th., 1994, Fluorescence quenching of polycyclic aromatic compounds by humic acid, *Analyst* 119:997.

Leung, V.W.H., Darvell, B.W., and Chan, P.C., 1988, A rapid algorithm for solution of the equations of multiple equilibrium systems - RAMESES, *Talanta* 35:713.

Manceau, A., Boisset, M.C., Sarret, G., Hazemann, J.L., Mench, M., Cambier, P., and Prost, R., 1996, Direct determination of lead speciation in contaminated soils by EXAFS spectroscopy, *Environ. Sci. Technol.* 30:1540.

Markovits, G., Klotz, P., and Newman, L., 1972, Formation constants for the mixed-metal complexes between indium(III) and uranium(VI) with malic, citric, and tartaric acids, *Inorg. Chem.* 11:2405.

Margane, J., Boenigk, W.1993: Reinigungsleistung thermischer Dekontaminationsverfahren für Schwermetalle, *Entsorgungspraxis* 30.

Miller, W.P., Martens, D.C., Zelazny, L.W., 1986, Effect of sequence in extraction of metals from soil, *J. Soil. Sci. Soc. Am.* 50:598

Morris, D.E., Allen, P.G., Berg, J.M., Chisholm-Brause, C.J., Conradson, S.D., Donnohoe, R.J., Hess, N.J., Musgrave, J.A., and Tait, C.D., 1996, Speciation of uranium in Fernald soils by molecular spectroscopic methods: Characterization of untreated soils, *Environ. Sci. Technol.* 30:2322.

Moulin, C., Decambox, P., Moulin, V., and Decaillon, J.G., 1995, Uranium speciation in solution by time-resolved laser-induced fluorescence, *Anal.Chem.* 67:348.

Moulin, C., Delorme, N., Berthoud, T., and Mauchien, P., 1988, Double beam thermal lens spectroscopy for actinides detection and speciation, *Radiochim.Acta* 44/45:103.

Mustre de Leon, J., Rehr, J.J., Zabinsky, S.I., and Albers, R.C., 1991, Ab initio curved-wave x-ray-absorption fine structure, *Phys. Rev. B* 44:4146.

Nevedov, V.I., Teterin, Y.A., Reich, T., and Nitsche, H., 1996, Determination of interatomic distances in uranyl compounds on the basis of $U6p_{3/2}$-level splitting, *Doklady Akademii Nauk* 348:643.

Nitsche, H., 1991. Solubility studies of transuranium elements for nuclear waste disposal: principles and overview, *Radiochim. Acta,* 52/53:3.

Okajima, S., Reed, D.T., Beitz, J.V., Sabau, C.A., and Bowers, D.L., 1991, Speciation of Pu(VI) in near-neutral solutions via laser photoacoustic spectroscopy, *Radiochim.Acta* 52/53:111.

Petiau, J., Calas, G., Petitmaire, D., Bianconi, A., Befatto, M., and Marcelli, A., 1986, Delocalized versus localized unoccupied 5f states and the uranium site structure in uranium oxides and glasses probed by x-ray-absorption near-edge structure, *Phys. Rev. B* 34:7350.

Petit-Maire, D., Petiau, J., Calas, G., and Jacquet-Francillon, N., 1989, Insertion of neptunium in borosilicate glasses, *Physica B* 158:56.

Pickering, I.J., Brown, G.E., and Tokunaga, T.K., 1995, Quantitative speciation of selenium in soils using x-ray absorption spectroscopy, *Environ. Sci. Technol.* 29:2456.

Pireaux, J.J., Martensson, H., Didriksson, R., Siegbahn, K., Riga, J., and Verbist, J., 1977 a, High-resolution ESCA study of uranium fluorides: UF_4 and K_2UF_6, *Chem. Phys. Lett.* 46:213.

Pireaux, J.J., Riga, J., Thibaut, E., Tenret-No'l, C., Cuadono, R., and Verbist, J., 1977 b, Shake-up satellites in the x-ray photoelectron spectra of uranium oxides and fluorides. A band structure scheme for uranium dioxide, UO_2, *Chem. Phys.* 22:113.

Plater, A.J., Ivanovich, M., and Dugdale, R.E., 1992, Uranium series disequilibrium in river sediments and waters: the significance of anomalous activity ratios, *Appl. Geochem.* 7:101.

Pompe, S., Bubner, M., Denecke, M.A., Reich, T., Brachmann, A., Geipel, G., Heise, K.H., and Nitsche, H., 1996, A comparison of natural acids with synthetic humic acid model substances: Characterization and interaction with uranium(VI), *Radiochim. Acta* 74:135.

Rajan, K.S., and Martell, A.E., 1964, Equilibrium studies of uranyl complexes. II., *J. Inorg. Nucl. Chem.* 26:1927.

Rajan, K.S., and Martell, A.E., 1965, Equilibrium studies of uranyl complexes. III. Interaction of uranyl ion with citric acid, *Inorg. Chem.* 4:462.

Rajan, K.S., and Martell, A.E., 1967, Equilibrium studies of uranyl complexes. IV., *J. Inorg. Nucl. Chem.* 29:523.

Rehr, J.J., Albers, R.C., and Zabinsky, S.I., 1992, High-order multiple scattering calculations of x-ray absorption fine structure, *Phys. Rev. Lett.* 69:3397.

Rüde, T.R., and Puchelt, H., 1994, Arsenbindungsformen in Böden, 72. Jahrestagung der DGM, Freiberg, 17.-25.09.94.

Ryzhkov, M.V., Gubanov, V.A., Teterin, Y.A., and Baev, A.S., 1991, Electronic structure and x-ray photoelectron spectra of uranyl compounds, *Soviet Radiochemistry* 33:19.

Ryzhkov, M.V., Gubanov, V.A., Teterin, Y.A., and Baev, A.S., 1992, Actinyl compounds of Np and Pu - x-ray photoelectron spectra and quantum-chemical calculations, *Soviet Radiochemistry* 34:69.

Sarret, G., Manceau, A., Hazemann, J.L., Gomez, A., and Mench, M., 1997, EXAFS study of the nature of zinc complexation sites in humic substances as a function of Zn concentration, *J. Phys. IV France* 7:799.

Shan, X.-Q., and Chen, B., 1993, Evaluation of sequential extraction for speciation of trace metals in model soil containing natural minerals and humic acid, *Anal. Chem.* 65:802.

Siegbahn, K., Nordling, C., Fahlman, A., Nordling, R., Hamrin, K., Hedman, Y., Yohansson, G., Bergmark, T., Karlsson, S.E., Lindgren, I., and Lindberg, B., 1967, Atomic, molecular and solid state structure studied by means of electron spectroscopy, *Nova Acta Regiae Societatis Scientiarum Upsaliensis Ser. IV* 20.

Siegbahn, K., Nordling, C., Johansson, G., Hedman, J., Heden, P.F., Hamrin, K., Gelius, U., Bergmark, T., Werme, L.O., Mann, R., and Baer, Y., 1969, *ESCA applied to free molecules*, North-Holland, Amsterdam.

Silva, R.J., Nitsche, H., 1995. Actinide Environmental Chemistry, *Radiochim. Acta,* 70/71:377.

SSRL, 1995, Molecular environmental science: Speciation, reactivity, and mobility of environmental contaminants; Report of DOE Molecular Environmental Science Workshop, July 5-8, 1995, Airlie Center, VA, USA, Stanford Synchrotron Radiation Laboratory, Stanford, California 94309, USA.

Stewart, B.D., McKleven, J.W., and Glinski, R.L., 1988, Determination of uranium and radium concentration in the waters of the grand canyon by alpha spectrometry, *J. Radioanal. Nucl. Chem.* 123:121.

Tessier, A., Campbell, P.G.C., 1988, Comments on the testing of the accuracy of an extraction procedure for determining the partioning of trace metals in sediments, *Anal. Chem.* 60:1475.

Tessier, A., Campbell, P.G.C., and Bisson, M., 1979, Sequential extraction procedure for the speciation of particulate trace metals, *Anal. Chem.* 51:844.

Teterin, Y.A. and Baev, A.S. , 1986, *X-ray photoelectron spectroscopy of light-actinide compounds*, ZNIIatominform, Moscow.

Teterin, Y.A., Baev, A.S., and Bogatov, S.A., 1994, X-ray photoelectron study of samples containing reactor fuel from "lava" and products growing on it which formed at Chernobyl NPP due to the accident, *J. Electron Spectrosc. Relat. Phenom.* 68:685.

Teterin, Y.A., Baev, A.S., Gagarin, S.G., and Klimov, V.D., 1985, Structure of photoelectron spectra of thorium compounds, *Radiokhimiya* 27:3.

Teterin, Y.A., Baev, A.S., Ivanvov, K.E., Mashirov, L.G., and Suglobov, D.N., 1996a, *Radiokhimiya* 38:395.

Teterin, Y.A., Baev, A.S., Mashirov, L.G., and Suglobov, D.N., 1984a, Structure of XPS spectra of $Cs_3NpO_2Cl_4$ and $Cs_2NpO_2Cl_4$ single crystals, *Dokladii Akademii Nauk SSSR* 276:154.

Teterin, Y.A., Baev, A.S., Mashirov, L.G., and Suglobov, D.N., 1984b, Dependence of the structure of XPS spectra of uranium and plutonium in $Cs_2AnO_2Cl_4$ (An=Pu, U) single crystals from their oxidation state, structure of these compounds and the character of the chemical bonds, *Dokladii Akademii Nauk SSSR* 277:131.

Teterin, Y.A., Baev, A.S., Vedrinsky, R.V., Gubinsky, A.L., Zelenkov, A.G., Kovtun, A.P., Kulakov, V.M., and Sachenko, V.P., 1981a, Structure of XPS spectra of low-energy electrons of oxides UO_2 and γ-UO_3, *Dokladi Akademii Nauk SSSR* 256:381.

Teterin, Y.A., and Gagarin, S.G., 1996, Inner valence molecular orbitals and the structure of x-ray photoelectron spectra, *Russian Chem. Rev.* 65:1.

Teterin, Y.A., Kulakov, V.M., Baev, A.S., Nevzorov, N.B., Melnikov, I.V., Streltsov, V.A., Mashirov, L.G., Suglobov, D.N., and Zelenkov, A.G., 1981b, A study of synthetic and natural uranium oxides by x-ray photoelectron spectroscopy, *Phys. Chem. Minerals* 7:151.

Teterin, Y.A., Kulakov, V.M., Baev, A.S., Zelenkov, A.G., Nevzorov, N.B., Melnikov, I.V., Streltsov, V.A., Mashirov, L.G., and Suglobov, D.N., 1980, A study of uranium oxidation state in uranium compounds by x-ray photoelectron spectroscopy, *Dokladii Akademii Nauk SSSR* 255:434.

Teterin, Y.A., Nefedov, V.I., Ivanov, K.E., Baev, A.S., Geipel, G., Reich, T., and Nitsche, H., 1995, X-ray photoelectron spectroscopy investigation of $UO_2(ClO_4)_2$-calcite interaction, *Doklady Akademii Nauk* 344:206.

Teterin, Y.A., Nefedov, V.I., Ivanov, K.E., Baev, A.S., Geipel, G., Reich, T., and Nitsche, H., 1996b, X-ray photoelectron spectroscopy investigation of $UO_2(ClO_4)_2$ with calcite and diabase minerals, *Zhuranl Neorganisheskoyi Khimii* 41:1884.

Thayer, J.S., 1995, *Environmental chemistry of the heavy elements: Hydrido and organo compounds* , VCH Verlag, Weinheim.

Thieme, M., Baraniak, L., Bernhard. G., Nindel, K., Schreyer, J., and Nitsche, H., 1995, Sequential extraction on radioactivity-containing sedimentary rocks of the Königstein region, Saxony, Annual Report, Research Center Rossendorf, Inst. Radiochem.

USDOE, 1989. Environmental Restoration and Waste Management, *DOE/S-0070,* U. S. Department of Energy, Washington, D.C.

USDOE, 1990. Basic Research for Environmental Restoration, *DOE/ER-0482T,* Office of Energy Research, U. S. Department of Energy, Washington, D.C.

Veal, B.W., Lam, D.J., Carnall, W.T., and Hoekstra, H.R., 1975, X-ray photoelectron study of hexavalent uranium compounds, *Phys. Rev. B* 12:5651.

Veal, B.W., Lam, D.J., Diamond, H., and Hoekstra, H.R., 1977a, X-ray photoelectron spectroscopy study of oxides of the transuranium elements Np, Pu, Am, Cm, Bk and Cf, *Phys. Rev. B* 15:2929.

Veal, B.W., Lam, D.J., and Diamond, H., 1977b, X-ray photoelectron spectroscopy of of 5f electrons in dioxides of neptunium and plutonium, *Physica* B86-88:1193.

Verbist, J., Riga, J., Pireaux, J.J., and Caudano, R., 1974, X-ray photoelectron spectra of uranium and uranium oxides. Correlation with the half-life of $^{235}U^m$, *J. Electron Spectrosc. Relat. Phenom.* 5:193.

Wachter, J.R., and Cremers, D.A., 1987, Determination of uranium in solution using laser-induced breakdown spectroscopy, *Applied Spectroscopy* 41:1042.

Waite, T.D., Davis, J.A., Payne, T.E., Waychunas, G.A., and Xu, N., 1994, Uranium(VI) adsorption to ferrihydrite: Application of a surface complexation model, *Geochim. Cosmochim. Acta* 58:5465.

Wang, Z.L., Cheng, C.K., Liu, X.N., Tang, F.X., and Pan, X.X., 1984, Determination of ultratrace quantities of uranium by laser-induced fluorescence spectrometry, *Anal.Chim.Acta* 160:295.

Waychunas, G.A., and Brown, G.E., 1990, Polarized x-ray absorption spectroscopy of metal ions in minerals, *Phys. Chem. Minerals* 17:420.

Westre, T.E., Di Cicco, A., Filipponi, A., Natoli, C.R., Hedman, B., Solomon, E.I., and Hodgson, K.O., 1995, GNXAS, a multipe-scattering approach to EXAFS analysis: Methodology and applications to iron complexes, *J. Am. Chem. Soc.* 117:1566.

WISMUT, 1994, Informationen zur Wismut, Wismut GmbH, Chemnitz

Wolery, T.J., 1992, EQ3/6, A software package for the geochemical modeling of aqueous systems. UCRL-MA-110662 Part I, Lawrence Livermore National Laboratory.

Zabinsky, S.I., Rehr, J.J., Ankudinov, A., Albers, R.C., and Eller, M.J., 1995, Multiple-scattering calculations of x-ray absorption spectra, *Phys. Rev. B* 52:2995.

MODELING THE EFFECT OF IONIC INTERACTIONS ON RADIONUCLIDES IN NATURAL WATERS

Frank J. Millero

Rosenstiel School of Marine and Atmospheric Science
University of Miami
Miami, Florida 33149

ABSTRACT

Dissolved nuclear wastes (Am, Cm, Pb, Np, Pu, Ra, Th, and U) can be affected by the composition of natural waters. Natural waters can have different values of pH, composition, ionic strength, temperature, and pressure. The solubility of these elements in natural waters can be determined experimentally or by modeling. Since direct measurements are not presently available for many metals of interest, it is important to develop methods that can be used to estimate the solubilities in natural waters of known composition. In this paper, one approach that can be used to estimate the solubilities using Pitzer's equations is discussed. For some metals (i.e., Pb and U) the effect of composition can be directly calculated using existing data. While, for others (i.e., Am and Cm) limited data is available and one must use model compounds (rare earths) as analogs. The solubility of Fe(III) in seawater as a function of pH is used to demonstrate how the model can be used. The model has also been used to estimate the speciation and solubility of a number of radionuclides in WIPP brines.

INTRODUCTION

The use of nuclear power promises to be the answer to our long term energy needs, if the problems of the disposal of nuclear wastes are to be solved. Since some radionuclides have long half-lives and high toxicity, their fate in the environment must be understood. This is essential if their impacts are to be properly assessed and effectively controlled. A number of workers have reviewed the chemical behavior of the actinide elements (Th, U, Np, Pu, Am, Cm) in the environment (Katz et al., 1986; Olofsson et al., 1984; Choppin and Stout, 1989; Choppin and Allard, 1985; Allard and Rydberg, 1983; Allard et al., 1984; Allard, 1986; Hobart, 1990; Grenthe et al., 1992). The potentially largest source of

Actinide Speciation in High Ionic Strength Media, edited by Reed *et al.*
Kluwer Academic / Plenum Publishers, New York, 1999

radionuclides comes from nuclear power plants. Release from explosions in Russia at Chernobyl in 1986 and the continuous contamination of the Irish Sea from the U.K. Windscale are examples of some sources. There are also a number of DOE sites with contaminated soils and groundwaters. The radionuclides of greatest concern are those with long life times (Pb, Ra, Th, U, Np, Pu, Am, Cm). It is these elements and their behavior in natural waters that are the concern of this paper.

Since the actinides can have a number of oxidation states, their behavior in natural waters is quite complicated. Plutonium for example can exist in four oxidation states (III, IV, V, and VI) under the same environmental conditions. The speciation of these different oxidation states are quite different in natural waters. This is demonstrated for the speciation of Pu in seawater in Figure 1 (Choppin, 1989; Choppin and Morse, 1989).

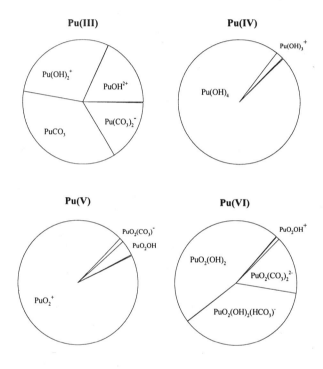

Figure 1. Plutonium Speciation in Seawater.

There presently is a lack of sufficient data on the chemistry of the actinides in natural waters at near neutral levels of pH (6 to 8). Most natural waters contain the same major cations (Na^+, Mg^{2+}, Ca^{2+}, K^+, Sr^{2+}) and anions (Cl^-, SO_4^{2-}, HCO_3^-, Br^-, CO_3^{2-}, $B(OH)_4^-$, F^-) that are in seawater. The ranges of ionic strength and temperature of these waters can vary, respectively, from 0 to 6m and 0 to 40°C. Higher temperatures (300 - 400°C) are found in hydrothermal systems.

The major anions in natural waters can form complexes with most divalent and trivalent metals; while, the major cations (Mg^{2+}, Ca^{2+}) can be competitors for both major and minor anions. The actinides as well as most divalent and trivalent cations can form strong complexes with anions in natural waters (Hobart, 1990). The most common inorganic ligands include OH^-, CO_3^{2-}, PO_4^{3-}, F^-, Cl^-, and SO_4^{2-}. Naturally occurring fulvic and humic acids can also affect the speciation of the actinides in natural waters (Choppin and Allard, 1985). A summary of the literature that is available for the speciation of the actinides with the major inorganic ligands is given in Table 1 (Hobart, 1990).

Table 1. The Sources of Stability Constants for the Formation of Ion Complexes of Some Radionucleotides.

Metal	F	Cl	OH	CO$_3$	Humics
Pb^{2+}		a	a	a	b
Ra^{2+}	c	c	c, d	c	
Pu^{4+}	e	e	f, d	g	h
Th^{4+}	e	e	d		h
U^{4+}	e	e	i, d	i	h
Np^{4+}	e	e	j, d	j	h
Am^{3+}	e	e	k, d	l	h
Cm^{3+}	e	e			h
UO$_2^{2+}$	e	e	d, m-r	m-r	h
PuO$_2^{2+}$	e	e	d, s-u	s-u	h
NpO$_2^{2+}$	e	e	v	v	h

a) Millero and Hawke (1992).
b) Turner et al. (1986); Campodaglio et al. (1990).
c) Langmuir and Riese (1985); Lowson (1985).
d) Baes and Mesmer (1976).
e) Ahrland (1986).
f) Kim et al. (1983).
g) Moskvin and Gel'man (1958).
h) Choppin and Allard (1985).
i) Ciavatta et al. (1983).
j) Moskvin (1971).

k) Bidoglio (1982).
l) Lundqvist (1982).
m) Ciavatta et al. (1979).
n) Scanlan (1977).
o) Maya (1982a,b).
p) Ferri et al. (1981).
q) Ciavatta et al. (1983).
r) O'Cinneide et al. (1975).
s) Gel'man et al. (1962).
t) Woods et al. (1978).
u) Sullivan et al. (1982).
v) Maya (1983).

The thermodynamic and kinetic behavior of dissolved metals in natural waters can be affected by the composition of the waters (Millero, 1990a, b). The composition of the water can affect the activity and speciation of an element due to ionic interactions with the major and minor components of the water (Millero, 1982). Other variables that can influence the activity and solubility of these elements include the temperature and redox conditions. Pressure is of little importance unless the nuclides are disposed in the deep oceans (Millero, 1979, 1983a). One can determine the solubility of a nuclide in a natural water of known composition at a given temperature and pressure experimentally or by modeling the system. Experimentally one could make direct measurements of the solubility in artificial solutions of expected composition as a function of pH. This approach is direct and will provide the best estimates of the solubility under a limited range of conditions. The solubility can also be estimated by using measurements of the solubility in pure water made under fixed redox conditions, pH, and temperature. Although direct stability constant measurements are available (see Table 1) for some of the actinides (Hobart, 1990), most of the measurements have been made in NaClO$_4$ media at various ionic strengths. Most natural waters, however, contain NaCl over a wide ionic strength. It is, thus, important to develop methods that can be used to estimate the stability constants in a wide range of natural waters using a consistent ionic interaction model that can account for changes in ionic strength and composition (Millero, 1992; Millero et al., 1995).

Recently, for example, the solubility of a number of elements that are part of the nuclear waste that will be buried in the WIPP (Waste Isolation Pilot Project) site have been estimated (Hobart et al., 1996). The site is made up of ancient brines that could become concentrated solutions if water gets into the system. The concentration of the nuclides in the resultant brines were needed to determine the probability of the future stability of the WIPP site to house the wastes. The solubility of metals in the brine at a given temperature and pressure can be effected by a number of factors.

1. The composition of the major components of the natural water (Na^+, Mg^{2+}, Cl^-, and SO_4^{2-}, etc.).
2. The composition of the minor components of the natural water (H^+, OH^-, CO_3^{2-}, Organic Ligands, etc.).
3. The pH and oxidation potential (Eh) of the metals and complexes.

The approach we used to solve this problem was to use the Pitzer equations (1973; 1979; 1991) combined with speciation models (Millero, 1992; Millero et al., 1995) to determine the activity of these elements in the WIPP brines. For some elements (e.g., Pb) the effect of composition can be made in a direct manner. For other elements (e.g., Am) one must use some model compounds (e.g., rare earths) as analogs. Computer codes that can be used to determine the speciation and activity of divalent and trivalent elements at 25°C (Millero and Hawke, 1992; Millero, 1992; Millero et al. 1995) were used. When limited data was available the activity of a given ion or ion pair was estimated by using known values for similar compounds. The solubilities in the brine were determined using the activity coefficients and measured solubilities of the least soluble salts in pure water.

In this paper I will outline how this approach can be used to determine the solubility of radionuclides in natural waters. The techniques used to determine the solubility of a metal in natural waters as a function of pH is demonstrated below for the solubility of Fe(III) in seawater.

MODELING THE IONIC INTERACTIONS

In determining the solubility of metals in natural waters the activity coefficients can be determined using the Pitzer model for ionic interactions (Pitzer, 1991). A brief overview of the application of this model for natural waters is discussed in this section. Although a number of computer codes are available to calculate the speciation of elements in natural waters (e.g., Morel and Morgan, 1972; Westall et al., 1976; Knowles and Wakeford, 1978; Nordstrom and Ball, 1984; Papelis et al., 1988), they are generally limited because the data base is confined to low ionic strengths. A quantitative treatment of the interactions of radionuclides in natural waters requires an appropriate, self consistent model describing the variation of activity coefficients with ionic strength and composition. The activity coefficients of ions in natural waters can be estimated by using the ion pairing model (Dickson and Whitfield, 1981; Millero and Schrieber, 1982; Millero and Hawke, 1992) and the specific interaction model (Harvie et al., 1984; Millero, 1982; Clegg and Whitfield, 1991). Millero and Hawke (1992) summarized the recent progress made in using these models to estimate the activity of ionic solutes and the speciation of metals. The most popular method used to account for the ionic interactions in natural waters is the ion pairing model. The application of this model to the major components of natural waters as a function of ionic strength has been made by a number of workers (Dickson and Whitfield, 1981; Millero and Schrieber, 1982). This model has also been used to examine the speciation of trace metals (Turner et al., 1981; Byrne et al., 1988; Millero and Hawke, 1992). These latter studies allow one to estimate reliable activity coefficients for a number of major and minor ions to 1 m. Byrne et al. (1988) extended the speciation calculations from 0 to 50°C for a number of trace metals. Millero and Hawke (1992) have provided equations that can be used to estimate the stability constants for divalent trace metals as a function of ionic strength. These equations are based upon the estimation of the activity coefficients of ions using the mean salt approach (Dickson and Whitfield, 1981; Millero and Schreiber, 1982). The limitations of these speciation calculations are related to the lack of reliable stability constants valid over a wide range of temperatures and ionic strengths. Extension to higher ionic strength, and other temperatures and pressures is complicated by the requirement for experimental data for the large number of ion pairs - 50 in the case of seawater for just the major components. The major problems with the model are the lack of association constant data, lack of conventions to assign values to the thermodynamic properties of the ion pairs and increasing complexity of the calculations as more species are involved in the solution.

The specific interaction model as formulated by Pitzer (1973, 1979, 1991) has been widely accepted and has been applied successfully in a number of areas such as association reactions between aqueous species, the solubility of minerals (Harvie and Weare, 1980) in multicomponent systems, and the solubility of atmospheric gases in natural waters (Clegg and Whitfield, 1991). In contrast to the ion pairing model, the ion interaction model treats strong electrolytes as completely dissociated, and the properties of the solutions are described in terms of interactions between free ions. By using this model, the definition of the conventional single ion activity coefficients results naturally from the derivation of the equations for the mean-ion activity coefficients and no additional assumptions are required. The model is based on a linear summation of the parameters obtained from single electrolyte solutions by fitting experimental osmotic and activity coefficients. A summary of the use of the Pitzer model to examine the interactions of ions in natural waters is given in Table 2 (Millero and Hawke, 1992).

The model was first used by Whitfield (1975a, b) to estimate the activity coefficients of a number of ions in a simple seawater solutions ($NaCl + MgSO_4$). Weare and co-workers (Harvie and Weare, 1980; Harvie et al., 1984; Felmy and Weare, 1986; Møller, 1988; Greenberg and Møller, 1989) and others (Millero, 1982; Pabalan and Pitzer, 1987; He and Morse, 1993; Campbell et al., 1993; Clegg and Whitfield, 1995; Millero and Roy, 1996) have extended the model. The present model (Millero and Roy, 1996) can be used to estimate reliable activity coefficients of the major components (H^+, Na^+, K^+, Mg^{2+}, Ca^{2+}, Sr^{2+}, Cl^-, OH^-, F^-, $B(OH)_4^-$, SO_4^{2-}, HCO_3^- CO_3^{2-}, $B(OH)_3$, and CO_2) of natural waters over a wide range of temperatures (0 - 50°C) to high ionic strengths (≤ 6 m). This model can serve as the backbone for the calculation of trace activity coefficients of metals in natural waters.

The extension of this model to trace metals has been made for solutions of Cl^- and SO_4^{2-} (Whitfield, 1975a, b). As pointed out by Whitfield (1975 a,b) the Pitzer major ion model can serve as a solid thermodynamic base for the determination of the speciation of trace constituents in natural waters over a wide range of ionic strength and composition.

Table 2. Summary of Work Published Using the Specific Interaction Model for Seawater Type Solutions.

Author	Year	Comments
Whitfield	1975	Major and Minor Cations, Cl and SO_4, 25°C
Harvie and Weare	1980	Na, K, Mg, Ca, Cl, SO_4, 25°C
Millero	1982	Major Cations and Acid Anions, 25°C
Harvie et al.	1984	H, Na, K, Mg, Ca, Cl, SO_4, OH, HCO_3, CO_3, CO_2, 25°C
Felmy and Weare	1986	Borate System, Major Seasalts, 25°C
Pabalan and Pitzer	1987	Na, K, Mg, Cl, SO_4, 25-250°C
Møller	1988	Na, Ca, Cl, SO_4, 25-250°C
Greenberg and Møller	1989	Na, K, Ca, Cl, SO_4, 0-250°C
Campbell et al.	1993	H, Na, K, Mg, Ca, Sr, Cl, HSO_4, SO_4, Br, 0-50°C
Clegg and Whitfield	1995	H, Na, K, Mg, Ca, Cl, HSO_4, SO_4 and ammonia in seawater, 0 to 40°C
Millero and Roy	1997	H, Na, K, Mg, Ca, Sr, Cl, SO_4, Br, HCO_3, CO_3, CO_2, $B(OH)_4$, $B(OH)_3$, 0-50°C

We have attempted to use this approach to examine the interactions of the major components of seawater with minor metals (Millero, 1992; Millero et al., 1995) and non-metals (Millero et al, 1989). The single activity coefficients calculated from the specific interaction model are used for the extension of the ion pairing model. The activity coefficients of ion pairs are treated in a similar manner as single ions. The specific interaction model accounts for the ionic interactions such as Na-SO_4, Na-Cl, K-Cl, K-SO_4, Ca-Cl, Ca-SO_4, Mg-Cl, Mg-SO_4 which are difficult to characterize experimentally using the ion-pairing concept and usually are the major components of natural waters. The strong cation-anion interactions of trace metals with the major and minor anions can be accounted for by using the ion pairing model. We have attempted to unite these two models in recent years (Millero, 1982, 1983b, 1984, 1982; Millero and Thurmond, 1983; Millero and Byrne, 1984; Hershey et al., 1986, 1989; Sharma and Millero, 1990).

Although the combination of the two models has been used to determine activity coefficients of the major components of natural waters, the extension of this approach to trace component speciation has not been widely attempted. Recently we (Millero, 1992; Millero et al. 1995) have used the Pitzer model (1991) combined with the ion pairing model to examine the speciation of divalent and trivalent metals in natural waters at 25°C as a function of ionic strength. The difficulty of combining the two models comes from a lack of knowledge of the activity coefficients of ion complexes and the resultant Pitzer parameters. The activity coefficients of the ion complexes (γ_{MX}) can be estimated from measurements of stability constants (K^*_{MX}) in a given ionic media (Millero, 1992; Millero et al., 1995)

$$\gamma_{MX} = (K_{MX}/K^*_{MX})\,\gamma_M\,\gamma_X \tag{1}$$

where K_{MX} is the thermodynamic stability constant and γ_M and γ_X are the activity coefficients of the metal (M) and ligand (X). To determine both K_{MX} and γ_{MX} it is necessary to know the stability constants for at least two concentrations in the media used for the measurements ($NaClO_4$). One needs to know the concentration dependence of γ_{MX} to determine Pitzer parameters for the ion complex (MX). When measurements have not been made as a function of concentration, it is necessary to estimate the activity coefficients of the ion complexes using known values for ion complexes of similar structure and charge (Millero and Schreiber, 1982; Millero and Hawke, 1992; Millero, 1992; Millero et al., 1996). By using the Pitzer model it is possible to account for the differences in the measured stability constants for the formation of complexes in different ionic media. This is shown for the formation of $(UO_2)_2(OH)_2^{2+}$ in various media (Baes and Mesmer, 1976) in Figure 2. Assuming that the experimental data is correct the

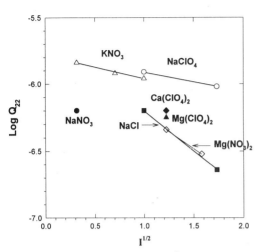

Figure 2. The $(UO_2)_2(OH)_2^{2+}$ hydrolysis constant in different media.

difference shown in this figure can only be accounted for by differences in the activity coefficients of the ions (UO_2^{2+} and OH^-) and ion complex $(UO_2)_2(OH)_2^{2+}$. To account for these differences one requires an ionic interaction model that can consider all the possible interactions that can occur in the solutions of interest.

An overview of the model we use to determine the activity coefficients of metals in natural waters is shown in Figure 3. The initial input required is the composition of the major cations (Na, Mg, Ca, K, Sr) and anions (Cl, SO_4, HCO_3, Br, $B(OH)_4$, F) in the solution and the temperature.

Since the model is designed for the calculation of trace components of the solution, it uses this information to determine the concentration of all the free ligands in the media (OH^-, HCO_3^-, CO_3^{2-}, $B(OH)_4^-$, etc.) at a given pH, TCO_2, phosphate, et al. The program determines the activity coefficients of minor divalent and trivalent metals in the solution including the speciation (fractions of the free ions and the various ion pairs) by a series of iterations. For the activity coefficients in solutions with low concentrations of H^+ (pH = 6 to 10), the model does not require the initial input of the proton concentration. This is also the case for the concentration of minor ligands (phosphate, et al.). The composition of the major components is used to determine the activity of water and all the activity coefficients of all the major and minor ions (e.g., Mg^{2+}) and ionic complexes ($MgCO_3$). The dissociation constants for the acids in the solution (carbonic, boric, water, phosphoric, silicic, hydrogen sulfide, ammonium, etc.) are also determined (Millero and Roy, 1997). We also have programs that can be used to determine other physical properties of the solution such as density, sound speed, heat capacity etc. (Millero, 1979).

After the calculations for the media are finished, the program requires the input of pH and TCO_2 or Total Alkalinity. The concentration of minor components such as phosphate, hydrogen sulfite, humics etc. can also be entered if one wishes to account for the interactions of these ligands with the trace metals. Iterations are made until the speciation is completed. Since the concentration of these ligands are normally quite low, they do not affect the activity coefficients of the major and minor components of the solution. If this is not the case, one must go back and redo the initial estimate of the activity coefficients and pK's of the acids.

The recent calculations on the WIPP brines (Hobart et al., 1996) were made at 25°C. For the major components of most natural waters, the calculation can now be expanded to

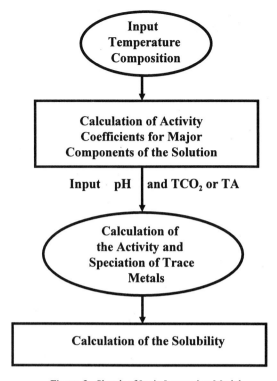

Figure 3. Sketch of Ionic Interaction Model.

temperatures from 0 to 50°C (Millero and Roy, 1997). This recent major component model has been shown to yield reliable dissociation constants for a number of acids in a seawater media. The results for the dissociation of carbonic acid are shown in Figure 4. The model predicts reliable pK's for a number of acids in natural waters from 0 to 50°C and I = 0 to 6 m (Millero and Roy, 1997). To extend the model for trace metals over a temperature range of 0 to 50°C, it is necessary to have Pitzer parameters for divalent and trivalent chlorides, perchlorates, and sulfates (when available) over the same temperature range. The chloride and sulfate data are needed to determined the activity coefficient of metal ions in natural waters whose major components are Cl^- and SO_4^{2-}. The perchlorate data are needed to interpret the stability constants determined in this media (Millero,1992; Millero et al., 1995). The model has been extended to the range of 0 to 50°C for these electrolytes by using heat capacity and enthalpy of dilution data in the literature (Criss and Millero, 1996).

Heat capacity and enthalpy data at 25°C have been shown (Millero, 1979) to provide reasonable estimates of activity coefficients from 0 to 50°C and 0 to 2 m. The Pitzer coefficients derived from the heat capacities ($\{\beta^{(0)J} = \partial^2\beta^{(0)}/\partial T^2 + (2/T)(\partial\beta^{(0)}/\partial T)\}$ etc.) and enthalpies ($\beta^{(0)L} = \partial\beta^{(0)}/\partial T$, etc.) of electrolytes can be used to estimate the effect of temperature on the osmotic and activity coefficients. From the definition of the effect of temperature on the Pitzer coefficients ($\beta^{(0)}$, $\beta^{(1)}$, $\beta^{(2)}$ and C^ϕ) used to fit the enthalpies and heat capacities, the temperature dependencies of the necessary Pitzer coefficients may be obtained. The integration of these equations between T_R (298.15) and T gives (Criss and Millero, 1996)

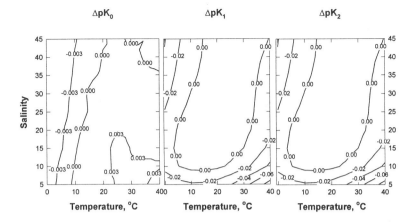

Figure 4. Difference in the measured and calculated dissociation constants for carbonic acids in seawater as a function of temperature and salinity (Millero and Roy, 1997).

$$\beta^{(0)} = \beta^{(0)}_R + a\,(1/T - 1/T_R) + b\,(T^2 - T_R^2) \tag{2}$$

where

$$a = (\beta^{(0)J}/3)\,T_R^3 - T_R^2\beta^{(0)L}_R \tag{3}$$

$$b = \beta^{(0)J}/6 \tag{4}$$

Similar equations can be derived for $\beta^{(1)}$, $\beta^{(2)}$ and C^ϕ. We recently (Criss and Millero, 1996) have determined the heat capacity Pitzer coefficients ($\beta^{(0)J}$, etc.) for a number of monovalent, divalent and trivalent ions. These heat capacity coefficients can be combined with literature values of $\beta^{(0)}_R$ and $\beta^{(1)}_R$ at 25°C to derive equations that can be used to estimate osmotic and activity coefficients of the electrolytes from 0 to 50°C (Criss and Millero, 1996; Millero, 1979).

SOLUBILITY OF Fe(III) IN SEAWATER

The solubility of a metal in a natural water can be estimated from direct measurements in a solution of similar ionic strength or from estimates of measurements made in water or dilute solutions. Both methods assume that the solid phase is the same. The first method assumes that the composition of the solution does not affect the solubility which is normally not the case. The second method attempts to account for the ionic interactions that occur in the water of known composition. An example of how one can estimate the solubility of radionuclides in natural waters can be demonstrated by considering the solubility of Fe(III). This metal was selected since it has a similar valence to many of the radionuclides and is strongly affected by hydrolysis. The procedure requires one to know the solubility of the least soluble mineral or salt in the natural water. This can be done using minimization techniques similar to those used by Harvie and Weare (1980).

The solubility of Fe(III) of various solid phases in seawater can be estimated from the calculated speciation of seawater (Zhu et al., 1992; Millero et al., 1995). The solubilities of the minerals $Fe(OH)_3$ as a function of pH are given by:

$$Fe(OH)_3 (s) + 3H^{3+} \rightarrow Fe^{3+} + 3H_2O \qquad (5)$$

At equilibrium, the thermodynamic equilibrium constant is given by:

$$K_{Fe(OH)3(s)} = a_{Fe} \, a_{H2O}^3/a_H^3 \qquad (6)$$

where $a_i = [i] \, \gamma_i$ are the activities of species i. If one uses the ion pairing model, the value of a_{Fe} is given by:

$$a_{Fe} = \gamma_{Fe} [Fe^{3+}] = \gamma_{Fe} [Fe(III)]/(1 + \Sigma K^*_{x,n} [X]^n + \Sigma K^*_j [H^+]^{-n}) \qquad (7)$$

where the cumulative stability constants $K^*_{x,n}$ are given by:

$$K^*_{x,n} = [Fe(X_i)_n]/[Fe^{3+}] [X]^n \qquad (8)$$

and K^*_j is the hydrolysis constant in the given ionic media. The values of $[Fe^{3+}]$ and $[Fe(III)]$ are the concentrations, respectively, of free and total iron and γ_{Fe} is the activity coefficient of free Fe^{3+}. The $[H^+]$ is defined on the free hydrogen ion molality scale.

The equilibrium solubility of $[Fe(III)]$ for the mineral can be determined for any natural water using

$$[Fe(III)] = K_{Fe(OH)3} \, \gamma_H^3 [H^+]^3/(\alpha_{Fe} \, a_{H2O}^3 \, \gamma_{Fe}) \qquad (9)$$

At a fixed ionic strength and temperature the solubility of the minerals as a function of pH can be determined from

$$[Fe(III)] = K^*_{Fe(OH)3} [H^+]^3/\alpha_{Fe} \qquad (10)$$

where $K^*_{Fe(OH)3} = K_{Fe(OH)3} \, \gamma_H^3/a_{H2O}^3 \, \gamma_{Fe}$ is the stability constants for the media. The fraction of free Fe in the solutions can be calculated from

$$\alpha_{Fe} = [Fe^{3+}]/[Fe(III)] = 1/(1 + \Sigma K^*_{x,n} [X]^n + \Sigma K^*_j [H^+]^n) \qquad (11)$$

The stability and hydrolysis constants at a given ionic strength in seawater can be determined from the equations given by Millero et al. (1995). The reliability of the model can be demonstrated by calculating the solubility of Fe(III) in seawater.

The solubility of $Fe(OH)_3(s)$ has been determined in seawater by Byrne and Kester (1976) and Kuma et al. (1992). The values of the log [Fe(III)] determined in these studies are shown as a function of pH in Figure 5 (Millero et al., 1995). Byrne and Kester (1976) used filtration (0.05 μm), ultrafiltration (150 to 350 D), and dialysis rates. Kuma et al. (1992), also, determined the solubility by dialysis rate measurements. From a pH of 3.3 to 7.0 the 0.05 μm filtration and the Ultrafiltration method gave similar results (open circles). The dialysis results of Kuma et al. (1992) from pH = 5.5 to 8.06 (solid circles) are in good agreement with the Byrne and Kester (1976) measurements determined by dialysis and acidification followed by filtration (open triangles). The measured solubilities between a pH of 7 to 9 determined by dialysis (solid circles and open triangles) are higher than the Ultrafiltration measurements (open circles). The solubility product for $Fe(OH)_3(s)$ at given salinity in seawater have be estimated from equation (10) where $K^*_{Fe(OH)3}(s) = 10^{4.5}$ is the solubility of $Fe(OH)_3(s)$ in average seawater. The speciation of Fe(III) in seawater is given in Figure 6 (Millero, 1997). Over most of the pH range of most natural waters the speciation and solubility is largely affected by the hydrolysis of Fe^{3+}. The formation of $Fe(OH)_4^-$ is not shown since it does not appear to affect the solubility of $Fe(OH)_3(s)$ in seawater.

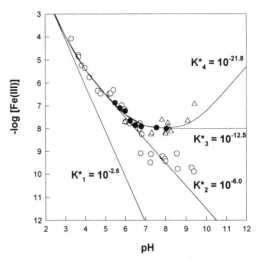

Figure 5. The Solubility of Fe(III) in Seawater (Millero et al., 1995).

To fit the solubility determined by dialysis, one needs to consider the formation of $FeOH^{2+}$, $Fe(OH)_2^+$ and $Fe(OH)_3$ (K^*_1, K^*_2, and K^*_3). The filtration (0.05 μm) solubilities of Byrne and Kester (1976), however, can be predicted by only considering the formation of $FeOH^{2+}$ and $Fe(OH)_2^+$ (K^*_1 and K^*_2). The differences are quite significant with changes in the solubility of Fe(III) going from 10 nM using the dialysis experiments to 0.4 nM for the filtration results at a pH = 8.

More recently Kuma et al. (1996) have measured the solubility of Fe(III) in open ocean and coastal waters as a function of pH with and without UV irradiation and after filtration through a 0.025

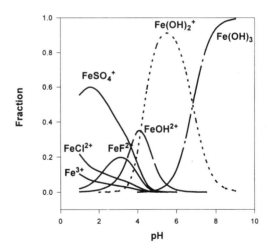

Figure 6. The speciation of Fe(III) in seawater (Millero, 1997).

μm filter. His results can be used to examine the effect that natural organic ligands have on the solubility of Fe(III) in seawater. The solubility of $Fe(OH)_3(s)$ without UV irradiation (Kuma et al., 1996) after 5 weeks of equilibration are shown in Figure 7 (Millero, 1997).

The solubilities in the coastal waters (1.7 nM at pH = 8.16) are much higher than in open-ocean waters (0.6 nM at pH = 8.16). This effect has been attributed to higher

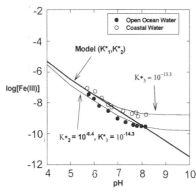
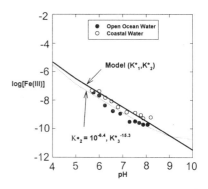

Figure 7. The solubility of Fe(III) in unaltered seawater (Millero, 1997).

Figure 8. The solubility of Fe(III) in UV irradiated seawater (Millero, 1997).

concentrations of organic ligands in the coastal waters. The model calculations for seawater (using K^*_1, K^*_2) are in better agreement (Millero, 1997) with the coastal values than the open ocean values. By adding a value of K^*_3, it is possible to get a reasonable fit of the coastal data; while, the open ocean data requires changes in the values of K^*_2 and K^*_3 (see Figure 7).

Model calculations (Millero, 1997) indicate that the curvature in the solubilities above pH = 8 can be accounted for by considering the formation of organic complexes rather than Fe(OH)$_3$. The results obtained by Kuma et al. (1996) for UV irradiated coastal and open ocean waters are shown in Figure 8 (Millero, 1997). The UV irradiation lowers the solubility in the coastal waters at the higher pH (1.7 nM to 0.6 nM at pH = 8.16); but, has a smaller effect on the open ocean values (0.28 nM to 0.19 nM at pH = 8.16). The UV irradiated coastal waters still have a higher solubility than the open ocean water values at the higher pH. This could be due to the incomplete destruction of the coastal organics by UV irradiation.

Recently a number of workers (Gledhill and van den Berg, 1994; Wu and Luther, 1994; Rue and Bruland, 1995; van den Berg, 1995) have used voltammetry to examine the concentration and strength of natural organic ligands to complex Fe(III) in seawater. They have found ligand concentrations of [L] = 0.4 - 13 nM and apparent stability constants of $K'_{FeL} = 10^{19}$ to 10^{23} defined by

$$K'_{FeL} = [FeL]/[Fe^{3+}]_F[L] \qquad (12)$$

where [L] is the total ligand not complexed by Fe^{3+}. The effect of an organic ligand on the speciation of Fe(III) can be determined from by adding K'_{FeL} [L] terms to the inorganic terms given in equation (11) depending upon the number of organic ligands present

$$\alpha_{Fe} = 1/(1 + \Sigma K^*_{x,n} [X]^n + \Sigma K^*_j [H^+]^{-n} + \Sigma K'_{FeL} [L]) \qquad (13)$$

and its affect on the solubility can be determined using equation (10) with and without the organic ligand by a series of iterations. The fraction of total Fe(III) complexed with the organic ligand in the solution can be determined from

$$\alpha_{FeL} = (\sum K'_{FeL} [L]) \, \alpha_{Fe} \qquad\qquad (14)$$

Since the concentrations of Fe(III) and L are the same order of magnitude, equations (9) and (10) must be solved by a series of iterations. The recent measurements of [L] and K'_{FeL} in seawater give increases in the solubility of 10 to 95% which is comparable to the decrease found after UV irradiation of the open ocean (32%) and coastal waters (65%).

The new measurements of Kuma et al. (1996) for UV irradiated coastal waters are in good agreement (see Figure 9) with the earlier measurements of Byrne and Kester (1976) and can be adequately represented by our speciation model by including the formation of $FeOH^{2+}$ and $Fe(OH)_2^+$ (K^*_1 and K^*_2). The UV irradiated and non UV irradiated open ocean waters solubility measurements (0.2 to 0.3 nM) of Kuma et al. (1996) are slightly lower than the UV irradiated coastal water as well as the earlier filtration measurements (0.4 nM) on Sargasso seawater by Byrne and Kester (1976).

The model calculations for the solubility at pH = 8.0 (free scale) considering only the formation of $FeOH^{2+}$ and $Fe(OH)_2^+$ yield a solubility of Fe(III) of 0.3 nM in seawater at 25°C. These results are in good agreement with measurements in open ocean surface waters (0.2 nM) and deep waters (0.6 nM) determined by various workers (Landing and Bruland, 1987; Martin and Gordon, 1988; Wu and Luther, 1995; Gledhill and van den Berg, 1994; Rue and Bruland, 1995; Kuma et al., 1996).

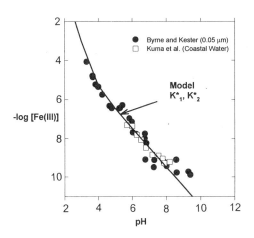

Figure 9. The solubility of Fe(III) in seawater (Millero, 1997).

Future measurements of the solubility of Fe(III) as well as a number of radionuclides in natural waters as well as ionic media such as NaCl should be made over a wide range of pH (2-9), temperatures (0 to 50°C) and ionic strength (0 to 6m) to elucidate the speciation and its affect on the solubility.

COMPOSITION OF WIPP BRINES

The composition of the expected brines have been considered by Brush (1990). A more thorough study of the WIPP brines has been made by Horita et al. (1991). The waters are largely Na-K-Mg-Ca-Cl-SO₄ brines that have been formed from seawater. Brush (1990) has suggested that three possible brines need to be considered. The composition of these brines are given in Table 3. The G-Seep brine was collected from the WIPP underground workings.

The SB-3 brine was defined by Brush and Anderson (1988), while brines A and B are standard brines defined to be in equilibrium with the minerals overlying the site

Table 3. The Composition (molality) of the Seawater and Brines in the WIPP Site[a]

Ion	Seawater	G-Seep	SB-3	Brine A	Brine B
Na^+	0.4862	3.688	3.424	1.543	4.707
Mg^{2+}	0.0547	0.528	0.837	1.249	0.0043
Ca^{2+}	0.0106	0.0064	0.0074	0.017	0.017
K^+	0.0106	0.293	0.425	0.665	0.0043
Sr^{2+}	0.00009				
Cl^-	0.6568	0.960	5.975	5.359	4.954
SO_4^{2-}	0.0293	0.257	0.141	0.034	0.034
HCO_3^-	0.0018				
Br^-	0.00087	0.014	0.027	0.0084	0.0086
CO_3^{2-}	0.00028	0.007	.0092	0.059	0.0044
F^-	0.00070				
$B(OH)_4^-$	0.00010	0.012	0.027	0.013	0.00034
pH_{NBS}		6.1	6.0	6.5	6.5
pH_F		7.10	7.16	7.56	7.22
pH_T	8.00	6.65	6.85	7.49	7.09
a_{H2O}	0.981	0.785	0.744	0.783	0.806
Density [b]	1.023	1.209	1.219	1.188	1.170
I	0.723	6.066	6.910	6.394	4.949

a) Converted to molality from the tabulations of Brush (1990). The values of Na^+ have been adjusted by balancing the equivalents.
b) Estimated using an additivity principle at 25°C for the apparent molal volumes (Millero, 1979).

(A) and entering from below the site (B). I have used the Pitzer equations (Hobart et al., 1996) to determine the activity coefficients in the major ions and activity of water in the brines at 25°C. The stoichiometric pK*'s of carbonic acid, boric acid, hydrogen sulfide, and water in the brines are given in Table 4. These dissociation constants are defined in terms of the total concentrations

$$K^*_{HA} = [HA]_T/[H]_T[A]_T \tag{15}$$

It is interesting to note that the pK's in Brine A are much lower than in Brine B due to the higher concentration of Mg^{2+} in Brine A. The interaction of Mg^{2+} with the acid anions lowers the activity coefficients of these anions and increases the dissociation of the acids.

To use these constants to estimate the anions of these acids (OH^-, CO_3^{2-}) that can complex metals in the brines, it is necessary to know the total proton concentration $[H^+]_T$ in the brines. Unfortunately measurements of pH of the waters were made using NBS buffers. The apparent activity obtained using NBS buffers is related to the total proton concentration by

$$a_H = f [H^+]_T \tag{16}$$

where the factor f includes the activity coefficient of the proton in the brine and a term related to liquid junction potential. The value of f can be determined experimentally by measuring the emf using electrodes without junctions in artificial brines with different amounts of HCl. The electrode emf can be fit to the Nernst equation

Table 4. Dissociation Constants for Acids in the WIPP Brines.[a]

Acid		Seawater	G-Seep	SB-3	Brines A	Brine B
H_2CO_3	pK_0	1.54	1.88	1.91	1.82	1.88
	pK_1	5.84	5.68	5.99	6.18	6.05
	pK_2	8.95	8.20	8.18	8.18	9.24
H_2O	pK_w	13.22	11.80	11.66	11.83	13.90
$B(OH)_3$	pK_{HB}	8.60	7.80	7.54	7.18	8.76
HF	pK_{HF}	2.55	2.34	2.17	2.11	3.30
HSO_4	pK_{HSO4}	0.99	0.89	0.85	0.58	1.05
H_3PO_4	pK_1	1.67	1.67	2.13	2.61	1.72
	pK_2	5.94	4.69	4.49	4.36	5.97
	pK_3	8.72	9.48	9.60	9.08	9.27
H_2S	pK_1	6.54	6.73	7.04	7.14	7.04
H_2SO_3	pK_1	1.49	1.85	2.13	2.19	2.22
	pK_2	6.16	5.22	4.99	4.42	6.49
NH_4	pK_{NH4}	9.2	9.67	9.97	10.06	9.95

a)   All the calculated pK's are stoichiometric values ($K^*_{HA} = [HA]_T/[H]_T[A]_T$) with the exception of HF and HSO_4^- where $[H]_F$ replaces $[H]_T$ (F = free and T = total). Equation (20) can be used to convert these stoichiometric constants to the free scale.

$$E = E^* + (RT/F) \ln[H^+]_T \tag{17}$$

where E* is the standard potential in the brine at a fixed ionic strength. This equation can be used to determine the $[H^+]_T$ before the addition of HCl. If the electrode has also been calibrated using an NBS buffer the emf obtained before the addition of the HCl can be used to determine the apparent activity and the resultant f factor for the electrode system (Millero, 1986). The total proton concentration in a brine can also be determined by using buffers such as TRIS to calibrate the electrode system (Millero, 1986; Millero et al., 1987). Since the Pitzer parameters are available in all the major brine salts, it is possible to determine the pK* of TRIS in any brine. A TRIS buffer made up in this brine can be used to determine the pH of an unknown brine. The pH is determined using the equation

$$pH(Brine) = pK^*(TRIS) + (E_{BRINE} - E_{TRIS})/k \tag{18}$$

where k = (RT/F) log10 = 55.16 mV at 25°C. Since the pK* of TRIS in various brines are quite similar, this method can be used for brines of similar composition without serious errors. To get an estimate of the f factor in the WIPP brines, I have determined the value in 5 and 6 M NaCl buffered with TRIS (Millero et al. 1987) using a glass and calomel electrode system in my laboratory. These results yield the following equation that can be used to estimate the free proton concentration in the WIPP brines

$$pH_F = pH_{NBS} + (0.18 I - 0.20) \tag{19}$$

which is valid from I = 5 to 7 (the subscript F is used to denote the free proton). This equation has been used to estimate the values of pH_F for the brines in Table 3. The concentration of the total proton in a brine is related to the free value by

$$pH_T = pH_F + \log(1 + K_{HSO4} [SO_4^{2-}]) \tag{20}$$

where K_{HSO4} is the stability constant for the formation of HSO_4^-. The values of K_{HSO4} are calculated from the Pitzer program. The values of pH_T have been used to estimate the concentration of various acidic anions at a given pH in the brine using the ionization constants given in Table 4.

The level of total carbonate in the brine can be estimated (if unmeasured) by assuming that the concentration is controlled by the solubility of $MgCO_3(s)$ in the brine. The pKsp for $MgCO_3$ in water is between 7.6 to 8.22 (Morse and Mucci, 1990). The values of pKsp* = $[Mg]_T [CO_3]_T$ in the various brines vary between 4.6 to 6.3. These values can be used to determine the maximum (0.06m) and minimum (2×10^{-4} m) concentrations of total carbonate in the solutions. If the total carbon dioxide (TCO_2) is known for the sample, the total carbonate can be calculated from

$$[CO_3]_T = TCO_2 K^*_1 K^*_2/([H]_T^2 + K^*_1 [H]_T + K^*_1 K^*_2) \tag{21}$$

where K^*_1 and K^*_2 are the stoichiometric dissociation constant in the media (Table 4). The concentration of the total hydroxide in the solution can be calculated from

$$[OH]_T = K^*_W/[H]_T \tag{22}$$

Where K^*_W is the stoichiometric dissociation constant in the media (Table 4). Calculations as a function of the total carbonate and hydroxide can be made to examine the effect they have on the total activity coefficients and speciation of elements in the brines.

SOLUBILITY OF RADIONUCLIDES IN WIPP BRINES

Recently the methods described above have been used to estimate the solubility of radionuclides in the WIPP brines (Hobart et al., 1996). Since little or no solubility data was available then, these estimations were the best that could be made at the time. The activity of free or uncomplexed nuclides in the WIPP brines have been determined using Pitzer equations that account for interactions with Cl^- and SO_4^{2-} and for anions with Na^+ and Mg^{2+} salts. For those metals where Pitzer parameters were not available, the estimates were made using the values for divalent (Ra = Pb) and trivalent (Am = Cm = La) metals of similar structure and charge. The activity coefficients for the ion complexes were estimated using known values for ions or ion complexes of similar valence. The stability constants at a given ionic strength needed to determine the fraction of free nuclides (α_M) have been estimated using literature stability constants at infinite dilution (Hobart, 1990) and estimates of the activity coefficients of the ions and ion complexes. Combining these fractions with estimates of the activity of the free metal ($\gamma(M)_F$), it was possible to estimate the total or stoichiometric activity coefficient ($\gamma(M)_T$) of the metal.

The speciation of elements is a strong function of the composition of the solutions. This is demonstrated in Figures 10 and 11 where a comparison is given for the speciation of Uranium in seawater and WIPP brine A with and without carbonate.

Speciation of UO₂(II)

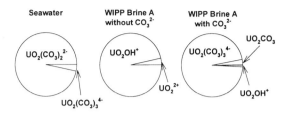

Figure 10. The Speciation of $UO_2(II)$ in Seawater and WIPP Brine A with (10 μm) and without carbonate.

Speciation of Pb(II)

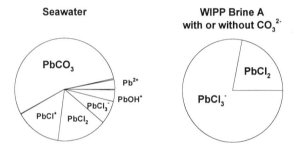

Figure 11. The Speciation of Pb(II) in Seawater and WIPP Brine A with and without carbonate.

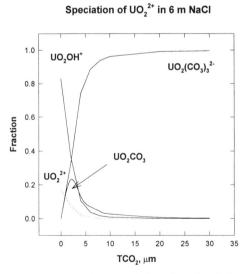

Figure 12. The changes in the speciation of Uranium in 6m NaCl with added carbonate.

The speciation of Uranium is strongly affected by the level of carbonate both in seawater (0.0022m) and in brines when $[CO_3^{2-}] \gg [UO_2^{2+}]$. This is more clearly demonstrated in Figure 12 where the speciation of Uranium in 6 m NaCl is given as a function of TCO_2. At low levels of carbonate, UO_2OH^+ is dominant while at concentrations of TCO_2 above 30 μm, $UO_2(CO_3)_3^{4-}$ is dominant at low levels of UO_2^{2+}. These results can be contrasted with the speciation of Pb. In seawater at levels of Cl⁻ of 0.5 m the carbonate speciation is important while in Brine A (5.3 m Cl⁻) carbonate concentrations as high as 1 m do not change the speciation. The large changes in the speciation of Uranium as a function of carbonate can lead to large variations in its solubility in the WIPP Brines (Hobart et al., 1996).

Recently, the solubility of Uranium, Plutonium and Americium have been determined (Nitsche, 1990) in the J-13 ground waters of the Yucca Mountain waters at a pH = 6, 7, 8.5. These solubilities were extrapolated to infinite dilution using the Pitzer equations. The resulting thermodynamic solubilities for these and other elements in WIPP brines are given in Table 5. By appropriate estimation of the activity coefficients, it is possible to

Table 5. Estimated Solid Phase and Thermodynamic Solubilities of Some Radionucleotides.

Element	Solid Phase	pKsp
Americium [a]	$Am(OH)(CO_3)$	24.39
Curium [b]	$Am(OH)(CO_3)$	24.39
Lead [c]	$PbCO_3$	13.2
Neptunium [d]	$Np(OH)_4$	32.32
Plutonium [a]	$Pu(OH)_4$	51.73
Radium [d]	$RaSO_4$	10.43
Thorium [d]	$Th(OH)_4$	52.3
Uranium [a]	UO_2CO_3	8.74
	$UO_2(OH)_2$	18.68

a) Estimated form the solubility measurements of Nitsche (1990) in J-13 ground waters from Yucca Mountains at pH = 6, 7 and 8.5.
b) Assumed to be equal to Am.
c) From Morse and Mucci (1990).
d) From Hobart et al. (1996).

estimate the solubilities in other ionic media. As a sample of how this can be done, the solubility of Pb(II) will be estimated in WIPP Brine A with and without carbonate. To make this estimate, it is necessary to select the compound that limits the solubility. For the sake of this calculation we will assume that solubility is controlled by (Hobart et al., 1996).

$$PbCO_3 (s) = Pb^{2+} + CO_3^{2-} \qquad (22)$$

The solubilities of Pb^{2+} in Brine A without carbonate can be estimated from

$$[Pb(II)] = [Ksp(PbCO_3)/\{\gamma_T(Pb^{2+})\,\gamma_T(CO_3)\}]^{0.5} \qquad (23)$$

where the subscript T is the total activity coefficient or concentration and Ksp is the thermodynamic solubility product. The value of pKsp in Table 5 combined with $\gamma_T(Pb^{2+})$ = 4.8 x10⁻⁶ and $\gamma_T(CO_3)$ = 0.0014 in Brine A gives a solubility of 0.003 m when $PbCO_3$ is the limiting solid. As discussed elsewhere, the addition of carbonate or considering $PbCl_2$(s) as the limiting solid (Hobart et al., 1996) can also affect the solubility of Pb(II) in the brines. These calculations point out the importance of making reliable pH and TCO_2 measurements, to be able to examine the interaction of OH⁻ and CO_3^{2-} ligands with radionuclides in natural waters. As discussed earlier the pH measurements should be made on the total scale to be consistent with the chemical model.

The solubilities for a number of radionuclides estimated in the WIPP brines have been estimated by using the Pitzer equations to estimate activity coefficients and thermodynamic data to determine the speciation of the most likely species (Hobart et al, 1996). The results for a cumulative probability of 0.5 are given in Table 6. Although

Table 6. Estimated Solution and Solid Phases and Solubility of Radionuclides in WIPP Brines[a]

Metal	Solution Species	Solid Phase		Estimated Solubility in WIPP Brines	
		Maximum	Minimum	Maximum	Minimum
Pb(II)	$PbCl_4^{2-}$	$PbCO_3$	$PbCl_2$	8.0×10^{-3}	1.64
Ra(II)	Ra^{2+}	$RaSO_4$	$RaCO_3$	1.0×10^{-8}	1.6×10^{-6}
Am(III)	$AmCl_2^+$	$Am(OH)_3$	$AmOHCO_3$	1×10^{-9}	1×10^{-9}
Cm(III)[b]	Cm^{3+}	$Cm(OH)_3$	$Cm(OH)(CO_3)$	1×10^{-9}	1×10^{-9}
Np(V)	$NpO_2CO_3^-$	NpO_2OH	$NaNpO_2CO_3 \cdot 3.5H_2O$	6.0×10^{-7}	
Np(IV)	$Np(OH)_5^-$	$Np(OH)_4$	NpO_2	6.0×10^{-9}	
Pu(V)	PuO_2^+	$Pu(OH)_4$	PuO_2	6.0×10^{-10}	
Pu(IV)	$Pu(OH)_5^-$	$Pu(OH)_4$	PuO_2	6.0×10^{-10}	
Th(IV)	$Th(OH)_4^\circ$	$Th(OH)_4$	ThO_2	1.0×10^{-10}	
U(VI)	$UO_2(CO_3)_2^{2-}$	$UO_3 \cdot 2H_2O$	UO_2	2.0×10^{-3}	
U(IV)	$U(OH)_4^\circ$	UO_2	U_3O_8	1.0×10^{-6}	

a) Hobart et al. (1996) in solutions with carbonate at a probability of 0.5.
b) Assuming speciation is similar to Am.

more recent measurements are now available for some of these elements in WIPP brines, the methods of estimating activity coefficients discussed in this paper provide a framework that can be used to examine the effect of composition, pH, TCO_2, etc. on the solubility in other natural waters. The addition of new Pitzer parameters for radionuclides to our present divalent and trivalent models can be added, and it will be possible to examine how changes in composition affect the solubility in natural waters of variable compositions.

Direct solubility measurements are of course needed to determine the reliability of these estimates. If the solubilities are determined in simple media, the effects of the composition on the solubilities in more complex media can be modeled using a model that includes ion complex formation (Millero, 1992).

CONCLUSIONS

The brief review given in this paper outlines a framework that can be used to determine the solubility of radionuclides in natural waters. It clearly demonstrates the need to make solubility measurements in NaCl solutions as a function of pH, ionic strength and temperature. With the addition of small amounts of other ligands (CO_3^{2-}, $B(OH)_4^-$, etc.) to this media, it will be possible to model the effect they have on a given radionuclide. These results can provide the fundamental stability constants needed to construct a model that can be used to determine the solubility of radionuclides in natural waters of known composition. By making the measurements in natural waters (rivers, lakes and coastal waters), it is possible to determine how natural organics affect the solubility of these elements.

ACKNOWLEDGMENTS

The author wishes to acknowledge the support of the Oceanographic Section of the National Science Foundation and the Office of Naval Research for supporting his marine physical chemical studies. The National Oceanographic and Atmospheric Administration and the Department of Energy are also recognized.

REFERENCES

Ahrland, S., 1986, in *The Chemistry of the Actinide Elements,* J.J. Katz, G.T. Seaborg, and L.R. Morss, Eds., Chapman and Hall, London, pp. 1480-1546.

Allard, B. and Rydberg, J., 1983, *Plutonium Chemistry*, ACS Symp. Ser. 216, American Chemical Society, Washington, D.C., 275-295.

Allard, B., Olofsson, U. and Tortenfelt, B., 1984, *Environ. Act. Chem. Inorg. Chim. Acta*, **94**:205-221.

Baes, C. F. and Mesmer, R.E., 1976, *The Hydrolysis of Cations,* John Wiley & Sons Inc., NY.

Bidoglio, G., 1982, Characterization of Am(III) complexes with bicarbonate and carbonate ions at ground-water concentration levels, *Radiochem. Radioanal. Lett.* **53**:45-60.

Brush, L.H., 1990, Test Plan for Laboratory and Modeling Studies of Repository and Radionuclide Chemistry for the Waste Isolation Pilot Program, SAND90-0266, Albuquerque, NM: Sandia National Laboratories.

Brush, L.H., and Anderson, D.R., 1989, Effects of Microbial Activity on Repository Chemistry, Radionuclide Speciation, and Solubilities in WIPP Brines, Southeastern New Mexico; March 1989, A.R. Lappin, R.L. Hunter, D.P. Garber, and P.B. Davies, Eds., SAND89-0462, Albuquerque, NM, Sandia National Laboratories. A-31 through A-50.

Byrne, R.H., and Kester, D.R., 1976, Solubility of hydrous ferric oxide and iron speciation in sea water, *Mar. Chem.* **4**:255-274.

Byrne, R.H., Kump, L.R., and Cantrell, K.J., 1988, The influence of temperature and pH on trace metal speciation in seawater, *Mar. Chem.* **25**:163-181.

Campbell, D.M., Millero, F.J., Roy, R., Roy, L., Lawson, M, Vogel, K.M., and Moore, C.P., 1993, The standard potential for the hydrogen - silver, silver chloride electrode in synthetic seawater, *Mar. Chem.* **44**:221-233.

Campodaglio, G., Coale, K.H, and Bruland, K.W., 1990, Lead speciation in surface waters of the eastern North Pacific, *Mar. Chem.* **29**:22-233.

Choppin, G.R., 1989, Soluble rare earths and actinide species in seawater, *Mar. Chem.*, **28**:19-26.

Choppin, G.R., and Allard, B., 1985, *Handbook on the Physics and Chemistry of the Actinides*, A.J. Freeman and C. Keller, Eds., Vol. 3, Chapt. 11, Elsevier, Amsterdam.

Choppin, G.R., and Stout, B.W., 1989, Actinides behavior in natural waters, *Sci. Tot. Environ.* **83**:203-216.

Choppin, G.R., and Morse, J.W., 1989, in *Environmental Research on Actinide Elements in Terrestrial Ecosystems*, Laboratory studies of Actinides in marine systems, pp.49-72.

Ciavatta, L., Ferri, D., Gnmaldi, M., Palombari, R., and Salvatore, F., 1979, Dioxouranium (VI) carbonate complexes in acid solution, *J. Inorg. Nucl. Chem.* **41**:1175-1182.

Ciavatta, L., Ferri, D., Grenthe, L., Salvatore, F., and Spahiu, K., 1983, Studies on metal carbonate equilibria 4. Reduction of the tris(carbonato) dioxouranium (VI) ion, $UO_2(CO_3)_3^{4-}$ in hydrogen carbonate solutions, *Inorg. Chem.* **22**:2088-2092.

Clegg, S.L., and Whitfield, M., 1991, Activity coefficients in natural waters, In: *Activity Coefficients in Electrolyte Solutions*, K.S. Pitzer, ed., CRS, Boca Raton, FL, pp. 279-434.

Clegg, S.L., and Whitfield, M., 1995, A chemical model of seawater including dissolved ammonia and the stoichiometric dissociation constant of ammonia in estuarine water and seawater from -2 to 40°C, *Geochim. Cosmochim. Acta.* **59**, 2403-2421.

Criss, C.M., and Millero, F.J. 1996, Modeling the heat capacities of aqueous 1-1 electrolyte solution with Pitzer's equations, *J. Phys. Chem.* **100**:1288-1294.

Dickson, A.G., and Whitfield, M., 1981, An ion-association model for estimating acidity constants (at 25°C and 1 atm total pressure) in electrolyte mixtures related to seawater (ionic strength < 1 mol kg^{-1} H$_2$O), *Mar. Chem.* **10**:315-333.

Felmy, A.R., and Weare, J.H., 1986, The prediction of borate mineral equilibria in natural waters: Application to Searles Lake, California. *Geochim. Cosmochim. Acta.* **50**:2771-2783.

Ferri, D., Grenthe, I., and Salvatore. F., 1981, Dioxouranium (VI) carbonate complexes in neutral and alkaline solutions, *Acta Chem. Scand.* **A35**:165-168.

Gel'man, A., Moskvin, A.L., and Zaitseva, V.P., 1962, The plutonyl carbonates, *Sov. Radiochem,* **4**:154-162.

Gledhill, M., and van der Berg, C. M. G., 1994, Determination of complexation of iron (III) with natural organic complexing ligands in sea water using cathodic stripping voltammetry, *Mar. Chem.* **47**:41-54.

Greenberg, J.P., and Møller, N., 1989, The prediction of mineral solubilities in natural waters: A chemical equilibrium model for the Na-K-Ca-Cl-SO$_4$-H$_2$O system to high concentration from 0 to 250°C, *Geochim. Cosmochim. Acta.* **53**:2503-2518.

Grenthe, I., Fuger, J., Koning, R., Lemire, R.J., Muller, A.B., Nguygen-Trung, C., and Wanner, H., 1992, *Chemical Thermodynamics of Uranium*, North Holland, Amsterdam.

Harvie, C.E., and Weare, J.H., 1980, The prediction of mineral solubilities in natural waters: the Na-K-Mg-Ca-SO$_4$-Cl-H$_2$O system from zero to high concentration at 25°C. *Geochim. Cosmochim. Acta.* **44**:981-997.

Harvie, C.E., Møller, N., and Weare, J.H., 1984, The prediction of mineral solubilities in natural waters: The Na-K-Mg-Ca-H-Cl-SO$_4$-OH-HCO$_3$-CO$_3$-CO$_2$-H$_2$O system to high ionic strengths at 25°C, *Geochim. Cosmochim. Acta.* **48**:723-752.

He, S., and Morse, J.W., 1993, The carbonic acid system and calcite solubility in aqueous Na-K-Ca-Mg-Cl-SO$_4$ solutions from 0 to 90°C, *Geochim. Cosmochim. Acta.* **57**:3533-3554.

Hershey, P.J., Fernandez, M., Milne, P.J., and Millero, F.J., 1986, The ionization of boric acid in NaCl, Na-Ca-Cl and Na-Mg-Cl solutions at 25°C, *Geochim. Cosmochim. Acta.* **50**:137-148.

Hershey, P. ., Millero, F.J., and Fernandez, M., 1989. The ionization of phosphoric acid in NaCl and NaMgCl solutions at 25°C, *J. Solution Chem.* **18**:875-892.

Hobart, D.E. 1990, Actinides in the Environment, Proceedings of the Robert A. Welch Foundation Conference on Chemical Research XXXIV: Fifty Years with Transuranium Elements, Houston, TX, Oct. 22-23.

Hobart, D.E. , Millero, F.J. , Chou, I.-M., Bruton, C.J., Trauth, K.M., and Anderson, D.R. 1996, Estimates of the solubilities of waste element radionuclides in waste isolation pilot plant brines: A preliminary report by the expert panel on source term solubilities, Sandia 96-0098, UC-721, Sandia National Laboratories, Livermore, CA.

Horita, J., Friedman, T.J., Lazar, B., and Holland, H.D., 1991, The composition of Permian Seawater, *Geochim. Cosmochim. Acta,* 55(2), 417-432.

Katz, J.J., Seaborg, G.T., and Morss, L.R., 1986, *The Chemistry of the Actinide Elements*, Chapman and Hall, London.

Kim, J., Lierse, Ch., and Baumgartner, F., 1983, Complexation of the plutonium (IV) ion in carbonate-bicarbonate solutions, in *Plutonium Chemistry*, ACS Symposium Series, No. 216, W.T. Carnall and G. R. Choppin, Eds., Amer. Chem. Soc., Washington, DC, pp. 317-334.

Knowles, G., and Wakeford, A.C., 1978, A mathematical deterministic river-quality model. Part 1: Formulation and description, *Water Res.* **12**:1149-1153.

Kuma, K., Nakabayashi, S., Suzuki, Y, and Matsunaga, K., 1992, Dissolution rate and solubility of colloidal hydrous ferric oxide in seawater, *Mar. Chem.* **38**:13-143.

Kuma, K., Nishioka, J., and Matsunaga, K., 1996, Controls on iron (III) hydroxide solubility in seawater: The influence of pH and natural organic chelators, *Limnol. Oceanogr.* **41**:396-407.

Landing, W.M., and Bruland,. K.W., 1987, The contrasting biogeochemistry of iron and manganese in the Pacific Ocean, *Geochim. Cosmochim. Acta.* **51**:29-43.

Langmuir, D., and Riese, A., 1985, The thermodynamic properties of Radium, Geochim. Cosmochim. Acta, **49**:1593-1601.

Lowson, R.T., 1985, The thermochemistry of Radium, *Thermochimica Acta.* **91**:185-212.

Lundqvist, R., 1982, Hydrophilic complexes of the actinides 1. carbonates of trivalent Americium and Europium, *Acta Chem. Scand.* **A36**:741-750.

Martin, J.H., and Gordon, R.M., 1988, Iron deficiency limits phytoplankton growth in the north -east Pacific subarctic, *Nature* **35**:177-196.

Martin, J.H., Fitzwater, S.E., and Gordon, R.M., 1990, Iron deficiency limits phytoplankton growth in Antarctic waters, *Global Biogeochem. Cycl.* **4**:5-12.

Maya, L., 1982a, Detection of hydroxo and carbonato species of dioxouranium (VI) in aqueous media by differential pulse polarography, *Inorg. Chem.* **21**:2895.

Maya, L., 1982b, Detection of hydroxo and carbonato species of dioxouranium (VI) in aqueous media by differential pulse polarography, Inorg. Chim. Acta. **65**:L13-L16.

Maya, L., 1983, Hydrolysis and carbonate complexation of dioxoneptunium (V) in 1.0 M $NaClO_4$ at 25 degree C, *Inorg. Chem.* **22**:2093-2095.

Millero, F.J., 1979, Effects of pressure and temperature on activity coefficients, Chapter 2, in **Activity Coefficients in Electrolyte Solutions**, Vol. II, R.M. Pytkowicz, ed., CRC Press, Boca Raton, FL, pp. 63-151.

Millero, F. J., 1982, Use of models to determine ionic interactions in the natural waters, *Thalassia Jugoslavica.* **18**:253-291.

Millero, F.J., 1983a, Influence of pressure on chemical processes in the sea, Chapt. 43, in **Chemical Oceanography**, 2nd Edition, Vol. 8, J.P. Riley and R. Chester, eds., Academic Press, London, England, pp. 1-88.

Millero, F.J., 1983b, The estimation of the pK*$_{HA}$ of acids in seawater using Pitzer's equations, *Geochim. Cosmochim. Acta.* **47**:2121-2129.

Millero, F.J., 1984, The activity of metal ions at high ionic strengths, in: C.J.M. Kramer and J.C. Duinker, eds., **Complexation of Trace Metals in Natural Waters**, Martinus Nijhoff/W. Junk, The Hague, pp. 187-200.

Millero, F.J., 1986, The pH of estuarine waters, *Limnol. Oceanogr.*, **31**:839-847.

Millero, F.J., 1990a, Effect of ionic interactions on the oxidation rates of metals in natural waters, in: **Chemical Modeling in Aqueous Systems** II, D. C. Melchior and R. L. Bassett, eds., ACS Press, Washington, D.C., Chapter 34, 447-460.

Millero, F.J. 1990b, Effect of speciation on the rates of oxidation of metals, in: **Metals Speciation, Separation and Recovery**, Vol. 2, J.W. Patterson and R. Passino, eds., Lewis Publishers, Inc., Chelsea, Michigan, pp. 125-141.

Millero, F.J. 1992, The stability constants for the formation of rare earth inorganic complexes as a function of ionic strength, *Geochim. Cosmochim. Acta.* **56**:3123-3132.

Millero, F.J., Solubility of Fe(III) in Seawater, *Earth Planet. Sci. Letters*, in press.

Millero, F.J., and Schreiber, D.R., 1982, Use of the ion pairing model to estimate activity coefficients of the ionic components of natural waters, *Am. J. Sci.*, **282**:1508-1540.

Millero, F.J., and Thurmond, V., 1983, The ionization of carbonic acid in Na-Mg-Cl solutions at 25°C, *J. Solution Chem.* **2**:401-412.

Millero, F.J., and Byrne, R.H., 1984, Use of Pitzer's equations to determine the media effect on the formation of lead chloro complexes, *Geochim. Cosmochim. Acta.* **48**:1145-1150.

Millero, F.J., and Hawke, D.J., 1992, Ionic interactions of divalent metals in natural waters, *Mar. Chem.* **40**:19-48.

Millero, F.J., and Roy, R., 1997, The carbonate and borate system in natural waters from 0 to 50°C. *Croatia Chemica Acta* **70**: 1-28.

Millero, F.J., Hershey, J.P., and Fernandez, M. 1987, The pK* of TRISH$^+$ in Na-K-Mg-Ca-Cl-SO_4 brines - pH scales, *Geochim. Cosmochim. Acta.* **51**:707-711.

Millero, F.J. Yao, W., and Aicher, J., 1995, The speciation of Fe(II) and Fe(III) in seawater, *Mar. Chem.* **50**, 21-39.

Millero, F.J., Hershey, J.P., Johnson, G., and Zhang, J., 1989, The solubility of SO_2 and the dissociation of H_2SO_3 in NaCl solutions, *J. Atmos. Chem.* **8**:377-389.

Møller, N., 1988. The prediction of mineral solubilities in natural waters: A chemical equilibrium model for the Na-Ca-Cl-SO_4-H_2O system, to high temperature and concentration, *Geochim. Cosmochim. Acta.* **52**:821-837.

Morel, F., and Morgan, J., 1972, A numerical method for computing equilibria in aqueous chemical systems, *Environ. Sci. Technol.* **6**:58-67.

Morse, J.W., and Mucci, A., 1990, Chemistry of Low-Temperature abiotic calcites: experimental studies on coprecipitation, stability and fractionation, *Rev. Aqua. Sci.* **3**:217-254.

Moskvin, A.I., 1971, Correlation of solubility products of actinide Compounds with properties of their resulting metal ions and anions of various acids, *Radiokhimiy*a 13(2), 293-5 (Russ.) Sov. *Radiochim.* 13.

Moskvin, A.I., and Gel'man, A.D., 1958, Composition and instability constant of oxalate and carbonate complexes of plutonium (IV*). J. Inorg. Chem. USSR* 3, 962-974.

Nordstrom, D.K., and Ball, J.W., 1984. Chemical models, computer programs and metal complexation in natural water,. In: *Complexation of Trace Metals in Natural Waters.* C.J.M. Kramer and J.C., ed., Martinus Nijhoff/W. Junk, The Hague, pp. 149-162.

Nitsche, H., 1990, Basic research for assessment of geologic nuclear waste repositories: what solubility and speciation studies of transuranium elements can tell us, paper presented at Intern. Symp. on Sci. Basis for Nuclear Waste Management, Boston Mass. Conference proceedings.

O'Cinneide, S., Scanlan, J.P., and Hynes, M.J., 1975, Equilibria in uranyl carbonate systems -I. The overall stability constant of $UO_2(CO_3)_3^{4-}$, *J. Inorg. Nucl. Chem.* **37**:1013-1018.

Olofsson, U., Bengtsson, M., and Allard, B., 1984, Scientific Basis for Nuclear Waste Management VII, *Proceedings of the Materials Research Society Symposia*, G.L. McVay, WD, Vol. 26, North-Holland, NY, 859-866.

Pabalan, R.T., and Pitzer, K., 1987, Thermodynamics of concentrated electrolyte mixtures and the prediction of mineral solubilities to high temperature for mixtures in the system Na-K-Mg-Cl-SO_4-OH-H_2O, *Geochim. Cosmochim. Acta*, **51**:2429-2443.

Papelis, C., Hayes, K.F., and Leckie, J. O., 1988, HYDRAQL: A program for the computation of chemical equilibrium composition of aqueous batch systems including surface-complexation modeling of ion adsorption at the oxide/solution interface, Tech. Rept. No. 306, Dept. of Civil Engineering, Stanford University.

Pitzer, K.S., 1973, Thermodynamics of electrolytes. I. Theoretical basis and general equations, *J. Phys. Chem.* **77**:268-277.

Pitzer, K.S., 1979, Theory: ion interaction approach, In: *Activity Coefficients in Electrolyte Solutions*, R.M. Pytkowicz, ed., Vol. I, CRC Press, Boca Raton, Fl., pp. 157-208.

Pitzer, K.S., 1991, Ion interaction approach: theory and data collection. In: *Activity Coefficients in Electrolyte Solutions*, K.S. Pitzer, ed., CRS, Boca Raton, FL, pp. 75-153.

Rue, E.L., and Bruland, K.W., 1995, Complexation of iron (III) by natural organic ligands in the Central North Pacific as determined by a new competitive ligand Equilibration /adsorptive cathodic stripping voltammetric method, *Mar. Chem.* **50**:116-138.

Scanlan, J.P., 1977, Equilibria in uranyl carbonate systems 2. Overall stability constant of $UO_2(CO_3)_2^{2-}$ and 3rd formation constant of $UO_2(CO_3)_3^{4-}$,. *J. Inorg. Nucl Chem.* **39**:635-639.

Sharma, V.K., and Millero, F.J., 1990, Equilibrium constants for the formation of Cu(I) halide complexes, *J. Solution Chem.* **19**:375-390.

Sullivan, J.C., Woods, M., Bertrand, P.A., and Choppin, G.R., 1982, Thermodynamics of plutonium (VI) interaction with bicarbonate, *Radiochim Acta.* **31**:45-50.

Turner, D.R., Varney, M.S., Whitfield, M., Mantoura R.F., and Riley J.P., 1986, Electrochemical studies of copper and lead complexation by fulvic-acid. 1. Potentiometric measurements and a critical comparison of metal -binding models, *Geochim. Cosmochim. Acta.* **50**:289-297.

Turner, D.R., Whitfield, M., and Dickson, A.G., 1981, The equilibrium speciation of dissolved components in freshwater and seawater at 25°C and 1 atm pressure, *Geochim. Cosmochim. Acta*, **45**:855-881.

van den Berg, C.M.G., 1995, Evidence for organic complexation of iron in seawater, *Mar. Chem.* **50**, 139-157.

Westall, J.C., Zachary, J.L., and Morel, F.M., 1976, MINEQL: A computer program for the calculation of chemical equilibrium composition of aqueous systems, Tech. Note No. 18, School of Engineering, Massachusetts Institute of Technology.

Whitfield, M., 1975a, An improved specific interaction model for seawater at 25°C and one atmosphere total pressure, *Mar. Chem.* **3**:197-213.

Whitfield, M., 1975b, The extension of chemical models for seawater to include trace components, *Geochim. Cosmochim. Acta.* **39**:1545-1557.

Woods, M., Mitchell, M.L., and Sullivan, J.C., 1978, Determination of the stability quotient for the formation of a tris (carbonato) plutonate (VI) complexes in aqueous solution, *Inorg. Nucl. Chem. Lett.* **14**:465-467.

Wu, J., and Luther, G.W., 1994, Complexation of Fe(III) by natural organic ligands in the North west Atlantic Ocean by a competitive ligand equilibration method and a kinetic approach, *Limnol. Oceanogr.* **50**:1119-177.

Yao, W., and Millero, F.J. 1996, Adsorption of phosphate on manganese dioxide in seawater, *Envir. Sci. Technol.*, **30**:536-541.

Zhu, X., Prospero, J.M., Millero, F.J., Savoie, D.L., and Brass, G.W., 1992, The solubility of ferric ion in marine mineral aerosol solutions at ambient relative humidities, *Mar. Chem.* **38**:91-107.

USE OF DISSOLVED AND COLLOIDAL ACTINIDE PARAMETERS WITHIN THE 1996 WASTE ISOLATION PILOT PLANT COMPLIANCE CERTIFICATION APPLICATION

Christine T. Stockman and Robert C. Moore

Sandia National Laboratories
PO Box 5800
Albuquerque, NM 87185-0733

BACKGROUND

The Waste Isolation Pilot Plant (WIPP) is a geologic repository operated by the U.S. Department of Energy (DOE) for disposal of transuranic radioactive wastes. The repository is located near Carlsbad in southeast New Mexico, and is approximately 650 meters underground in the Salado Formation. The geologic formations immediately above and below the Salado are the Rustler and Castile Formations, respectively. The Rustler is considered important because it contains the most transmissive units above the repository; the most significant of these is considered to be the Culebra Dolomite members. The Castile contains areas of pressurized brine (brine pockets); it is not known whether any such pockets are located directly under the repository. The 16 square miles surrounding the shafts, surface facilities, and the underlying subsurface are controlled by the DOE.

In October 1996, the DOE submitted a compliance certification application (CCA) to the U.S. Environmental Protection Agency (EPA) in accordance with the requirements of Title 40 of the *Code of Federal Regulations* (40 CFR) Parts 191 and 194. The containment requirements in 40 CFR 191.13(a) specify that the disposal system is to be designed to provide a reasonable expectation that radionuclide releases to the accessible environment during 10,000 years are not likely to exceed specified limits (which are based on the radionuclide inventory in the repository). The demonstration of having a reasonable expectation is to be based on a performance assessment (PA), which is defined in 40 CFR 191.12 as,

> ...an analysis that: (1) Identifies the processes and events that might affect the disposal system; (2) examines the effects of these processes and events on the performance of the disposal system; and (3) estimates the cumulative releases of radionuclides, considering the associated uncertainties, caused by all significant processes and events. These estimates shall be incorporated into an overall probability distribution of cumulative release to the extent practicable.

The accessible environment is defined as:

Actinide Speciation in High Ionic Strength Media, edited by Reed *et al.*
Kluwer Academic / Plenum Publishers, New York, 1999

(1) the atmosphere, (2) land surfaces, (3) surface waters, (4) oceans, and (5) all of the lithosphere that is beyond the controlled area.

Many of the papers in this volume present detailed descriptions of the chemical analyses and methodologies that have been used to evaluate the maximum dissolved and colloid concentrations of actinides within the WIPP repository as part of the performance assessment. This paper describes the program for collecting experimental data and provides an overview of how the PA modeled the release of radionuclides to the accessible environment, and how solubility and colloid parameters were used by the PA models.

RELEASE MECHANISMS

Computer simulations predict that there will be essentially no release of radionuclides to the accessible environment in 10,000 years if the repository is left undisturbed. However, the controlling regulations, 40 CFR191 and 40CFR194, require the evaluation of inadvertent intrusions, such as exploratory drilling for oil, and it is these intrusions and their possible releases of contaminated brine that are of concern. Figure 1 shows a conceptual view of the repository and possible release paths.

While searching for natural resources (oil) in the deeper formations of the region, drillers may penetrate the repository. A pressurized brine pocket may or may not be present in these lower formations, leading to uncertainty as to whether a brine pocket is encountered. If the repository is breached, the drill bit will cut through the waste and the drilling fluids will carry the solid radioactive materials directly to the surface and into the drilling mud pit. In addition, if the repository is at least half filled with brine and under high pressure at the time of drilling, contaminated brine may be forced up the drill hole and into the mud pit. As specified in the EPA regulations, it is assumed that no natural resources will be found and that present-day standards for plugging and abandoning the drill holes will be used. After 200 years, the cement borehole plugs are assumed to degrade to a material with a permeability equivalent to silty sand, and contaminated brine may flow up the borehole and into the Culebra dolomite. From this point, the contaminated brine may travel beyond the controlled area and be used for livestock and/or agricultural purposes.

Thus there are three mechanisms for release of radionuclides to the accessible environment: direct release (DR) of solids during drilling, direct release of contaminated brine during drilling, and long-term release (LR) of contaminated brine through the Culebra dolomite. For the last two mechanisms, it is necessary to determine the concentration of radionuclides that may be mobilized into the brine phase. This was accomplished using solubility and colloid parameters experimentally determined by the actinide source term group.

MODELS AND CODES USED IN THE ASSESSMENT

The performance assessment required the calculation of complementary cumulative distribution functions (CCDFs), which are ordered sets of points that span the cumulative normalized releases from the waste isolation system for all combinations of future histories of the repository over the 10,000-year regulatory period. The cumulative normalized releases are calculated by summing over time the releases of all regulated isotopes normalized by their release limits. The CCDF is compared with the quantitative release limits specified in 40 CFR § 919.13(a) to determine compliance.

Because of the large number of complex calculations required to produce CCDFs, components and subsystems of the WIPP were modeled in separate steps. Several computer

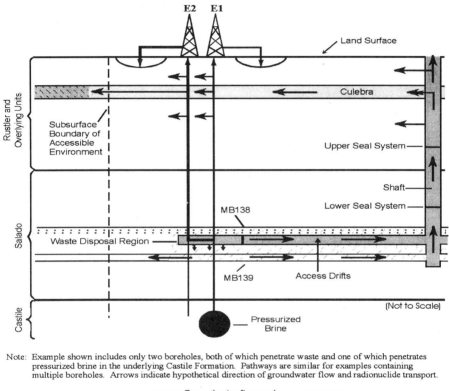

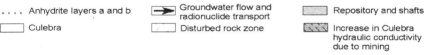

Note: Example shown includes only two boreholes, both of which penetrate waste and one of which penetrates pressurized brine in the underlying Castile Formation. Pathways are similar for examples containing multiple boreholes. Arrows indicate hypothetical direction of groundwater flow and radionuclide transport.

· · · · Anhydrite layers a and b

☐ Culebra

➤ Groundwater flow and radionuclide transport

☐ Disturbed rock zone

▨ Repository and shafts

▨ Increase in Culebra hydraulic conductivity due to mining

CCA-012-2

Figure 1. Conceptual view of WIIPP repository and possible human intrusion. (DOE, 1996, Vol.1, Chapter 6, p. 79)

codes were used to simulate the relevant features of the disposal system and calculate the consequences for different scenarios.

Ten codes were used to model different aspects of brine flow and radionuclide transport for the three release mechanisms. Figure 2 is a flow diagram of the codes used in performance assessment. For long-term release, BRAGFLO was used to model flow of gas and brine within the Castile and Salado formations and up to the Culebra (Vaughn, 1996). NUTS and PANEL were used to model transport of radionuclides within the BRAGFLO brine flow. (Stockman et al., 1997) SECOFL was used to model fluid flow within the Culebra using transmissivity fields generated by GRASP_INV, and SECOTP was used to model transport within the SECOFL flow fields. (Jow et al, 1998) For brine direct release, the BRAGFLO flow fields were read by BRAGFLO (version 4.01), which calculated the volume of brine released during the exploratory drilling. (O'Brian et al., 1998) PANEL was used again for the direct brine release to obtain the inventory of radionuclides per volume of brine as a function of time. CUTTINGS_S (Berglund, 1998) was used to calculate the volume of solid waste brought to the surface and EPAUNI (Sanchez et. al. 1997) was used to calculate the radionuclide inventory per waste volume. The output from all of these codes was brought together by the complementary cumulative distribution function grid flow code (CCDFGF) (Helton et al., 1996), to calculate releases to the accessible environment from the three mechanisms and produce the CCDF. Of these codes, only NUTS and PANEL used solubility and colloid concentration parameters to determine mobilized actinide concentrations within the brine. The other transport code, SECOTP, used the conservative assumption of no precipitation, and thus did not need these parameters.

Both NUTS and PANEL assumed, at each time step, instant mobilization of radioisotopes up to their solubility limits if inventory was sufficient, or up to their inventory limits if inventory was insufficient (below effective solubility limits). The total inventory contained within the repository was assumed to be homogeneously and uniformly distributed throughout. NUTS assigned portions of the inventory to each grid block on the basis of that block's volume fraction of the repository as a whole. PANEL did the same, but it treated an entire waste panel as its one and only grid block. Thus both codes required input of the repository inventory of each modeled isotope and the solubility for each modeled element. Although both codes were originally designed to mobilize and transport dissolved species, they may also be used to mobilize and transport any species that may be characterized by a maximum concentration and inventory. Actinides within the WIPP are expected to be mobilized into the brine phase as dissolved species and four types of colloidal species: actinide intrinsic colloids, humic colloids, mineral fragment colloids, and microbe colloids. Thus a method for representing these five species within the calculations had to be chosen.

Colloids may transport differently than dissolved species because of a number of physical processes such as diffusion, filtration, and sorption. (Stumm, 1992, Morel and Herring, 1993, Vold and Vold, 1993) In calculations where these processes are important, such as the SECOTP calculations of transport within fractured dolomite, colloids may transport very differently from dissolved species. Consequently, transport of dissolved and colloidal actinides were calculated separately in SECOTP. For transport through the degraded waste panels and boreholes, however, these processes would change with time and be very difficult to quantify. Consequently, the project decided not to take any credit for diffusion, filtration, or sorption within the Salado formation. Because no physical mechanisms which could transport the species differently were modeled in the NUTS and PANEL calculations, all five species were "lumped" together. A single "effective solubility" was assigned for each actinide; this was the sum of the solubility and the maximum steady-state colloid concentrations. After transport through the Salado formation, the actinides were "broken out" into their dissolved and colloidal components for transport through the Culebra formation.

RADIOISOTOPES INCLUDED IN THE PERFORMANCE ASSESSMENT

Of the 135 radioisotopes reported in the Waste Isolation Pilot Plant Transuranic Waste Baseline Inventory Report (DOE, 1995), 47 are regulated by 40 CFR 191. Sanchez et al. (1997) compiled a projected WIPP inventory of these isotopes and reported their results in curies and EPA units (an EPA unit is defined as the inventory of that isotope in curies divided by the EPA release limit for that isotope in curies as specified in 40 CFR Part 191, Appendix A, Table 1). The repository is considered to comply with the EPA regulations if there is less than a 0.1 probability that the cumulative release to the accessible environment is greater than 1 EPA unit, and less than a 0.001 probability that the cumulative release is greater than 10 EPA units.) Of the 135 isotopes listed, only 24 have more than 0.001 EPA units of inventory at any time within the 10,000-year regulatory period. Consequently, only those have a direct potential to affect calculated releases. In addition to those, however, there are several unregulated short-lived isotopes that (1) have significant inventory, and (2) decay to regulated isotopes. The radioisotopes that could affect regulated releases indirectly have long been included in the list of 30 radioisotopes treated by PANEL. The two lists, i.e., the 24-member list of top EPA-unit isotopes and PANEL's list of 30 isotopes, were combined. Because of overlap, the amalgamated list included a total of 33 distinct isotopes, and these were considered for inclusion in the 1996 CCA PA. Table 1 lists these isotopes in order of maximum EPA units over the 10,000-year regulatory period. All of the 33 radionuclides except ^{14}C, ^{137}Cs, ^{147}Pm, ^{147}Sm, ^{90}Sr, and ^{232}U belong to the following decay chains[1]:

^{238}Pu ↘
^{242}Pu → ^{238}U → ^{234}U → ^{230}Th → ^{226}Ra → ^{210}Pb
^{243}Cm ↘
^{243}Am → ^{239}Pu → ^{235}U → ^{231}Pa → ^{227}Ac
^{244}Cm ↘
^{252}Cf → ^{248}Cm → ^{244}Pu → ^{240}Pu → ^{236}U → ^{232}Th → ^{228}Ra
^{245}Cm → ^{241}Pu → ^{241}Am → ^{237}Np → ^{233}U → ^{229}Th

Short-lived isotopes like ^{238}Pu decay quite rapidly, dropping to less than an EPA unit by 1100 years, while other isotopes like ^{230}Th that initially have low EPA units grow to over an EPA unit in 10,000 years. Some isotopes at the bottom of the table have short half-lives and are not regulated by the EPA, but they also have significant curies and are the parents of regulated isotopes, so they were included in the calculations.

The number of isotopes had to be reduced for the computationally intensive codes because each isotope that is included increases the calculation time. The first three columns in Table 1 show which isotopes were included in the calculations done by EPAUNI (solid direct release), PANEL (long-term and direct release), and NUTS (long-term release). For the EPAUNI calculations of direct solid release, waste was categorized into more than 500 waste streams based on the source and inventory of the waste. After all of the waste

---

[1] These decay chains were simplified by leaving out the many very short-lived and therefore unregulated intermediates. Leaving out short-lived intermediates that lie between regulated isotopes does not affect the rate of decay and ingrowth of the remaining long-lived isotopes. This is demonstrated by the verification of PANEL's simplified decay chains against ORIGEN's full decay chains in PANEL's software quality assurance documents.

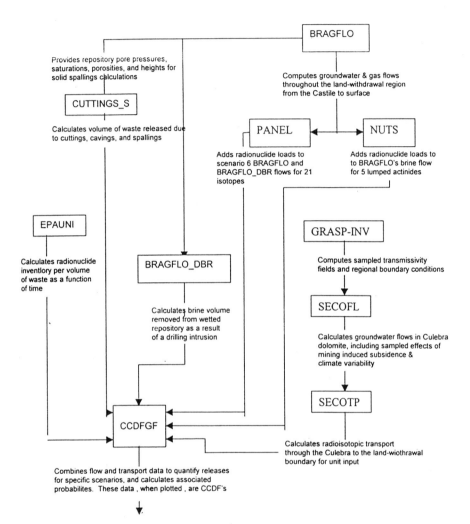

Figure 2. Flow diagram of the computer codes used for performance assessment of the WIPP (Garner et al., 1996).

Table 1. The 33 Isotopes Considered for Modeling in the 1996 CCA PA [Based on the data reported in the Analysis Report for EPAUNI (Sanchez et al., 1997)]

SOLID	PANEL	NUTS			Half life	Release	0 years	0 years	100 years	350 years	10,000 yr. yryears
DR	DR+LR	LR		Isotope	(years)	Limit (Ci)	Ci	EPA Units[2]	EPA Units	EPA Units	MAX EPA
x	x	x	t,p	Pu-238	8.77E+01	344	1.94E+06	5.64E+03	2.55E+03	3.55E+02	1.32E-22	5.64E+03
x	x	x	t,p	Pu-239	2.41E+04	344	7.95E+05	2.31E+03	2.31E+03	2.29E+03	1.73E+03	2.31E+03
x	x	x	t,p	Am-241	4.32E+02	344	4.88E+05	1.42E+03	1.24E+03	8.31E+02	1.78E-04	1.42E+03
x	x	c	t,p	Pu-240	6.54E+03	344	2.14E+05	6.22E+02	6.16E+02	6.02E+02	2.16E+02	6.22E+02
x			t,p	Cs-137	3.00E+01	3440	9.31E+04	2.71E+01	2.69E+00	8.31E-03	0.00E+00	2.71E+01
x			t,p	Sr-90	2.91E+01	3440	8.73E+04	2.54E+01	2.35E+00	6.10E-03	0.00E+00	2.54E+01
x	x	c	t,p	U-233	1.59E+05	344	1.95E+03	5.67E+00	5.67E+00	5.67E+00	5.44E+00	5.67E+00
x	x	x	t,p	U-234	2.45E+05	344	7.51E+02	2.18E+00	3.28E+00	4.07E+00	4.10E+00	4.10E+00
	x	x	t,p	Th-230	7.70E+04	34	3.06E-01	8.90E-03	3.40E-02	1.19E-01	3.55E+00	3.55E+00
	x	c	t,p	Pu-242	3.76E+05	344	1.17E+03	3.40E+00	3.40E+00	3.40E+00	3.34E+00	3.40E+00
	x	c	t,p	Th-229	7.34E+03	344	9.97E+00	2.90E-02	8.20E-02	2.12E-01	3.40E+00	3.40E+00
	x		t,p	Np-237	2.14E+06	344	6.49E+01	1.89E-01	2.32E-01	3.14E-01	4.83E-01	4.83E-01
			t,p	Ra-226	1.60E+03	344	1.14E+01	3.31E-02	3.20E-02	2.94E-02	2.77E-01	2.77E-01
			t,p	Pb-210	2.23E+01	344	8.75E+00	2.54E-02	3.20E-02	2.97E-02	2.77E-01	2.77E-01
	x		t,p	U-238	4.47E+09	344	5.01E+01	1.46E-01	1.46E-01	1.46E-01	1.46E-01	1.46E-01
	x		t,p	U-236	2.34E+07	344	6.72E-01	1.95E-03	3.78E-03	8.28E-03	1.16E-01	1.16E-01
	x		t,p	Am-243	7.37E+03	344	3.25E+01	9.45E-02	9.36E-02	9.16E-02	3.69E-02	9.45E-02
	x		t,p	U-235	7.04E+08	344	1.75E+01	5.09E-02	5.12E-02	5.15E-02	7.06E-02	7.06E-02
	x		t,p	Cm-243	2.91E+01	344	2.07E+01	6.02E-02	5.29E-03	1.21E-05	0.00E+00	6.03E-02
			t	U-232	6.89E+01	344	1.79E+01	5.20E-02	1.99E-02	1.79E-03	0.00E+00	5.20E-02
			t	C-14	5.72E+03	344	1.28E+01	3.72E-02	3.69E-02	3.58E-02	1.11E-02	3.72E-02
	x		t,p	Th-232	1.41E+10	34	1.01E+00	2.94E-02	2.94E-02	2.94E-02	2.94E-02	2.94E-02
			t	Ac-227	2.18E+01	344	5.05E-01	1.47E-03	1.44E-03	1.69E-03	1.28E-02	1.28E-02
			t,p	Pa-231	3.28E+04	344	4.67E-01	1.36E-03	1.46E-03	1.72E-03	1.28E-02	1.28E-02
	x		p	Cm-248	3.39E+05	344	3.72E-02	1.08E-04	1.08E-04	1.08E-04	1.06E-04	1.08E-04
	x		p	Cm-245	8.53E+03	344	1.17E-02	3.40E-05	3.52E-05	3.66E-05	1.85E-05	3.66E-05
	x		p	Pu-244	8.26E+07	344	1.51E-06	4.39E-09	4.48E-09	4.68E-09	1.26E-08	1.26E-08
			p	Sm-147	1.06E+11	344	4.55E-10	1.32E-12	1.32E-12	1.32E-12	1.32E-12	1.32E-12
			p	Pm-147	2.62E+00		8.10E-04					
			p	Ra-228	5.75E+00		1.00E+00					
			p	Cf-252	2.64E+00		1.72E-04					
x	x		p	Cm-244	1.81E+01		7.44E+03					
x	x	c	p	Pu-241	1.44E+01		3.94E+05					
0.9996	0.9948	0.9947	:fraction of total EPA units at closure								

x = included in calculation,
c = combined with a transported isotope
t = one of top 24 isotopes
p = one of PANEL isotopes

---

[2] An EPA unit is a normalized unit obtained by dividing the curies of an isotope by the EPA release limit (in curies) for that isotope.

inventories were examined, the top 8 isotopes in Table 1 were chosen (indicated with an x in the table), along with two short-lived parents of these 8 isotopes because these isotopes dominate the inventory in all waste streams and account for 99.96% of the EPA units at the time of closure of the repository in 2033 A.D. For the faster-running code PANEL, it was possible to use more isotopes. Twenty-one isotopes were used for the PANEL brine release calculations, totaling 99.48% of the EPA units at closure. In the NUTS calculations, only five isotopes were directly modeled, as indicated by the x in the third column, but five additional isotopes were indirectly modeled by "lumping" their inventory with modeled isotopes with similar decay and transport properties. The resulting 10 isotopes account for 99.47% of the EPA units at closure. For the SECOTP, ^{238}Pu was dropped from the list of five lumped isotopes because the amount of it calculated to be delivered to the Culebra was small.

BRINE COMPOSITION

Actinide solubility and maximum actinide concentrations on colloids may vary significantly with oxidation state, pH, carbonate concentration, and brine composition. The pH and carbonate concentration within the repository were expected to be well controlled by the MgO backfill (Wang, 1996), leaving brine composition and oxidation state as the only major variables influencing solubility. Oxidation state is described in Section 6.3. Brine composition is described here.

The choice of brine compositions for modeling actinide solubility was determined by the brines present within the WIPP formations and the changes that would take place in their compositions when they enter the repository. There are three distinct brine compositions (Table 2) in the three WIPP formations (Salado, Castile, and Culebra sub-unit of the Rustler).

Under all scenarios, brine will flow from the surrounding Salado Formation through the disturbed rock zone (DRZ), and into the repository in response to the pressure difference between the repository at closure and the surrounding formation. In scenarios where a borehole is drilled into the repository but not into an underlying brine pocket, brine may flow down the borehole from the Rustler and Dewey Lake Formations. In scenarios where a pressurized Castile brine pocket is penetrated, brine from the Castile Formation may flow up the borehole into the repository. In this case there is expected to be very little Culebra brine flow down the borehole before the repository is filled by Castile brine flowing up from the brine pocket.

Table 2. Major Chemical Components of WIPP Brines (Brush, 1990)

Element or Chemical Property	Salado (Brine A)	Castile (ERDA-6)	Culebra (Air Intake Shaft)
HCO_3- (mM)	--	43	1.1
B^{3+} (mM)	20	63	2.8
Br– (mM)	10	11	0.37
Ca^{2+} (mM)	20	12	23
Cl– (mM)	5350	4800	567
K^+ (mM)	770	97	8.3
Mg^{2+} (mM)	1440	19	23
Na^+ (mM)	1830	4870	600
SO_4^{2-} (mM)	40	170	77
pH	6.5	6.17	7.7
Total Dissolved Solids (mg/liter)	306,000	330,000	42,600

The composition of the more dilute brines of the Rustler and Dewey Lake Formations is expected to change rapidly when it enters the repository due to fast dissolution of host Salado Formation minerals (about 93.2% halite and about 1.7% each of polyhalite, gypsum, anhydrite, and magnesite; Brush, 1990). If the dilute brines dissolve only the surfaces of the repository, they will attain Castile-like compositions, but if they circulate through the Salado Formation after saturating with halite, they may attain compositions within the range for Salado brine. The actual brine within the repository may be described as a mixture of the two concentrated brine "end members" -- Salado and Castile. The brine ratio in this mixture is, however, difficult to quantify, since it is both temporally and spatially variable. Only in the undisturbed scenario is the mixture well defined as 100% Salado brine over the 10,000-year time period. In the borehole intrusion scenarios, it is expected that the fraction of Salado brine within the mixture will be high in areas of the repository distant from the borehole and much lower near the borehole.

As seen in Table 2, Salado and Castile brines have different compositions, most notably in the Mg^{2+} concentration and ionic strength. These differences can cause significant differences in the actinide solubilities. Even after equilibrating with the MgO backfill, Salado brine has higher Mg^{2+} concentration and ionic strength than Castile. The higher ionic strength stabilizes highly charged actinide species, resulting in higher solubilities in some cases.

Since radioisotope transport up the borehole is required for significant release, it is the solubility of radioisotopes near the borehole that is most important. Given these uncertainties, and NUTS's requirement for time-independent solubilities, calculation of brine mixing was not attempted in the CCA calculations. Instead, actinide solubilities calculated in Castile brine were used for scenarios where a borehole hit a pressurized brine pocket and solubilities calculated in Salado brine were used for scenarios where it did not. This simplification should bracket the range of behavior of the repository and should therefore suffice for CCDF calculations.

PARAMETERS USED TO CALCULATE THE "EFFECTIVE SOLUBILITIES OF THE ACTINIDES

The parameters required for constructing the "effective solubility" were: (1) modeled solubilities for four oxidation states in each end-member brine, (2) a distribution of the uncertainty in the model solubility values, (3) the scheme for assigning sampled dominant oxidation states, (4) colloidal concentrations or proportionality constants for the 6 actinides and the 4 oxidation states for each of four colloid types, and (5) upper limits on the actinide concentrations that may be carried on two colloid types.

Solubility Modeling

There are numerous models for accurately calculating equilibrium concentrations in low ionic strength aqueous solutions; however, few can adequately describe solution behavior in concentrated electrolytes. The activity coefficient formalism of Pitzer (1991) has been parameterized for concentrated aqueous electrolyte systems by Harvie-Møller-Weare (HMW, Harvie et al., 1984), who have demonstrated the reliability of this formalism for predicting mineral solubility using chemical equilibrium in electrolyte solutions from zero to high ionic strength. The database developed by Harvie et al. (1984) includes all parameters necessary to predict equilibrium in the Na^+-K^+-H^+-Ca^{2+}- Mg^{2+} - OH^-- SO_4^{2-}- HSO_4^--Cl^- - HCO_3^-- CO_3^{2-} - CO_2 - H_2O system. This database serves as the reference database for the parameterization of the WIPP model.

The Pitzer ion-interaction model (Pitzer, 1973, 1991) is a semiempirical model for calculating activity coefficients for high ionic strength, and has been proven accurate for calculating solubilities in multicomponent electrolyte solutions (Harvie et al., 1984). In the Pitzer model, the excess free energy is represented by combining a modified Debye-Hückel equation for the dilute solution region with a virial expansion for higher ionic concentrations. The model can be reliably extended to multicomponent systems using parameters derived from binary and ternary systems (Harvie and Weare, 1980; Harvie et al., 1984).

The Pitzer model has been successfully used with marine evaporite systems (Harvie et al., 1980; Eugster et al., 1980; Brantley et al., 1984) mineral precipitation in lakes (Spencer et al., 1985; Felmy and Weare, 1986), and actinide solubility in high ionic strength brines (Novak et al., 1996; Al Mahamid et al., 1997). Pitzer (1991) gives a complete description of the formalism and model parameters for many ionic species.

Parameterization of the Pitzer Model and Solubility Calculations

Pitzer parameters or data used to calculate Pitzer parameters for actinides in the +III oxidation state in high ionic strength brines have been reported by Fanghanel et al., 1994, Felmy et al., 1995, Felmy et al., 1990, Felmy et al, 1989, Novak et al, 1995, Rai et al, 1992, Rai et al 1995, Rai et al, 1992, and Rao et al , 1997. Data or Pitzer parameters for actinides in the +IV oxidation state have been reported by Felmy and Rai, 1992, Felmy et al, 1991, Felmy et al, 1997, Rai et al, 1995, and Rai et al, 1996. Data or Pitzer parameters for actinides in the +V oxidation state have been reported by Novak and Roberts, 1994, Al Mahamid et.al., 1997, Roy et al, 1992, Fanghanel et al, 1995, and Neck et al, 1995. Stability constants for actinide-organic ligands have been reported by Pokrovsky ete al, 1998, Novak et al, 1996, and Martell and Smith, 1977. Data for much of the organic ligand work performed for WIPP has either already been published in the literature (including this volume) or will be published in the near future. The ligands investigated in this work included acetate, lactate, oxalate, citrate, and ethylenediamine tetraacetic acid (EDTA). These compounds were chosen for their aqueous solubility and because they were expected to be present in WIPP waste. Empirical solubility experiments to test and challenge the WIPP solubility model are under way at Argonne National Laboratory and at Pacific Northwest National Laboratory (PNNL). The papers in this volume describe some of the experimental methods and procedures used to determine the data.

Experiments were conducted with one actinide for each of the four possible oxidation states and, through the oxidation state analogy (in which actinides in the same oxidation state tend to exhibit similar behavior), bounding models were developed to represent each oxidation state of the actinides. The actinides for which data were obtained americium, thorium, neptunium, and uranium in the +III, +IV, +V, and +VI oxidation states, respectively. For example, the solubility of Pu in the +III and +IV states was represented by the solubility of Am(III) and Th(IV). Sufficient data have been collected to parameterize the model for +III, +IV, and +V actinides. Data for +VI are still being developed. Since a complete model for +VI actinides was unavailable, literature data from empirical solubility studies for U(+VI) were used to determine a +VI actinide solubility. The use of solubility bounding analogs for each oxidation state instead of actual values for each element in each oxidation state required fewer experiments and greatly simplified the solubility model. Because the analogs provide a bounding case for each oxidation state, the model always predicts conservative values for actinide solubilities that were used in WIPP performance assessment calculations.

The experimental data were used to calculate Pitzer parameters and standard chemical potentials using the NONLIN V2.0m software, which is a Gibbs energy minimization routine developed by A. Felmy at PNNL. The complete parameterized database was used in conjunction with the FMT V2.0 aqueous speciation computer code developed by C. Novak at Sandia National Laboratories for use on a DEC/Alpha platform to predict maximum actinide equilibrium solubilities in Salado and Castile brines. Organic ligands were not included in solubility calculations. Utilizing the results of organic ligand experiments at Florida State University, divalent cations (e.g., Mg^{2+}, Ca^{2+}, Fe^{2+}, Ni^{2+}) ions present in large quantities in the disposal room have been shown to compete with the actinides for complexation with the ligands. The effect is that actinide concentrations will not be significantly enhanced by the presence of organic ligands (U.S. Department of Energy, 1996). Calculations for both Salado and Castile brines were performed assuming the brines were equilibrated with MgO backfill. Sufficient MgO backfill will be emplaced with the waste to react quickly with any microbially produced CO_2, preventing a drop in pH and a rise in dissolved carbonate species. The results for actinide solubility calculations for each oxidation state are given in Table 3.

Table 3. Modeled Solubilities for Actinides in the Four Oxidation States and Two Brines [pH and $f(CO_2)$ controlled by the $Mg)OH)_2$ - $MgCO_3$ buffer] (DOE, 1996, appendix SOTERM)

	+IV	+V	+VI
Salado	4.4e–06	2.3e–06	8.7e–06
Castile	6.0e–09	2.2e–06	8.8e–06

Distributions

After the model solubilities were obtained from Novak (1997), solubility distributions were calculated by Bynum (1996), who compared calculated solubilities with those observed experimentally and prepared a distribution of the divergence of model solubilities from experimental results. Figure 3 shows the distributions of the observed solubilities around the model solubilities and is plotted as the log of the molar concentration. It can be seen that the observed solubilities are up to 2 orders of magnitude lower than those calculated by the model and in some cases are up to 1.4 orders of magnitude above the model solubilities. These values were used as the total uncertainty in mode predicted solubility. Uncertainties in the model and the data were not treated as separate, but were "lumped" into parameter uncertainty. The information was entered into the parameter database as the actual points on the cumulative distribution.

Oxidation States

The solubilities of actinides are dependent on the actinide oxidation-state distributions (Weiner, 1996), which are expected to be determined by reactions of the actinides with the major components of the waste. Microbially mediated reactions with the organic waste and anoxic corrosion of the steel waste containers, which results in the production of dissolved Fe(+II), are expected to have the largest impact on the oxidation-state distribution, but because the propensity of plutonium to exist in multiple oxidation states, and lack of redox data in concentrated brines, it is impossible to define a single redox state for the repository.

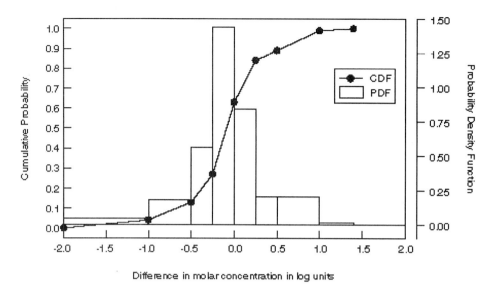

Figure 3. Distribution of actinide log solubilities (Bynum, 1996).

It is expected that the redox state of the repository may range from "reducing" to "extremely reducing," and experiments (Weiner, 1996) have shown that for most of the actinides, high oxidation states will not persist. The most likely persistent oxidation states for the six actinides are shown in Table 4.

For U, Np, and Pu, two oxidation states are possible under the anticipated repository conditions. While both may coexist, one will dominate the solubility. A bounding case was created through independent sampling of each oxidation state at its thermodynamic maximum solubility.

Colloid Parameters

Colloid parameters were obtained from Papenguth (1996a-d). These were of two types: the steady-state colloid actinide concentrations for mineral fragments and actinide intrinsic colloids, and proportionality constants for humic and microbe colloids. Proportionality constants indicated the amount of actinides on the humic or microbe colloids versus the concentrations in solutions. There were also concentration caps for those colloid types.

For the steady-state colloid actinide concentrations, Papenguth (1996a) provided a range of 2.6^{-10} to 2.6^{-8} moles per liter for Am, Pu, U, Th, and Np sorbed on mineral fragments. For actinide intrinsic colloids, Papenguth (1996b) provided a concentration of 10^{-9} M for Pu. The other actinide intrinsic colloids were not stable but agglomerated and settled out, so 0 M was used.

For humic and microbe colloids, Papenguth (1996c,d) provided proportionality constants and concentration upper limits or caps as shown below in Table 5. For humic colloids, the cap was based on the maximum steady-state number of sorption sites on the suspended humic colloids. For microbe colloids, the cap was based on the actinide concentration that was toxic to the microbes. Note that for microbe colloids, the proportionality constant is provided by the actinide element, but the humic proportionality constant is provided by the oxidation state and brine composition. A comparison of the WIPP data to literature data was not possible due to a lack of data for colloid actinide interactions under WIPP conditions in the literature.

Table 4. Possible Oxidation States of Actinides under
Repository Conditions (DOE, 1996)

Actinide	Oxidation State
Am	+III
Cm	+III
Np	+IV, +V
Pu	+III, +IV
Th	+IV
U	+IV, +V

Table 5. Proportional Constants and Concentration Upper Limits or Cap for Microbial
and Humic Colloids used in PA (DOE, 1996).

Microbial Colloids			
Actinide	Proportionality Constant		Cap
Am	3.6		c
Pu	0.3		6.8e-5
U	2.1e-3		2.1e-3
Th	3.1		1.9e-3
Np	12		2.7e-3
Humic Colloids			
Total Actinide by Oxidation State	Proportionality Constant[d]		Cap[b]
	Salado	Castile	Both Brines
+III	0.008 to 0.19	0.065	1.1e-5
+IV	6.3	6.3	1.1e-5
+V	5.3e-5 to 9.1e-4	4.3e-4 to 7.4e-3	1.1e-5
+VI	0.008 to .12	0.062 to 0.51	1.1e-5

[a] Moles of mobile microbial actinide (An) / moles dissolved An.

[b] Cap on total moles mobile An / liter

[c] High cap entered so that only inventory limit would cap Am mobilized on microbes (^{241}Am
inventory limit = 538 moles / 3e+4m^3 = 1.8e-5 M)

[d] Moles of mobile humic-bound An / moles dissolved An.

Combining the Source Term Parameters into an Effective Solubility

Before use by NUTS and PANEL, the uncertain dissolved and colloidal parameters were sampled, resulting in 100 sample sets of parameters. For each sample set, the parameters were then combined into "effective solubilities" for each of the actinides as described below. Parameters that were sampled, and values derived from them, have been indicated by italics. Parameters read from the parameter database are in boldface type.

Dissolved Solubility = **Model Solubility** * $10^{Sampled\ from\ Solubility\ Distribution}$

Humic Colloid Concentration = Dissolved Solubility * **Proportionality Constant**
        if *Dissolved* * **Prop. Const.** < **Humic Cap**, otherwise

Humic Colloid Concentration = **Humic Cap**

Microbe Colloid Concentration = Dissolved Solubility * **Proportionality Constant**
        if the *Total Mobile* < **Microbe Cap**, otherwise
Microbe Colloid Concentration = **Microbe Cap**

Mineral Colloid Concentration = **Database Concentration**

Intrinsic Colloid Concentration = **Database Concentration**

Total Mobile = Dissolved + Humic + Microbe + Mineral + Intrinsic

LOGSOLM = log_{10}(Total Mobile)

where LOGSOLM is the log of the "effective solubility" in moles/liter used by NUTS and PANEL. Table 7 shows LOGSOLM for each brine and oxidation state calculated using median values for all sampled parameters.

Table 6. Median "Effective Log Solubilities" for Each Brine and Oxidation State (DOE, 1996)

Brine	Am(+III), Cm(+III)	Pu(+III)	Pu(+IV)	U(+IV)	U(+VI)	Th(IV)	Np(IV)	Np(V)
Salado	-5.64	-6.14	-4.80	-4.84	-5.10	-4.59	-4.17	-4.52
Castile	-6.47	-6.77	-7.19	-7.16	-4.96	-7.05	-6.85	-4.54

For actinides with more than one oxidation state, the above procedure was performed for each oxidation state, and the final total mobile concentration is set based on the oxidation state parameter:
Total mobile = Total mobile(lower oxidation state) if OXSTAT ≤ 0.5
        = Total mobile(higher oxidation state) if OXSTAT > 0.5
where OXSTAT is the oxidation-state parameter that is sampled uniformly from 0 to 1.

RESULTS

The results of the sampling and the construction of the total dissolved and colloid concentrations resulting from one set of calculations in the compliance certification application are shown in Figure 4. One hundred realizations were binned into half order of

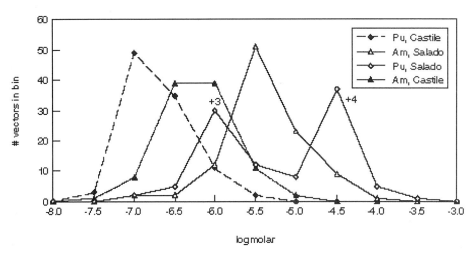

Figure 4. Total maximum log concentration (dissolved + colloid) Replicate 1. (DOE, 1996)

magnitude solubilities. The figure shows the number of vectors in each half order of magnitude bin. The "effective solubilities" are shown for only Am and Pu, since they account for over 99% of the EPA units of the inventory throughout the 10,000-year regulatory period.

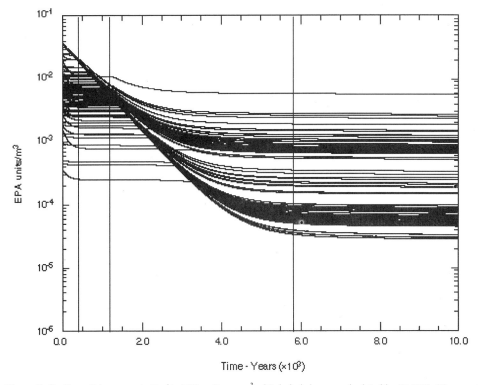

Figure 5. Radionuclide concentration in EPS units per m^3 of Salado brine, as calculated by PANEL (Garner et al., 1988).

Figures 5 and 6 show the radionuclide concentrations in EPA units per cubic meter of brine as a function of time as calculated by PANEL for the direct brine release calculations. These concentrations are the sum of the concentrations of all 21 isotopes included in the PANEL calculation. In each of these figures, one can see four "regions." (1) In the first several hundred years, there is a drop in the EPA unit concentrations of some realizations as the EPA units of ^{238}Pu decays to below that of ^{241}Am. (2) There is then a constant concentration seen for the period where ^{241}Am controls the total EPA unit concentration and is solubility limited. For the realizations that sampled a higher Am solubility, this period is shorter than for the realizations that sampled a lower Am solubility. (3) After the Am changes from solubility to inventory limited, the EPA unit concentration drops until (4) the ^{239}Pu solubility controls the EPA unit concentration. In region 4, the higher concentrations are constant but the lower concentrations show a slow decrease with time, because the sampled ^{239}Pu solubility is low enough that other isotopes which are inventory limited and have intermediate half-lives contribute to the total EPA unit concentrations. The spread in concentrations seen in region 2 reflects the spread in Am solubility, and the higher Am solubility in Salado brine than in Castile brine is clearly seen. Similarly, the region 4 spread mainly reflects the spread in Pu(+III) and Pu(+IV) solubilities. As can be seen, the solubility of both oxidation states of Pu is quite low in Castile brine, but in Salado brine there is a bimodal distribution showing the higher solubility of Pu(+IV) and lower solubility of Pu(+III).

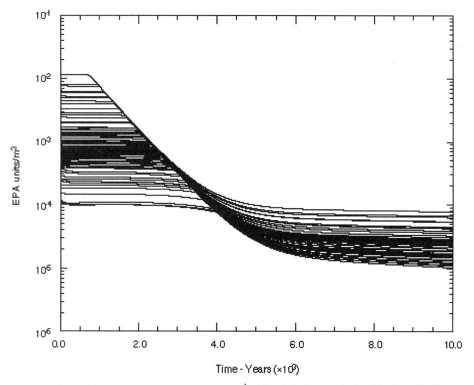

Figure 6. Radionuclide concentration in EPS units per m^3 of Castile brine, as calculated by PANEL (Garner et al., 1988).

This work was supported by the United States Department of Energy under Contract DE-ACO4-94AL85000.

REFERENCES

Note: Some of the work cited here (analysis packages) was prepared specifically for the Waste Isolation Pilot Plant and is available in the Sandia National Laboratories record center or the EPA public docket. Solubility and colloid work by Bynum, Papenguth, Weiner, and Wang also appear elsewhere in this volume.

Berglund, J.W., 9/6/98, Analysis Package for the Cuttings and Spallings Calculations (Tasks 5 & 6) of the Performance Assessment Analysis Supporting the Compliance Certification Assessment (WPO# 40521)

Brantley, S.L., D.A. Crerar, N.E. Moller, and J.H. Weare. 1984. "Geochemistry of a ModernMarine Evaporite: Bocana de Virrila," *Peru. J. Sed Petrol.* **54**: 462-477.

Brush, L.H. 1990. Test Plan for Laboratory and Modeling Studies of Repository and Radionuclide Chemistry for the Waste Isolation Pilot Plant, SAND90-0266, Sandia National Laboratories, Albuquerque, NM.

Bynum, R. V. 1996. Estimation of Uncertainties for Predicted Actinide Solubilities, WPO# 40512, Sandia National Laboratories, Albuquerque, NM.

DOE (U.S. Department of Energy). 1995. Waste Isolation Pilot Plant Transuranic Waste Baseline Inventory Report; DOE/CAO-95-1121; Revision 2. Department of Energy,

DOE(U.S. Department of Energy); DOE/CAO-1996-2184; Title 40 CFR 191 Compliance Certification Application, October 1996; DOE, Carlsbad Area Office, Carlsbad, NM. Washington, DC.

Eugster, H.P., C. E. Harvie, and J. H. Weare. 1980. "Mineral Equilibria in a Six-Component Seawater System, Na-K-Mg-Ca-Cl-SO4-H2O at 25°C," *Geochim. Cosmochim. Acta* **44**:1335.

Fanghanel, Th., J.I. Kim, P. Paiviet, and W. Hauser 1994. "Thermodynamics of Radioactive Trace Elements in Concentrated Electrolyte Solutions< Hydrolysis of Cm^{3+} in NaCl Solutions", *Radiochimica Acta* **66/67** 81-87.

Fanghanel, Th., V. Neck and J.I. Kim 1995 "Termodynamics of Neptunium(V) in Concentrated Salt Solutions:II. Ion Interaction (Pitzer) Parameters for Np(V) Hydrolysis Species and Carbonate Complexes" *Radiochimica Acta* **69** 169-176.

Felmy, A.R., and J.H. Weare. 1986. "The Prediction of Borate Mineral Equilibria in Natural Waters: Application to Searles Lake, California" *Geochim. Cosmochim. Acta* **50** 2771-2783.

Felmy, A.R., and D. Rai. 1992. "An Aqueous Thermodynamic Model for a High Valence 4:2 Electrolyte Th^{4+}-SO_4^{2-} in the System Na^+-K^+-Li^+-NH_4 -SO_4^{2-} -HSO_4 -H_2O to High Concentration." *Journal of Solution Chemistry* **21**(5) 407-423.

Felmy, A.R., D. Rai, and M.J. Mason. 1991. "The Solubility of Hydrous Thorium(IV) Oxide in Chloride Media: Development of an Aqueous Ion-Interaction Model." *Radiochimica Acta* **55** 177-185.

Felmy, A.R., D. Rai, J.A. Schramke, and J.L. Ryan. 1989. "The Solubility of Plutonium Hydroxide in Dilute Solution and in High-Ionic-Strength Chloride Brines," *Radiochim. Acta* **48**: 29-35.

Felmy, A.R., D. Rai, S.M. Sterner, M.J. Mason, N.J. Hess, and S.D. Conradson 1998 in preparation.

Felmy, A.R., D. Rai, and R.W. Fulton. 1990. "The Solubility of $AmOHCO_3$(c) and the Aqueous Thermodynamics of the System Na^+-Am^{3+}-HCO_3 -OH^--H_2O." *Radiochimica Acta* **50** 193-240.

Felmy, A.R., D. Rai, M.J. Mason and R W. Fulton 1995. "The Aqueous Complexation of
Nd(III) with Molybdate: The Effects of Both Monomeric Molybdate and
Polymolybdate Species", *Radiochimica Acta* **69** 177-193

Felmy, A.R., D. Rai, J.A. Schramke, and J.L. Ryan. 1989. "The Solubility of Plutonium
Hydroxide in Dilute Solution and in High-Ionic-Strength Chloride Brines."
Radiochimica Acta **48** 29-35.

Garner, J., A. Shinta, and C. Stockman, 9/6/98, Analysis Package for the Salado Transport
Calculations (Task 2) of the Performance Assessment Analysis Supporting the
Compliance Certification Assessment (WPO# 40515)

Harvie, C.E., and J.H. Weare. 1980. "The Prediction of Mineral Solubilities in Natural
Waters: The Na-K-Mg-Ca-Cl-SO$_4$-H$_2$O System from Zero to High Concentration at
25°C," *Geochim. Cosmochim. Acta* **44**: 981-997.

Harvie, C.E., J.H. Weare, L.A. Hardie, and H.P Eugster. 1980. "Evaporation of Seawater:
Calculated Mineral Sequences." *Science* **208**: 498-500.

Harvie, C.E., N. Møller, and J.H. Weare. 1984. "The Prediction of Mineral Solubilities in
Natural Waters: The Na-K-Mg-Ca-H-Cl-SO$_4$-OH-HCO$_3$-CO$_3$-CO$_2$-H$_2$O System to
High Ionic Strength at 25°C," *Geochim. Cosmochim. Acta* **48**: 723-751.

Harvie, C.E., J.P. Greenberg, and J.H. Weare. 1987. "A Chemical Equilibrium Algorithm
for Highly Non-ideal Multiphase Systems: Free Energy Minimization," *Geochim.
Cosmochim. Acta* **51** (5): 1045-1057.

Helton, J.C., J.D. Johnson, and L.N. Smith. 9/6/96, Analysis Package for the
Complementary Cumlative Distribution Function (CCDF) Construction (Task 7) of
the Performance Assessment Calculations Supporting the Compliance Certification
Application (CCA), AP-AAD, (WPO#040524)

Jow, H.N, J.L. Ramsey, and M.G. Wallace, 9/6/98, Analysis Package for the Culebra Flow
and Transport Calculations (Task 3) of the Performance Assessment Analysis
Supporting the Compliance Certification Assessment (WPO# 40516),

Morel, F.M and J.G. Hering 1993 *Principles and Applications of Aquatic Chemistry*. John
Wiley & Sons New York, New York

Neck, V., Th. Fanghanel, G. Rudolph and J.I. Kim 1995 "Thermodynamics of
Neptunium(V) in Concentrated Salt Solutions:Chloride Complexation and Ion
Interaction (Pitzer) Parameters for NpO2+ Ion" *Radiochemica Acta* **69** 39-47.

Novak, C.F., and K.E. Roberts. 1994. *Thermodynamic Modeling of Neptunium(V)
Solubility in Na-CO$_3$-HCO$_3$-Cl-ClO$_4$-H-OH-H$_2$O Electrolytes*. SAND94-0805C.
Albuquerque, New Mexico: Sandia National Laboratories.

Novak, C.F., M. Borkowski, and G.R. Choppin. 1996. "Thermodynamic Modeling of
Neptunium(V)-Acetate Complexation in Concentrated NaCl Media," *Radiochimica
Acta* (in press).

Novak, C.F., I. Al Mahamid, K.A. Becraft, S.A. Carpenter, N. Hakem, and T. Prussin. 1997.
"Measurement and Thermodynamic Modeling of Np(V) Solubility in Dilute Through
Concentrated K$_2$CO$_3$ Media," manuscript in preparation.

O'Brian, D. Solutions, and D.M. Stoelzel, 9/6/98, Assessment Analysis Supporting the
Compliance Certification Assessment (WPO# 40520) Analysis Package for the
BRAGFLO Direct Release Calculations (Task 4) of the Performance

Papenguth H. W. 1996a. Parameter Record Package for Mobile-Colloidal Actinide Source
Term. Part 1. Mineral Fragment Colloids, WPO# 35850, Sandia National
Laboratories, Albuquerque, NM.

Papenguth H. W. 1996b. Parameter Record Package for Mobile-Colloidal Actinide Source
Term. Part 2. Actinide Intrinsic Colloids, WPO# 35852, Sandia National
Laboratories, Albuquerque, NM.

Papenguth H. W. 1996c. Parameter Record Package for Mobile-Colloidal Actinide Source Term. Part 3. Humic Substances, WPO# 35855, Sandia National Laboratories, Albuquerque, NM.

Papenguth H. W. 1996d. Parameter Record Package for Mobile-Colloidal Actinide Source Term. Part 4. Microbes, WPO# 35856, Sandia National Laboratories, Albuquerque, NM

Pitzer, K.S. 1973. Thermodynamics of Electrolytes - I. Theoretical Basis and General Equations: *J. Phys. Chem.* **77**: 268-277.

Pitzer, K.S., ed. 1991. *Activity Coefficients in Electrolyte Solutions*, 2nd Ed., CRC Press, Boca Raton, Florida.

Pokrovsky, O.S., M.G. Bronikowski, R.C. Moore, and G.R. Choppin. 1998. "Interaction of Neptunyl(V) and Uranyl(VI) with EDTA in NaCl Media: Experimental and Pitzer Modeling", *Radiochimica Acta* **80**, 23-29.

Rai, D., A.R. Felmy, and R.W. Fulton. 1992a. "Solubility and Ion Activity Product of $AmPO_4 \cdot xH_2O(am)$." *Radiochimica Acta* **56** 7-14.

Rai, D., A.R. Felmy, and R.W. Fulton. 1994. "The Nd^{3+} and Am^{3+} Ion Interactions with SO_4^{2-} and their Influence on $NdPO_4(c)$ Solubility." *Journal of Solution Chemistry.* **24**(9) 879-895

Rai, D., A.R. Felmy, R.W. Fulton, and J.L. Ryan. 1992 "Aqueous Chemistry of Nd in Borosilicate-Glass/Water Systems." *Radiochimica Acta* **58/59** 9-16.

Rai, D., A.R. Felmy, D.A. Moore, and M. Mason 1995 "The Solubility of Th(IV) and U(VI) Hydrous Oxides in Concentrated NaHCO3 and Na2CO3 Solutions" Materials Resources Symposium Proceedings **353** 1143-1150

Rai, D., A.R. Felmy, S.M. Sterner, D.A. Moore, M.J. Mason, and C..F. Novak. 1995 The Solubility of Th(IV) and U(VI) Hydrous Oxides in Concentrated NaCl and MgCl2 Solutions" Presented at the Fifth International Conference on the "Chemistry and Migration Behavior of Actinides and Fission Products in the Geosphere", Saint-Malo, France, September 10-15, 1995

Rao, L., D. Rai, A.R. Felmy, R.W. Fulton, and C.F. Novak 1998 in preparation.

Roy, R.N., K.M. Vogel, C.E. Good, W.B. Davis, L.N. Roy, D.A. Johnson, A.R. Felmy, and K.S. Pitzer. 1992. "Activity Coefficients in Electrolyte Mixtures: HCl + ThCl₄ + H₂O for 5°–55°C." *Journal of Physical Chemistry* **96** 11065-11072.

Spencer R.J., H. Eugster, and B.F. Jones, "Geochemistry of the Great Salt Lake, Utah II: Pleistocene-Holocene Evolution," *Geochim. Cosmochim. Acta* **49**: 739-747.

Stumm, W. 1992. *Chemistry of the Solid-Water Interface*, John Wiley & Sons, Inc. New York, New York.

Sanchez, L. C., and J. Liscum-Powell, J. S. Rath, and H. R. Trellue. 1997. WIPP PA Analysis Report for EPAUNI: Estimating Probability Distribution of EPA Unit Loading in the WIPP Repository for Performance Assessment Calculations, Version 1.01, WPO# 43843, Sandia National Laboratories, Albuquerque, NM.

Siegel, M. D. 1996. Solubility Parameters for Actinide Source Term Look-Up Tables, WPO# 35835, Sandia National Laboratories, Albuquerque, NM.

Stockman, C.T., A. Shinta, and J. Garner. 1997. Analysis Package for the Salado Transport Calculations (Task 2) of the Performance Assessment Analysis Supporting the Compliance Certification Assessment (WPO# 40515), Sandia National Laboratories, Albuquerque, NM.

Vaughn, P.M., 9/5/96, Analysis Package for the Salado Flow Calculations (Task 1) of the Performance Assessment Analysis Supporting the Compliance Certification Assessment (WPO# 40514)

Vold, R.D. and M.J. Vold 1993 *Colloid and Interface Chemistry*. Addison-Wesley Publishing Co., London. England

Wang, Y. 1996. "Define Chemical Conditions for FMT Actinide Solubility Calculations," WPO# 30819, Sandia National Laboratories, Albuquerque, NM.

Weiner, R. F. 1996. Parameter Package for the Oxidation State Distribution of Actinides in the WIPP, WPO #35194, Sandia National Laboratories, Albuquerque, NM.

CHEMICAL SPECIES OF PLUTONIUM IN HANFORD SITE RADIOACTIVE TANK WASTES

G. S. Barney[1] and C. H. Delegard[2]

[1]B&W Hanford Company
Richland, Washington, 99352
[2]Pacific Northwest National Laboratory
Richland, Washington, 99352

INTRODUCTION

Large quantities of radioactive wastes have been generated at the Hanford Site over its operating life. Wastes with the highest activities are stored in 177 large (mostly 4 million liter capacity) underground concrete tanks with steel liners. The wastes consist of processing chemicals, cladding chemicals, fission products, and actinides. Waste solutions from chemical processing facilities were made alkaline by adding sodium hydroxide before sending the wastes to the underground tanks. This was done to prevent corrosion of the steel liners by acidic wastes. The wastes generally consist of sludge layers generated by precipitation of dissolved metals from aqueous waste solutions during addition of sodium hydroxide, salt cake layers formed by crystallization of salts after evaporation of the supernate solution, and aqueous supernate solutions that exist as separate layers or as liquid contained interstitially between sludge or salt cake particles. Because the mission of the Hanford Site was to provide plutonium for defense purposes, the amount of plutonium lost to the wastes was relatively small. The best estimate of the amount of plutonium lost to all the waste tanks is about 500 kg. Given uncertainties in the measurements, some estimates are as high as 1000 kg (Roetman et al., 1994).

Knowledge of the identity of plutonium chemical species in these wastes allows a better understanding of its behavior during waste storage, retrieval, and processing. Plutonium chemistry in the wastes is important to criticality and environmental concerns, and in processing the wastes for final disposal.

Plutonium has been found to exist mainly in the sludge layers of the waste along with precipitated metal hydrous oxides. This is expected because of the known low solubility of plutonium hydrous oxides in alkaline aqueous solutions. Tank supernate solutions do not contain high concentrations of plutonium even though some wastes contain high concentrations of complexing agents. Most solutions also contain significant concentrations of hydroxide which compete with other potential complexants. The sodium nitrate and sodium phosphate salts that form most of the salt cake have little interaction

Actinide Speciation in High Ionic Strength Media, edited by Reed *et al.*
Kluwer Academic / Plenum Publishers, New York, 1999

83

with plutonium in the wastes and contain relatively small plutonium concentrations. For these reasons we will consider plutonium species in the sludges and aqueous solutions only.

The low concentrations of plutonium in waste tank aqueous solutions and in the solid sludges prevent identification of chemical species of plutonium by ordinary analytical techniques. Spectrophotometric measurements are not sensitive enough to identify plutonium oxidation states or complexes in these waste solutions. Identification of solid phases containing plutonium in sludge solids using x-ray diffraction (XRD) or microscopic techniques would be extremely difficult. Because of these technical problems, plutonium speciation must be extrapolated from known behavior observed in laboratory studies of synthetic waste or of more chemically simple systems.

ORIGIN AND COMPOSITION OF WASTE

The U.S. Manhattan Engineering District created the Hanford Site in 1943 to produce plutonium for nuclear weapons. Plutonium production operations on the Site ceased in 1990. Three separation (reprocessing) methods were used to produce pure plutonium product from nitric acid solutions of dissolved irradiated uranium metal fuel (Gerber, 1996; Kupfer et al., 1997; Benedict et al., 1981). All relied on changes in the plutonium oxidation state to achieve separation and decontamination from uranium, fission products, and other materials. The Bismuth Phosphate Process (1944-1956) was based on cyclic batchwise carrier precipitations with $BiPO_4$ and LaF_3. Separations were achieved because Pu(IV) is carried and Pu(VI) is not carried. Alternate precipitations with plutonium in the tetravalent and hexavalent states achieved a high degree of separation from uranium and fission products. The REDOX Process (1951-1967) used continuous solvent extraction in methyl isobutyl ketone; Pu(VI) is extractable, Pu(III) is not. The PUREX Process (1956-1972, 1983-1990) used extraction in tributyl phosphate; Pu(IV) is extractable, Pu(III) is not.

Each process generated radioactive wastes containing spent process chemicals, dissolved aluminum- or zirconium-based fuel cladding, solvents and solvent decomposition products, fission products, uranium and plutonium process losses, and other transuranium elements (e.g., americium, neptunium). The wastes were adjusted to pH >8 and then discharged to storage in mild steel-lined reinforced concrete underground tanks. Because the Bismuth Phosphate Process did not recover uranium, uranium-rich waste also was discharged. The valuable uranium subsequently was reclaimed by tributyl phosphate solvent extraction in the Uranium Recovery Process (1952-1958).

The costs of tank storage drove each evolving process to optimize the use of process chemicals. Waste volumes of 64 liters/kg of uranium in the Bismuth Phosphate Process in 1945 were decreased to 1.3 liters/kg of uranium in the PUREX Process. Other methods were used to gain tank storage space. The isotopes $^{134,137}Cs$, $^{89,90}Sr$, and ^{60}Co were removed from dilute Bismuth Phosphate Process waste solutions by nickel ferrocyanide, calcium/strontium phosphate, and nickel sulfide coprecipitation, respectively (Waste Scavenging Process, 1954-1958). Discharge of the resulting decontaminated solution to the ground freed about 140 million liters, or 45 percent of the existing tank capacity.

The waste volume was further decreased by evaporation of water to concentrate liquors and form sodium nitrate and phosphate salt cakes. Practical limits in waste volume reduction, imposed by fission product isotope heat loading on the tanks, were addressed by removal of $^{134,137}Cs$ (ion exchange) and $^{89,90}Sr$ (solvent extraction) and their separate encapsulation (Waste Fractionation Process, 1967-1985). Organic complexing agents [ethylenediaminetetraacetate (EDTA), N-2-hydroxyethylethylenediaminetriacetate (HEDTA), citrate, and glycolate] were used in this process to sequester iron and aid in waste dissolution. Additional volume reduction by vacuum evaporators created solutions as high as 13 M in sodium and crystallized sodium nitrate and other salts [$NaNO_2$, Na_2CO_3, $NaAl(OH)_4$, and $Na_3PO_4 \cdot 12H_2O$]. Laboratory and equipment decontamination wastes, Plutonium Finishing Plant (PFP) plutonium scrap recovery wastes, ion exchange resins, and diatomaceous earth and portland cement liquid sorbents also have been added to some tanks.

About 330,000 metric tons of radioactive waste currently are stored in the 177 underground tanks on the Site (Gephart and Lundgren, 1996). The waste compositions are dominated by the facts that the plutonium separation processes were conducted in nitric acid and the wastes were made alkaline with sodium hydroxide prior to their disposal. Water and sodium nitrate thus constitute about 75 wt% of the tank waste; sodium nitrate-rich crystalline salt cake comprises about 42 percent of the waste volume. Water soluble sodium salts are found in liquors (about 33 percent of the waste volume) and water-insoluble compounds are found in liquor-saturated sludge (about 25 vol%). Plutonium primarily is found in the sludge with other metal hydrous oxides; lower plutonium concentrations are found in the liquors (Table 1). The sodium salt cakes have little interaction with plutonium and thus contain about 10^{-6} to 10^{-7} moles plutonium per kg dry salt cake, about a factor of 100 below the concentrations observed in dry sludge.

Table 1. Composition of Hanford Site tank solutions and sludges.

Component	Concentration in Solutions (M) (Tank Number)					Concentration in Sludges (mole/dry kg) (Tank Number)			
	BiPO₄ Sludge Liquor (107-T)	Complexant Concentrate (102-AN)	Double Shell Slurry (101-AW)	Waste Fract. (106-C)	Dilute Waste (108-AP)	BiPO₄ Sludge (107-T)	REDOX Sludge (104-S)	PUREX Sludge (101-AZ)	PFP Sludge (102-SY)
Al	0.0008	0.56	1.0	0.01	0.04	1.4	6.6	1.6	3.0
Bi	0.00008	NR*	<0.0006	0.005	NR	0.13	<0.0003	NR	NR
Ca	0.0001	0.01	0.0008	NR	NR	0.04	0.16	0.09	0.08
Cr	0.003	0.006	0.003	NR	NR	0.02	0.07	0.05	0.40
Fe	0.0004	0.0009	<0.0008	NR	<0.0001	1.2	0.05	2.7	0.26
Mn	0.000001	0.0007	0.0005	NR	NR	0.009	0.03	0.08	0.08
Na	2.3	10.4	10.0	4.0	0.76	12.9	8.0	6.0	9.7
Ni	0.0001	0.006	NR	NR	NR	0.01	0.001	0.12	0.06
Si	0.003	<0.0007	<0.005	0.05	NR	0.49	0.07	0.32	0.09
U	0.001	<0.0008	0.0009	0.002	NR	0.25	0.04	0.04	0.06
Zr	0.000009	NR	<0.0002	NR	NR	0.002	0.0006	0.59	NR
F^-	0.03	0.10	<0.04	0.009	0.03	1.4	0.01	0.20	0.06
NO_2^-	0.28	1.8	2.2	0.21	0.08	0.58	0.86	1.6	1.3
NO_3^-	1.4	3.6	3.5	1.1	0.24	2.7	4.7	1.3	3.5
PO_4^{3-}	0.06	0.05	0.02	0.04	0.003	2.7	0.005	0.01	0.28
SO_4^{2-}	0.11	0.14	0.01	0.05	0.005	0.24	0.04	0.31	0.67
CO_3^{2-}	0.15	1.1	0.21	0.75	0.14	0.56	0.11	0.77	0.92
OH^-	NR	0.21	5.1	0.01	0.16	NR	NR	NR	1.8
TOC	0.07	2.2	0.21	0.94	0.03	0.18	0.22	1.3	1.2
Pu	<6.8x10⁻⁷	3.5x10⁻⁷	6.9x10⁻⁸	1.3x10⁻⁵	3.6x10⁻⁹	1.8x10⁻⁵	2.6x10⁻⁵	4.8x10⁻⁴	1.9x10⁻⁴
pH	10.2	13.2	NR	10.3	13.2	11.4	12.9	13.6	NR
Ref.	1	2	3	4	5	1	6	7	8

* NR - not reported.

[1]Valenzuela & Jensen (1994); [2]Jo et al. (1996); [3]Baldwin et al. (1995); [4]Castaing (1993); [5]Baldwin & Stephens (1996); [6]DiCenso et al. (1994); [7]Hodgson (1995); and [8]Delegard (1995a).

The broadly representative solution and sludge compositions shown in Table 1 reveal the waste heterogeneity, the pervasiveness of sodium nitrate (even in sludge), and the low plutonium concentrations in the wastes. Low-solubility crystalline phases identified in sludge include the oxides $Fe(Cr,Fe)_2O_4$, UO_3, and hematite and maghemite (both Fe_2O_3 phases); the oxyhydroxides goethite (α-FeOOH) and boehmite (γ-AlOOH); the aluminum hydroxides gibbsite, bayerite, and nordstrandite [all $Al(OH)_3$]; the aluminosilicates cancrinite (nominally $NaAlSiO_4$) and albite ($NaAlSi_3O_8$); the phosphates $Na_2Fe_2Al(PO_4)_3$, $FeFe_2(PO_4)_2(OH)_2$, $Pb_5OH(PO_4)_3$, and hydroxyapatite [$Ca_5OH(PO_4)_3$]; and the salts sodium oxalate ($Na_2C_2O_4$) and $Na_7F(PO_4)_2 \cdot 19H_2O$. Sludges also contain amorphous phases: iron, aluminum, chromium, zirconium, and manganese hydrous oxides; aluminum phosphate; aluminosilicates; and sodium aluminosilicates (Temer and Villarreal, 1995, 1996; Lumetta et al., 1996; Delegard et al., 1994).

Plutonium concentrations in waste solutions range from about 10^{-9} to 10^{-5} M and from 10^{-6} to 10^{-4} moles per kilogram in sludge (Figure 1). The error bars in this figure show the standard deviations from the average of multiple analyses and the numbers above the bars are the number of analyses included in the average. Most of the plutonium in the waste tanks is concentrated in the sludge layers. Low concentrations in solution for all the tank samples analyzed thus far indicate that there are no strong complexing agents in the tanks that greatly influence dissolution of plutonium under tank storage conditions. Tank 106-C

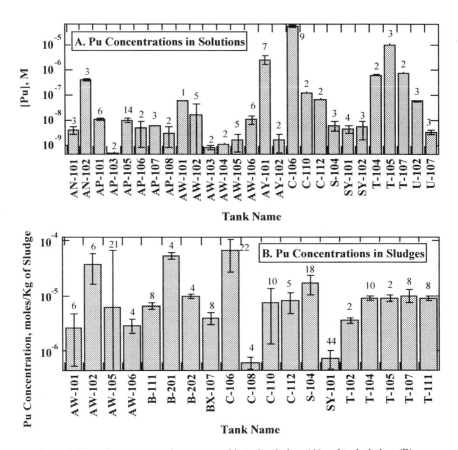

Figure 1. Plutonium concentrations measured in tank solutions (A) and tank sludges (B).

has the highest plutonium concentration in solution (5×10^{-5} M), probably because of a low hydroxide concentration (10^{-5} M) and a high carbonate concentration (0.75 M). Both of these factors increase plutonium solubility and inhibit adsorption of plutonium on sludge solids, as will be discussed in the sections to follow. Concentrations of plutonium in the supernates would be expected to be much higher than those given in Figure 1 if the plutonium species were predominately in the (VI) oxidation state, as will be discussed later. Because the (III) oxidation state is known to be unstable under the alkaline conditions of the tanks, the oxidation state of plutonium in the wastes will be (IV) or (V).

PLUTONIUM SOLUTION CHEMISTRY IN ALKALINE MEDIA

The Hanford Site tank waste is alkaline (pH 8 to greater than 5 M NaOH). However, the chemistry of plutonium in alkaline media is considerably less investigated than that in acid. Published research on plutonium chemistry in alkaline media is reported primarily in (1) studies of Hanford Site tank waste, (2) related tank waste studies for the US Department of Energy's Savannah River Site, (3) studies for alkaline nuclear waste repository systems, and (4) investigations related to the discovery of heptavalent plutonium conducted at the Institute of Physical Chemistry (IPC) of the USSR Academy of Sciences (Krot et al., 1977). The chemistry of the transuranium elements and technetium in strongly alkaline media was recently reviewed by the IPC (Peretrukhin et al., 1995). However, the review does not cover the significant developments of the last few years. Following is a brief summary of the chemistry of plutonium in alkaline solution as it applies to Hanford Site tank waste solution. The studies may be divided between those conducted for strongly concentrated ((0.5 M hydroxide) and moderately concentrated (< 0.5 M hydroxide) alkaline media.

Plutonium in Strongly Alkaline Media

The earliest references to the chemistry of plutonium in strongly alkaline media are those of the Manhattan Project. Studies indicated that Pu(VI) and particularly (IV) form compounds of low solubility in aerated NaOH solution and do not readily interconvert between their tetra- and hexavalent oxidation states (Connick et al., 1949). Plutonium(III) hydroxide will reduce water in 1 M NaOH solution (Cunningham, 1954) and thus should not be stable in tank waste. The solubility of Pu(VI) hydroxide in 5 M NaOH was reported to be about 5×10^{-4} M (Hindman, 1954). In a later study, the increasing solubility of Pu(VI) hydroxide in potassium hydroxide solution (from about 10^{-5} to 10^{-4} M as potassium hydroxide concentration increased from 0.11 to 10 M) was attributed to formation of an anionic hydroxo complex (Pérez-Bustamante, 1965). The discovery of heptavalent neptunium and plutonium, achieved by the oxidation of the respective hexavalent actinides by ozone in NaOH solution (Krot and Gel'man, 1967), was followed by a program of experimental work at the IPC on the chemistry of this alkali-stabilized actinide oxidation state (Krot et al., 1977).

Plutonium Oxidation States in Solution. The oxidation-reduction potentials of Pu[(VII)-(VI); (VI)-(V); (V)-(IV); (IV)-(III)] in 0.1 to 14 M NaOH were studied using polarographic techniques (Peretrukhin and Alekseeva, 1974). The results generally agreed with, and expanded on, slightly earlier findings (Bourges, 1972 and 1973). Further investigations of the oxidation / reduction reactions of plutonium have since been conducted by electrochemical means (Peretrukhin and Spitsyn, 1982; Peretrukhin et al., 1994). The studies showed Pu(VII) oxidizes water to form oxygen in 0.1 to 15 M NaOH solution and Pu(III) reduces water in 1 M NaOH solution (Peretrukhin and Spitsyn, 1982) in agreement with earlier findings (Cunningham, 1954). Thus, neither Pu(III) nor Pu(VII) should be stable indefinitely in strongly alkaline waste. The Pu(VI)/(V) couple was found to be reversible (Peretrukhin and Alekseeva, 1974; Bourges, 1973). By implication, the respective plutonium species should have similar structures. In contrast, the Pu(V)/(IV) couple was found to be irreversible implying different structures. The polarographic measurements show that all dissolved valent states are stable to disproportionation in sodium hydroxide solution (Peretrukhin and Alekseeva, 1974; Peretrukhin and Spitsyn, 1982; Peretrukhin et al., 1994).

Other tests confirmed that Pu(V) solution, stored for over a month in 12 M NaOH, is stable. However, in 4 M sodium hydroxide, Pu(V) was found to disproportionate after three days' storage to form a brown-green precipitate, postulated to be hydrous oxides of Pu(IV) (Bourges, 1973). Tests from the opposite direction (i.e., reproportionation) confirmed this behavior. When freshly precipitated Pu(IV) was mixed with an equimolar (2×10^{-3} M) amount of dissolved Pu(VI) in sodium hydroxide solution, Pu(V) formation was incomplete. The fraction of plutonium present as Pu(V), about 4 percent in 4 M NaOH, increased to about 12 percent in 6 M NaOH. The extent of disproportionation thus decreases with increasing sodium hydroxide concentration, becoming negligible by 12 M NaOH. Based on these observations, the formal potential [that is, the potential at 1 M Pu(V)] of the Pu(V)(aq)/$PuO_2 \cdot xH_2O$(s) couple was calculated to be 0.44 V at 4 M NaOH (Bourges, 1972).

Later experiments showed that above at least 5 M NaOH, the solubility of $PuO_2 \cdot xH_2O$(s) in aerated solution is controlled by oxidative dissolution to form the Pu(V) complex $[PuO_2(OH)_4]^{3-}$ according to the half-reaction

$$PuO_2 \cdot xH_2O_{(s)} + 4OH^- = [PuO_2(OH)_4]^{3-} + e^- + xH_2O$$

(Delegard, 1985). The published equilibrium constant expression may be rearranged in terms of electron activity to evaluate the formal potential of the Pu(V)(aq)/$PuO_2 \cdot xH_2O$(s) couple as a function of sodium hydroxide concentration. At 5 M NaOH, the formal potential is about 0.35 V, similar to the formal potential, 0.44 V, for the same reaction reported at 4 M NaOH (Bourges, 1972). The potential of the Pu(V)(aq)/$PuO_2 \cdot xH_2O$(s) couple depends on Pu(V) concentration, decreasing 0.059 V per decade decrease in Pu(V) concentration. At 10^{-6} M Pu, representative of plutonium concentrations at saturation, the Pu(V)(aq)/$PuO_2 \cdot xH_2O$(s) potentials would be about 0.35 V lower than at 1 M Pu(V).

Recent tests, based on Pu(V) and Pu(VI) spectrophotometry, confirmed that Pu(V) disproportionation to form $PuO_2 \cdot xH_2O$(s) and Pu(VI)(aq) decreases with increasing NaOH concentration but increases with increasing temperature (Budantseva et al., 1997). Plutonium(V) does not disproportionate in 8 M NaOH but disproportionates almost completely in 1 M NaOH. In the presence of additional salts typical of Hanford tank waste (about 5 M sodium as nitrate, nitrite, aluminate, and others), Pu(V) is stable in 4 M NaOH; without the additional salts, partial disproportionation occurs in 4 M NaOH within one minute. The formal potential of the Pu(V)(aq)/Pu(IV)(aq) couple was evaluated (Budantseva et al., in press) based on the known $PuO_2 \cdot xH_2O$(s) solubility-limited Pu(IV) concentration in hydrazine-bearing alkaline solution and the published Pu(VI)/Pu(V) potential. By these determinations, the potential at 4 M NaOH is 0.26 V, decreasing to 0.17 V at 8 M NaOH. The potentials are about 1.1 V higher than the corresponding values found by polarography (Peretrukhin and Alekseeva, 1974) but only about 0.2 V lower than the potential of the Pu(V)(aq)/$PuO_2 \cdot xH_2O$(s) couple measured by Bourges (1972) and derived from the results of Delegard (1985). Because Pu(V) disproportionation to form Pu(VI) and $PuO_2 \cdot xH_2O$(s) is observed at NaOH concentrations less than about 6 to 8 M NaOH, the Pu(V)/Pu(IV) polarographic results (Peretrukhin and Alekseeva, 1974) must be in error.

The formal potentials of the adjacent plutonium oxidation state couples in alkaline solution are presented in Figure 2. The Pu(V)(aq)/PuO$_2$•xH$_2$O(s) and Pu(V)(aq)/Pu(IV)(aq) potentials cross the Pu(VI)/Pu(V) plot with a downward slope indicating Pu(V) is unstable to disproportionation at NaOH concentrations lower than the crossing point. The nitrate/nitrite couple (Bratsch 1989), which may control the redox potential in nitrate/nitrite-laden tank waste solution, lies in the PuO$_2$•xH$_2$O(s) region of stability at 1 M NaOH; i.e., plutonium should be present as Pu(IV) solids in tank waste but possibly present as Pu(V) in solution. Higher sodium hydroxide concentrations, higher ionic strength (or salt effects), and lower temperatures favor Pu(V).

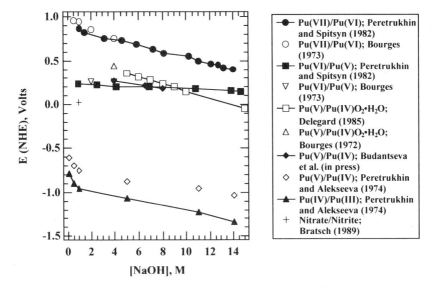

Figure 2. Formal potentials of dissolved plutonium species versus sodium hydroxide concentrations.

Solubility of Plutonium Compounds. Studies were conducted to determine the effects of Hanford Site tank waste solution components (sodium hydroxide, nitrate, nitrite, aluminate, carbonate, phosphate, sulfate, fluoride, EDTA, HEDTA, citrate, and glycolate) on the solubilities of plutonium compounds precipitating in NaOH solution (Delegard and Gallagher, 1983). Hydroxide, aluminate, and nitrate were found to increase plutonium compound solubility significantly, while the organic agents (EDTA, HEDTA, citrate, and glycolate) had no discernible effect.

Subsequent solubility tests investigated the effects of varying sodium hydroxide, aluminate, nitrate, nitrite, and carbonate concentrations (Delegard, 1985). The green plutonium solid phase formed when Pu(IV) nitrate solution was added to sodium hydroxide solution was identified with XRD to be poorly crystalline PuO$_2$•xH$_2$O. Brown solids were observed initially when Pu(VI) nitrate was added to NaOH solution. However, with months of aging, the solids became green PuO$_2$•xH$_2$O. The chemical reduction of Pu(VI) has been attributed to α-radiolytic reactions (Peretrukhin et al., 1995).

The plutonium concentrations increased then decreased with time as the solid phase aged and increased sharply with increasing NaOH concentration. As shown in Figure 3, the observed solubilities were identical for tests begun with Pu(IV) nitrate and Pu(IV) hydrous oxide and the Pu(VI) data were trending to those of Pu(IV) (Delegard, 1985 and 1987). The Pu(V) spectrum was observed in 15 M NaOH supernates; Pu(V) also was found in 10 M NaOH. Direct identification of plutonium valence at lower hydroxide (and plutonium) concentrations was not possible because the spectra were too weak; however, indirect evidence for Pu(V) or Pu(VI) was found at 5 M NaOH. These findings showed that Pu(IV) hydrous oxide dissolves in aerated 5 to 15 M NaOH solution by being oxidized to form Pu(V) solution species. The plutonium hydrous oxide solubilities observed in aerated pure NaOH solution (Figure 3) increase about a factor of 10 with each 2 to 3 M increase in hydroxide concentration.

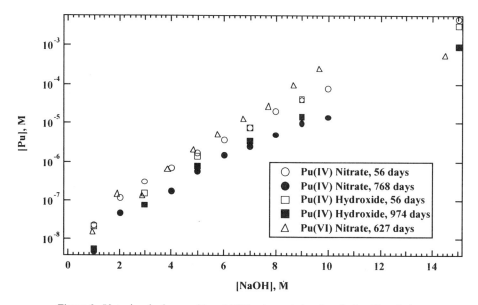

Figure 3. Plutonium hydrous oxide solubilities in aerated sodium hydroxide solution.

Sodium nitrate and nitrite in sodium hydroxide solution were found to increase the plutonium solution concentration because of their contribution to the chemical activity of sodium hydroxide. However, in mixed (1 to 4 M) $NaNO_2$ and (4 to 7 M) NaOH solution, measured plutonium concentrations and electrode potentials were lower than those expected in pure sodium hydroxide solution of the same sodium hydroxide chemical activity (Delegard, 1985). Subsequent analysis of these data showed that the $\log\{[Pu_{measured}]/[Pu_{expected}]\}$ is proportional to the difference in the measured and predicted electrode potentials (Delegard, 1996). The slope of a plot of these values indicates a one-electron change, thus corroborating the one-electron oxidative dissolution of Pu(IV) hydrous oxide to form Pu(V) dissolved species as proposed in the original study.

Plutonium compound solubility in simulated highly alkaline radioactive waste solutions also has been investigated at the Savannah River Site (Karraker, 1993; Hobbs and Karraker, 1996). Parametric tests identified and quantified the effects of waste components (sodium hydroxide, nitrate, nitrite, aluminate, carbonate, sulfate, and chloride), temperature (25 °C and 60 °C), and time on plutonium compound solubility. Evaporation tests determined the influence of evaporative concentration of simulated waste solutions on plutonium compound solubility.

A simplified solubility equation based on hydroxide concentration was derived from initial test data, from unpublished Savannah River Site work, and from selected Hanford

Site studies for solutions containing sodium hydroxide, nitrate, nitrite, and other salts (Hobbs and Edwards, 1993). Plutonium concentrations in salt-laden sodium hydroxide solution are about 10 times higher than shown in Figure 3 for pure sodium hydroxide solutions. The possible complexation of plutonium and the influence of the additional salts on hydroxide chemical activity likely account for the higher observed plutonium concentrations.

Similar effects of added salts on Pu(IV) solubility were found in other studies (Budantseva et al., 1997). Simulated Hanford Site waste solutions containing fixed sodium nitrate, nitrite, aluminate, carbonate, sulfate, fluoride, EDTA, HEDTA, citrate, glycolate, and acetate concentrations but varying sodium hydroxide concentrations were prepared. At 1.9 M NaOH, plutonium concentrations were about 100 times higher than shown in Figure 3; at 7.9 M NaOH, about a tenfold enhancement in solubility was observed because of complexing or changes in hydroxide activity.

The Savannah River Site simulated waste evaporation tests also gave higher plutonium concentrations than are found in pure sodium hydroxide solution for a given hydroxide concentration (Karraker, 1993; Hobbs and Karraker, 1996). Plutonium concentration increased about a factor of 10 with each 3 to 4 M increase in sodium hydroxide concentration. The plutonium solid phase observed in the evaporation tests was $PuO_2 \cdot xH_2O$.

The effects of oxidation state and sodium hydroxide concentration are evident in the plutonium compound solubility results obtained in recent IPC research and in earlier data (Figure 4). All tests were conducted at 6-hour to 3-day equilibration times. The short times decreased possible oxidation state changes caused by radiolysis and solid phase aging. Plutonium compound solubilities increase with increasing oxidation state and with increasing hydroxide concentration.

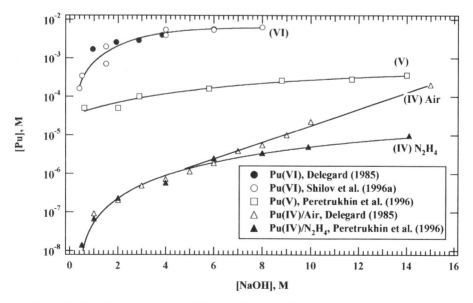

Figure 4. Plutonium compound solubilities for different oxidation states versus sodium hydroxide concentrations.

The stability of Pu(IV) at lower hydroxide concentrations is shown by the similarity of the solubility data for Pu(IV) hydrous oxide in the presence of hydrazine reductant and in the presence of air. The added reductant affected Pu(IV) solubility only above 6 M NaOH, presumably by preventing Pu(IV) solids from oxidizing to form dissolved Pu(V) as observed in aerated 10 and 15 M NaOH solution. Disproportionation also was observed below about 6 M NaOH in the Pu(V) solubility tests and below about 8 M NaOH in the stability tests (Peretrukhin et al., 1996; Budantseva et al., 1997). Therefore, in pure sodium hydroxide solution, Pu(IV) solution species likely occur below 6 M NaOH and Pu(V) solution species above 6 M NaOH.

Plutonium Complexation. The increasing plutonium hydrous oxide solubility with increasing sodium hydroxide concentration found for all plutonium oxidation states is strong evidence of anionic hydroxide complexes [see Bourges, 1973 for Pu(V), Pu(VI), and Pu(VII)]. Postulated hydroxide complexes vary with oxidation state and hydroxide concentration but most are octahedrally coordinated by oxygen. Heptavalent complexes proposed for alkaline solution are $[PuO_4(OH)_2]^{3-}$ (Krot et al,. 1977) or $[PuO_5OH]^{4-}$ or $[PuO_6]^{5-}$ in greater than 10 M NaOH (Tananaev et al. 1992). Recent x-ray absorption investigations of Np(VII) in 2.5 M NaOH show the existence of a central trans dioxo species, likely coordinated to four hydroxide ligands and a single water molecule, $[NpO_2(OH)_4H_2O]^-$ (Clark et al., 1997). The analogous Pu(VII) complex is likely. The Pu(VI) complex $[PuO_2(OH)_4]^{2-}$ is proposed on spectrophotometric and electrochemical evidence above 1 M hydroxide (Tananaev, 1989). Alkali metal salts $MPuO_2(OH)_2 \cdot xH_2O$ and $M_2PuO_2(OH)_3 \cdot xH_2O$ have been prepared (Tananaev, 1992); the latter is the solid phase used in the Pu(V) compound solubility tests (Peretrukhin et al., 1996). Because of the existence of these compounds, and by electrochemical correlations, the Pu(V) dissolved species $[PuO_2H_2O(OH)_3]^{2-}$ and $[PuO_2(OH)_4]^{3-}$ have been proposed (Peretrukhin et al., 1995). The latter complex also was proposed on solubility and electrochemical bases (Delegard, 1985). The Pu(V) and Pu(VI) complexes should have the same number of hydroxide ligands (i.e., four) because their electrode potential (Figure 2) is practically independent of sodium hydroxide concentration (Bourges, 1973). The Pu(IV) hydroxide complex is less well established; $[Pu(OH)_6]^{2-}$ is suggested based on the dependence of $PuO_2 \cdot xH_2O(s)$ solubility on sodium hydroxide concentration in hydrazine-bearing solution (Peretrukhin et al., 1996).

Thus, the plutonium solution species expected in pure NaOH solution are $[Pu(V)O_2(OH)_4]^{3-}$ above about 6 M and $[Pu(IV)(OH)_6]^{2-}$, or other Pu(IV) complex, below 6 M. The transition from Pu(IV) to (V) solution species likely occurs at lower NaOH concentration in wastes containing non-complexing dissolved salts that contribute to the hydroxide chemical activity.

A Pu(IV) carbonate complex, $[Pu(OH)_4(CO_3)_2]^{4-}$, is postulated based on Pu(IV) hydrous oxide solubility in carbonate media at pH 12 and 13 and the thenoyltrifluoroacetone extractability of the acidified dissolved plutonium (Yamaguchi et al., 1994). As shown in Figure 5, these data are consistent with solubilities observed at 0.25 to 1 M carbonate and 3 to 5 M hydroxide in controlled laboratory tests, and in high carbonate Hanford Site tank waste solutions. The dissolved plutonium oxidation state analyses described by Yamaguchi et al. (1994) may have been compromised by nitrite. However, limited spectrophotometric evidence suggests that a Pu(IV) carbonate complex exists in mixed hydroxide and carbonate solution (Delegard, 1995b).

A 1:1 Pu(V):aluminate complex has been proposed based on Pu(IV) hydrous oxide solubility in 5 to 10 M NaOH and 0.3 to 2 M $NaAl(OH)_4$ (Delegard, 1985). The oxidation state of the postulated complex, suggested by the observation of Pu(V) spectra in concentrated NaOH solution without aluminate, was not confirmed. A red Pu(V) peroxide complex exists, but only in the cold at 1 M NaOH (Musikas, 1971). The complex is increasingly stable to decomposition at higher (12 to 16 M) NaOH concentration (Shilov et al., 1996b). However, peroxide itself, though formed continuously by radiolysis, is unstable to disproportionation in tank waste and the corresponding Pu(V) peroxide complexes are unlikely. Plutonium fluoride complexes in highly alkaline solution also are unlikely based on studies in the potassium hydroxide/fluoride aqueous system (Delegard, 1987).

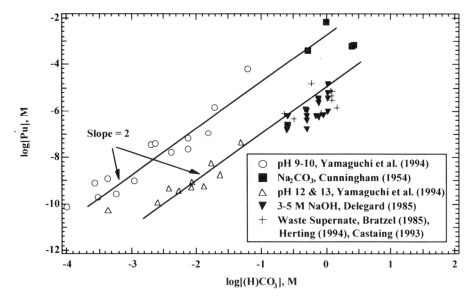

Figure 5. Enhancement of plutonium dissolution caused by carbonate complexation.

Oxidation/Reduction Reagents and Reactions. Oxidation reactions with freshly precipitated Pu(IV) hydrous oxide in strongly alkaline solution were recently reported (Shilov et al., 1996a). Oxidizing agents tested were air, oxygen, ozone, permanganate, persulfate, hypochlorite, hypobromite, ferricyanide, ferrate, and chromate; oxidation of Pu(V) with oxygen also was studied. Related studies for peroxide have been published (Shilov et al., 1996b). All reagents except air, oxygen, and chromate fully or partially oxidize Pu(IV) hydrous oxide. Ozone, persulfate, and hypochlorite are the most effective. Oxygen can oxidize Pu(V) to Pu(VI). The rates and extents of oxidation increase with sodium hydroxide concentration and temperature. Nitrate and nitrate/nitrite in combination have no net effect on the plutonium oxidation state though nitrite by itself acts as a mild reducing agent by preventing oxidative dissolution to Pu(V) (Delegard, 1985 and 1996). These data show that Pu(IV) hydrous oxide is not appreciably oxidized, beyond that observed in pure sodium hydroxide, by oxidizing agents present in the waste (mixed nitrite/nitrate, chromate, air).

Reducing agents in sodium hydroxide solution tested with Pu(V) and Pu(VI) were dithionite, sulfite, thiourea dioxide, hydrazine, and hydroxylamine (Shilov et al., 1996a). At lower hydroxide concentrations, reduction was complete for Pu(VI) or Pu(V) to Pu(IV) hydrous oxide for all reductants listed except sulfite. Hydroquinone and ascorbate also reduce Pu(VI). Hydroxylamine is the most effective reductant. Lower hydroxide concentrations favor reduction.

The stability of Pu(VI) in NaOH solutions of individual and combined Hanford Site tank waste components (nitrate, nitrite, aluminate, carbonate, sulfate, fluoride, EDTA, HEDTA, citrate, glycolate, and acetate) was determined (Budantseva et al. 1997). Reduction occurred at a few percent per hour in 4 M NaOH solution containing HEDTA at 50°C,. In 4 M NaOH, the other components individually had no effect although nitrite addition to the NaOH/HEDTA solution further increased the reduction rate. In a simulated waste solution containing all components, Pu(VI) also was reduced.

These data further sustain the finding that the higher plutonium oxidation states are stabilized by increasing hydroxide concentrations. Aside from HEDTA and nitrite, no waste component examined was an effective reductant and none was an effective oxidant except air for Pu(V). Certain organic agents (formate, alcohols) present in tank wastes also may act as reducing agents. Their effects have not been studied.

Radiolysis Induced Chemical Reactions. On average, the radioactive Hanford Site tank waste contains about $3.7x10^{10}$ Becquerel (1 curie) of combined beta and gamma activity per liter (alpha activity is negligible). Most of the activity arises from decay of ^{90}Sr and ^{137}Cs. The radioactive decay energy deposits almost entirely in the waste to yield radiolytic dose rates up to 7 Gray/h (700 Rad/h) or 60,000 Gray/y. Radiolysis occurring in the tank waste generates reductants (such as e_{aq}^-, H, and H_2O_2) and oxidants (OH) from water. Secondary reactions of water radiolysis products with waste solutes, such as nitrate and nitrite with e_{aq}^- to form NO_3^{2-} and NO_2^{2-} radical ions, also occur (Meisel et al., 1993; Pikaev et al., 1996). Both primary and secondary radiolysis products can influence plutonium oxidation state distribution.

The behavior of Pu(VI) in irradiated, aerated alkaline media has been investigated to absorbed doses up to 58,000 Gray (Pikaev et al., 1996). Radiation-induced chemical reduction occurs readily for $2x10^{-4}$ M Pu(VI) in 1.3 to 6.9 M NaOH to form Pu(V). The Pu(V) disproportionates to Pu(VI) and $PuO_2 \cdot xH_2O(s)$. The net yield, about 1.4 ions per 100 eV, is near that predicted based on Pu(VI) reduction by water radiolysis products e_{aq}^-, H, and H_2O_2 and reoxidation of Pu(V) by OH. Nitrate and nitrite do not significantly alter the net results of these reactions. In $NaOH/NaNO_3$, the Np(V) peroxide complex is formed by reaction of radiolytic peroxide with Np(V). However, the corresponding Pu(V) peroxide complex is unstable (Musikas, 1971) and is not observed under similar conditions. The organic solutes EDTA and formate both enhance the radiolytic reduction yield of Np(VI) and Np(V) because they scavenge radiolytic OH oxidant and because the resulting organic products reduce neptunium. Similar effects might be expected for plutonium. However, free radicals from EDTA solution radiolysis apparently do not react with Pu(VI).

Other radiolysis studies of Pu(VI) stability (50,000 Gray dose in 2 to 6 M NaOH solution) also found reduction to Pu(IV) hydrous oxide (Karraker, 1995). At 8 to 10 M NaOH, where Pu(V) is stable, reduction did not occur. Tests with 60,000 Gray irradiation in mixed $NaOH/NaNO_3/NaNO_2$ solutions were not markedly different from the tests with sodium hydroxide alone. The solubility of Pu(IV) hydrous oxide was measured in the presence and absence of gamma irradiation (Karraker, 1994). For pure sodium hydroxide solutions, irradiation increased the plutonium concentration. With nitrate and nitrite present, irradiation had no distinct effect on Pu(IV) hydrous oxide solubility.

Plutonium in Moderately Alkaline Media

Investigations of plutonium chemistry in moderately alkaline media have focused on environmental conditions (fresh and sea waters) and nuclear waste repository conditions (groundwaters and brines). Few systematic studies simulate Hanford Site tank waste conditions.

Solubility of Plutonium Compounds and Hydrolysis. Studies of plutonium compound solubilities in aqueous solution were reviewed (Puigdomènech and Bruno, 1991). In the absence of complexing ions such as carbonate or phosphate, solubilities of the respective oxidation state solid phases are limited by hydrolysis. The solubility of $Pu(OH)_3$ was found to vary, near the detection limit, between about 10^{-9} and 10^{-10} M in the pH range of 8.5 to 13 (Felmy et al., 1989). Because additions of metallic iron powder and exclusion of air were necessary to maintain Pu(III), its existence in tank waste is unlikely.

The plutonium concentration above Pu(IV) hydrous oxide decreases from 10^{-3} to 10^{-10} M as pH increases from 2 to 9; that is, about a factor of 10 per pH unit (data summarized by Puigdomènech and Bruno, 1991). From pH 9 to 13, an average plutonium concentration of 7×10^{-11} M was observed in low carbonate (3×10^{-5} M) cement-equilibrated water above Pu(IV) hydrous oxide (Ewart et al., 1992). These results are similar to other data (Lierse, 1985).

Plutonium(V) is much more soluble. The plutonium concentration above a compound presumed to be PuO_2OH was 2.8×10^{-4} M at pH 8.3 (Gevantman and Kraus, 1949). The plutonium concentrations above Pu(VI) hydroxide are near 10^{-8} M in the pH range 9 to 12.5 (Lierse, 1985). Slightly increased solubility is shown at higher pH. This trend must curve sharply upward at yet higher pH to tie with Pu(VI) compound solubilities observed in stronger sodium hydroxide solutions (Figure 3).

Solubility of Plutonium Compounds and Complexation. Plutonium concentrations increase dramatically in the presence of carbonate or, at lower pH, bicarbonate. Initial work in this field was performed under the Manhattan Project. For example, Pu(IV) hydrous oxide solubility in 0.53 to 2.7 M Na_2CO_3 at room temperature was found to range irregularly from 4×10^{-4} to 6.6×10^{-3} M (versus about 10^{-10} M expected in the absence of carbonate). Plutonium(VI) carbonate was even more soluble, exceeding 0.3 M in 0.5 M Na_2CO_3 (Cunningham, 1954).

Plutonium(III), reportedly unstable in carbonate media and rapidly air oxidized to Pu(IV) (Charyulu et al., 1991), should not exist in tank waste. Blue Pu(III) can, however, be produced by electrochemical reduction of Pu(IV) in bicarbonate solution (Wester and Sullivan, 1983) and is stable in the absence of oxygen (Varlashkin et al., 1984).

Plutonium concentrations observed above $PuO_2 \cdot xH_2O$ in bicarbonate solutions of pH 9.4 to 10.1 increase with the square of the total bicarbonate concentration (Yamaguchi et al., 1994). At low (10^{-4} M) total carbonate, plutonium concentration is 7×10^{-11} M, identical to that observed by Ewart and colleagues (1992). At 0.1 M total carbonate and pH 10, about 2×10^{-5} M Pu was observed. Extrapolating to 1 M carbonate, about 2×10^{-3} M plutonium should occur. This is similar to concentrations found by Manhattan Project researchers in molar concentrations of sodium carbonate (Figure 5). Based on the solubility data, the 2:1 Pu(IV):carbonate complex, $[Pu(OH)_2(CO_3)_2]^{2-}$, was postulated (Yamaguchi et al., 1994). Plutonium concentrations in carbonate-rich pH 10 waste solutions (Table 1) are substantially lower than predicted by these published data, however.

The solubility of PuO_2CO_3 as a function of total carbonate concentration has been determined and the complex $[PuO_2(CO_3)_3]^{4-}$ (an analogue of the well known uranyl triscarbonato complex) was proposed (Robouch and Vitorge, 1987). At 10^{-2} M carbonate, about 10^{-3} M plutonium was observed. About 0.2 M Pu(VI) was obtained in 1.0 M carbonate at pH 9.5 (Clark et al., 1993). These data indicate that under oxidizing bicarbonate conditions, and in the absence of sorption [which is known to be low for Pu(VI) carbonate complexes], plutonium should be completely dissolved in tank waste and not be found in the sludge phase. Because plutonium is found primarily in the sludge phase, combined oxidizing and high bicarbonate conditions apparently do not exist in tank waste. However, if alkaline-side processing to dissolve plutonium from sludge is desired, oxidizing bicarbonate leaching conditions may be successful provided plutonium is not deeply occluded in sludge particles or irreversibly adsorbed on the particles.

Plutonium(V) disproportionates to Pu(IV) and Pu(VI) in carbonate media from pH 9.3 to at least pH 11.4 but becomes stable at pH 12.6 (Varlashkin et al., 1984). Nevertheless, Pu(V) carbonate complexes can be observed around pH 8 at low (10^{-5} M) Pu(V) where disproportionation is less likely and rates are lower (Bennett et al., 1992). The formal oxidation potential for Pu(VI)/Pu(V) is about 0.35 V (versus NHE) in 2 M Na_2CO_3 at pH less than 13 and decreases to 0.31 V in 2 M Na_2CO_3 / 2 M NaOH (Varlashkin et al., 1984), about 0.10 V higher than found in the absence of carbonate (Peretrukhin and Spitsyn, 1982). Nitrite reportedly is effective in maintaining Pu(IV) in (bi)carbonate solution (Yamaguchi et al., 1994). The potential of the nitrate/nitrite couple is 0.017 V in 1 M OH⁻ (Bratsch, 1989); at pH 10, the potential is 0.253 V. Therefore, if the plutonium oxidation state is controlled by the abundant nitrate/nitrite, Pu(IV) carbonate should be stable in alkaline carbonate tank waste solution.

Carbonate complexation of Pu(IV) yields to hydrolysis as pH increases. Thus, Pu(IV) dissolved in $NaHCO_3$ solution as a carbonate complex will hydrolyze and precipitate as the pH is raised above 11.4 (Varlashkin et al., 1984). As shown in Figure 5, the postulated Pu(IV) complex formed at pH 12 and 13 is about 190-times less soluble than the complex observed at pH 9.4 to 10.1 (proposed to be $Pu(OH)_2(CO_3)_2]^{2-}$); the solid phase in both cases is Pu(IV) hydrous oxide (Yamaguchi et al., 1994). Other studies suggest the existence of the $[Pu(CO_3)_5]^{6-}$ limiting complex that transforms to $[Pu(CO_3)_4]^{4-}$ as pH and carbonate concentration decrease (Capdevila et al., 1996).

Similarly, absorption spectra for Pu(VI) and Pu(V) carbonate complexes change markedly from pH 12.6 to molar sodium hydroxide in concentrated sodium carbonate solution, reflecting changes in the mixed hydroxide/carbonate complexes (Varlashkin et al., 1984). Spectra for both Pu(V) and (VI) tend to, but do not match, spectra observed in highly alkaline solution without carbonate. The technical literature on actinide carbonate complexation was reviewed recently (Clark et al., 1995). Because of the confounding influences of hydroxide and (bi)carbonate concentrations in the published research, the review authors declined to identify specific Pu(IV) hydroxide/carbonate complexes.

The effects of tank waste components [nitrate, nitrite, (bi)carbonate, phosphate, fluoride, sulfate, and EDTA] on plutonium solubility were studied by parametric tests in the pH range 10 to 13 (Delegard, 1997). Carbonate/bicarbonate concentration and pH had the largest influence and gave plutonium concentrations, solid phase ($PuO_2 \cdot xH_2O$), and solution oxidation state [Pu(IV); no Pu(VI)] similar to those observed previously (Yamaguchi et al., 1994). At 0.005 M (bi)carbonate, 0.1 M EDTA increased plutonium concentration about a factor of 10 above that observed at 0.001 M EDTA. The remaining waste components had no significant effect on $PuO_2 \cdot xH_2O$ solubility.

Plutonium Solid Phase

The solid phase most often observed (usually by XRD) in plutonium solubility studies in neutral or alkaline media is Pu(IV) hydrous oxide. Broad diffraction peaks are attributed to small crystallite size or poor crystallinity. This compound also is loosely designated as Pu(IV) hydroxide although a discrete $Pu(OH)_4$ compound has not been identified (Lloyd and Haire, 1978). With time, the initially amorphous Pu(IV) gel dehydrates to form PuO_2 crystallites; the rate of crystallization increases with temperature and with lower salt concentrations (Haire et al., 1971). The crystallization is opposed by self-radiolysis to give a steady-state solid phase in aged (up to 1,300 days) pH 3 to 5 solution (Rai and Ryan, 1982). However, steady-state plutonium solubility is not observed in strongly alkaline solutions aged almost 1,000 days (Figure 3).

In alkaline solutions with plutonium compounds comprising the only possible solid phase(s), Pu(IV) hydrous oxide is observed to form in experiments begun with Pu(IV) (Rai and Ryan, 1982; Delegard, 1985, 1987, 1997; Yamaguchi et al., 1994; Hobbs and Karraker, 1996), by air or water oxidation in tests begun with Pu(III) (Cunningham, 1954; Charyulu et al., 1991; Delegard, 1987), by disproportionation of Pu(V) in 0.6 to 6 M NaOH (Bourges, 1972; Peretrukhin et al., 1996; Budantseva et al., 1997), and by slow radiolytic reduction of Pu(VI) in 1 to 15 M NaOH (Delegard, 1985; Peretrukhin, et al. 1995). Only in the presence of strong oxidants or reductants are pure plutonium phases other than $PuO_2 \cdot xH_2O$ stable. Tank wastes, without exception, are nitrate/nitrite mixtures subject to water radiolysis products and are expected to stabilize the Pu(IV) oxidation state in the solid phase.

FORMATION OF SOLID PHASES CONTAINING PLUTONIUM

Most of the high-level radioactive waste solutions generated from chemical operations at the Hanford Site were acidic solutions containing metals dissolved in nitric acid. Small concentrations of dissolved plutonium (typically $< 10^{-5}$ M plutonium) that were not recovered in processing also were present in these waste solutions. Before sending the waste to underground tanks, the solutions were made alkaline with excess sodium hydroxide to prevent corrosion of the carbon steel tanks. Metals that were insoluble in alkaline solutions (including plutonium) were precipitated during sodium hydroxide addition. The hydrous oxides formed when the solutions were made alkaline are the major solid components of sludge in the tanks.

Dissolved plutonium in acidic waste solutions could precipitate during the sodium hydroxide addition by the following mechanisms: (1) formation of a crystalline plutonium hydrous oxide phase, (2) formation of a non-crystalline, colloidal, plutonium hydrous oxide polymer, (3) formation of a crystalline solid solution phase where plutonium enters into the crystalline lattice of another hydrous oxide, or (4) adsorption of plutonium species onto the surface of other hydrous oxides that are present in great excess in the waste solutions. The latter two processes are considered to be coprecipitation mechanisms. Each of these potential precipitation mechanisms will be discussed below.

Concentrations of plutonium were very small compared to the other metals dissolved in the waste solutions from the various processing operations (Braun et al., 1994). For example, typical mole ratios of iron/plutonium and aluminum/plutonium were 12,000 and 28,000, respectively for Bismuth Phosphate waste; 1,400 and 130,000, for REDOX waste; 63,000 and 1,400,000, for PUREX aluminum clad wastes; 3,300 and 13,000, for PUREX zirconium clad waste; and 100 and 7,400, for plutonium finishing wastes (this does not include iron added to the waste as a criticality control). In addition to iron and aluminum, other dissolved metals such as uranium, zirconium, manganese, bismuth, chromium, and nickel were present in the acid waste and precipitated as hydrous oxides during neutralization (see Table 1).

Precipitation of Pure Plutonium Phases

Formation of individual crystals of $PuO_2 \cdot xH_2O$ or hydrous plutonium oxide polymer was unlikely during hydroxide addition to the waste because initial plutonium concentrations were thousands of times lower than the major metal ion concentrations in the waste solutions. To precipitate as a pure plutonium phase, one of two conditions would have to exist. Either the interaction between dissolved plutonium and precipitated iron, aluminum and other hydrous oxides would have to be negligible or the plutonium would have to be precipitated before the iron and aluminum precipitated from waste solutions.

Many examples exist of strong adsorption of dissolved plutonium species onto hydrous oxide surfaces including aluminum and iron hydrous oxides. These sorption reactions show that plutonium strongly bonds to precipitated hydrous oxide surfaces. Sorptions reactions are discussed in a later section.

Prediction of which metal ions precipitated first during hydroxide addition shows that pure $PuO_2 \cdot xH_2O$ or hydrous polymer was not likely formed in the waste from dissolved plutonium. In a nitric acid solution containing Pu(IV) as the only dissolved hydrolyzable metal ion, neutralization of acid with sodium hydroxide, as required for waste solutions sent to the tanks, would produce plutonium polymer along with plutonium hydroxide. However, because of the low solubility of iron and aluminum hydroxides and their much higher initial concentrations in the waste solution, these compounds will precipitate first as the waste is made alkaline and plutonium will be coprecipitated with them. Equilibrium solubilities of amorphous $PuO_2 \cdot xH_2O$ (Rai et al., 1980), amorphous $Fe(OH)_3$ (Lindsay, 1979), and crystalline $Al(OH)_3$ (gibbsite) (Lindsay, 1979) are compared for a range of pH values in Table 2. Assuming no significant precipitation rate differences, these data show that iron will precipitate before any plutonium can precipitate as plutonium hydrous oxide during waste neutralization.

Amorphous iron is much less soluble than amorphous plutonium hydroxide. Because the wastes contain thousands of times more iron and aluminum than plutonium, a pure plutonium phase (plutonium hydroxide or polymer) will not be formed. Instead, the plutonium will be coprecipitated on the iron and aluminum hydroxide that are precipitated first.

Table 2. Comparison of plutonium, iron, and aluminum hydroxide solubilities.

pH	Log of Solubilities (M)		
	$PuO_2 \cdot xH_2O$, Amorphous	$Fe(OH)_3$, Amorphous	$Al(OH)_3$, Crystalline
3	-3.6	-5.5	-1
4	-4.4	-7.1	-3
5	-5.3	-8.0	-6
6	-6.0	-9.2	-7
7	-6.8	-10.0	-7
8	-7.6	-10.3	-7
9	-8.4	-10.0	-6

The rate of iron hydroxide precipitation during waste neutralization was rapid compared to the rate of plutonium hydroxide precipitation (or plutonium hydroxide polymer formation). As long as the initial molar ratio of $[OH^-]/[Fe^{3+}]$ is greater than three, amorphous $Fe(OH)_3$ will precipitate immediately (Yariv and Cross, 1979). During neutralization of the acidic wastes, this ratio greatly exceeded three. Toth et al. (1981) have shown that plutonium hydroxide polymer forms slowly at the plutonium concentrations known to exist in the acidic wastes. The rate of polymer formation decreases with decreasing plutonium concentration. For example, only about 30 percent of the plutonium in a 0.05 M plutonium solution was polymerized after one hour at an acid concentration of 0.092 M. Because the plutonium concentration in the waste is about 10,000 times less than 0.05 M, polymer formation would be expected to be much slower.

Typical concentrations of plutonium in the acid wastes range from 10^{-5} to 10^{-6} M. At these low plutonium concentrations, incipient polymerization of plutonium hydroxide requires a relatively high pH. Also, polymerization is not instantaneous, but requires an induction period before the polymer is formed (Toth et al., 1981). Although polymer formation is unlikely, any polymer formed has a strong affinity to adhere to other particles suspended in solution or on nearby surfaces (Dran et al., 1994).

Some plutonium existed as solids in the waste before neutralization. These were in the form of oxides or other insoluble precipitates (fluorides, phosphates, etc.). Small amounts of solid plutonium oxide were added to the tanks, mainly from the PFP. These are oxides that were not dissolved during plutonium scrap processing and ended up in the acid waste as suspended solids. These solids were disposed of in underground tanks 102-SY and 118-TX.

Hobbs (1995) has shown that plutonium is effectively coprecipitated with iron and uranium hydroxides from simulated acidic PUREX waste solutions by making the solutions 0.6 M in sodium hydroxide. The concentrations of plutonium in the resulting supernate solutions were more than a factor of 100 lower than blank solutions that contained only plutonium nitrate before neutralization. The precipitated plutonium was therefore either adsorbed on the precipitated hydroxides or incorporated into a crystalline phase of these compounds. Because XRD measurements of the dried precipitates show that they were mostly amorphous, adsorption seems more likely. Also, plutonium is not likely to be incorporated into an iron compound crystalline structure because of the great difference in the ionic radii of Fe(III) and Pu(IV).

Herting (1995) has studied coprecipitation of plutonium from actual waste tank supernate solutions using iron hydroxide. The initial concentration of plutonium in these supernate solutions was 2×10^{-6} M. First, sodium hydroxide solution was added to samples of supernate solution from tank 107-AN to make the solutions 1.5 M and 3.0 M in free hydroxide. This addition removed about 22 to 27 percent of the plutonium from solution. Addition of a 0.3 M ferric nitrate solution removed an additional 46 to 57 percent of the initial plutonium by coprecipitation with ferric hydroxide. These results show that coprecipitation of plutonium with iron hydroxide occurs under actual tank waste conditions.

Coprecipitation of plutonium with iron and zirconium hydroxides has been used as a separation technique in analytical methods for measurement of plutonium concentrations and identification of plutonium oxidation states in solution. These methods depend on the preferential adsorption of plutonium on the hydrous oxide surfaces generated when base is added to the acidic solutions. Plutonium and other actinide elements were separated from other Hanford Site tank waste components by coprecipitation with iron hydroxides (Maiti and Kaye, 1992). Separation of plutonium from platinum solutions was successfully accomplished by coprecipitation of plutonium with ferric hydroxide (Johnson and Fowler, 1967). Plutonium was removed from basic solutions using ferric hydroxide generated by adding sodium ferrate (Na_4FeO_5) to the solutions (Stupin and Ozernoi, 1995). The ferrate was slowly reduced by water and the resulting ferric hydroxide precipitate adsorbed the dissolved plutonium. Krot et al. (1996) showed that Pu(V) and Pu(VI) could be effectively coprecipitated from alkaline solutions using hydroxides of Fe(III), Cr(III), Co(III), Co(II), and Mn(II) as carriers. The carriers were generated homogeneously from solution by reducing and/or hydrolyzing soluble precursors (e.g., FeO_4^{2-} and CrO_4^{2-}). The plutonium was reduced and coprecipitated as Pu(IV) during the formation of the precipitates. The pH ranges where Pu(III), Pu(IV), and Pu(VI) coprecipitate with iron and zirconium hydroxides were determined by Novikov and Starovoit (1969). They found that Pu(IV) coprecipitated at the lowest pH range, followed by Pu(III) and Pu(VI) as the pH was increased. Separation of the different oxidation states by this method was explained by the relative tendencies of these species to hydrolyze.

The vast majority of liquids added to the tanks were aqueous solutions of soluble nitrates resulting from sodium hydroxide addition to nitric acid waste streams. The concentrations of plutonium in these aqueous solutions are very low because of the strong sorption reactions of dissolved plutonium with precipitated metal hydrous oxide surfaces.

Also, the solubility of plutonium hydrous oxide is very small. Maximum expected concentrations of dissolved plutonium, based on solubility studies, are about 10^{-10} to 10^{-2} M, depending on the hydroxide and carbonate concentrations in solution (see Figure 5). The concentration of plutonium in tank 106-C supernate solution was determined by measurement to be 5×10^{-5} M (see Figure 1). This value is lower than expected for the solubility of $PuO_2 \cdot xH_2O$ because of the high carbonate and relatively low hydroxide concentrations in the supernate solutions, as discussed in more detail later. This lower concentration is likely caused by plutonium sorption on sludge solids. The dissolved plutonium species are probably anionic hydroxide and carbonate complexes of Pu(V) and Pu(IV), respectively.

Sorption of plutonium on the large surface areas of the solid metal hydrous oxides precipitated during neutralization is the most important coprecipitation mechanism. The hydrous oxides of iron and aluminum, in particular, develop large, amorphous surfaces as they are formed that attract heavy metal ions (Laitinen, 1960). The sorption capacity (in terms of available sorption sites) for iron hydrous oxide gel has been estimated to be about 500 μmoles/mmole iron (Kinniburgh and Jackson, 1981). This ratio greatly exceeds the molar ratios of plutonium/iron in the waste tank sludges which range from about 0.016 to 10 μmoles/mmole iron.

Coprecipitation as Solid Solutions

Coprecipitation of plutonium as a micro component in the crystalline lattice of solid crystalline phases precipitated during waste neutralization would require that (1) the substituted macro ion must have about the same ionic radius as plutonium and (2) the crystal lattices of the micro and macro components must be isomorphous or isodimorphous (able to form a common crystalline lattice structure). Because the initial precipitates of the macro ions were amorphous solids (mostly iron, aluminum, bismuth, and zirconium hydroxides), formation of solid solutions was not possible (Baes and Mesmer, 1976). Partial crystallization will slowly occur during years of storage of the wastes and this crystallization could result in solid solution formation. This is not important for most of the sludge wastes containing mostly iron and aluminum solids because of the large difference in ionic radii and charge for Pu(IV) and Fe(III) or Al(III). Shannon and Prewitt (1969) give values of crystal ionic radii of 1.10 Å for octacoordinate Pu(IV), 0.74 Å for hexacoordinate Fe(III), and 0.67 Å for hexacoordinate aluminum. Obviously, the large plutonium ion would be unlikely to substitute for iron or aluminum in the same crystalline lattice.

The hydrolysis of metal ions has been studied extensively (Baes and Mesmer, 1976). It has been shown that as these hydrolysis reactions occur, the ions tend to aggregate through hydroxyl bridging, forming dimers, trimers, or extensive polymeric networks that can reach colloid dimensions. These polymers can then condense with the loss of water to form oxygen-bridged polymeric species (Thiyagarajan et al., 1990). More than one metal hydroxide (hydroxides of other metal ions present in the initial solution mixtures) can be bonded to the polymer. For example, if uranium is present during plutonium or thorium hydroxide polymerization, the uranium is bonded through hydroxyl bridges to the polymer (Toth et al., 1981; and Toth et al., 1984). This type of bonding is similar to that proposed for surface complexation models of sorption of hydrolyzable metal ions on oxide surfaces (Schindler, 1981; and Davis et al., 1978). Surface complexation models have been successfully used to predict sorption of uranyl ions onto amorphous iron hydrous oxide, goethite, and hematite solids (Hsi and Langmuir, 1985). Structures of uranyl complexes at the surface of clay minerals have been identified by Chisholm-Brause et al. (1994). These studies indicate that the trace levels of plutonium present during neutralization of acidic wastes will be bonded to other hydrous oxides that are present in much greater abundance in the sludge waste.

Sorption/Desorption Reactions

Most plutonium in the high-level waste tanks exists as a dilute solid mixture with coprecipitated hydrous oxides in sludge layers. The plutonium is strongly sorbed onto the surfaces of these hydrous oxides. Sorption has been shown to be an important mechanism in the removal and fixation of plutonium from aqueous solution by a wide variety of metal hydrous oxides, including those present in tank sludges. These hydrous oxides include both pure phases and minerals that contain a mixture of phases. Plutonium sorption on goethite and hematite from basic solutions containing high dissolved salt concentrations was very strong (170 to 1,400 mL/g) according to work reported by Ticknor (1993). Adsorption of plutonium from waste solutions at Rocky Flats onto ferrite ($FeO \cdot Fe_2O_3$) crystalline solids at a pH of 10 to 12 was effective in removing at least 99.9 percent of the plutonium (Boyd et al., 1983). Sorption of Pu(IV) from solutions containing high concentrations of carbonate onto alumina, silica gel, and hydrous titanium oxide was also very strong (Pius et al., 1995). Carbonate lowered the sorption distribution coefficient for these adsorbents, but even at 0.5 M carbonate the coefficients were 60 mL/g, 1,300 mL/g, and 15,000 mL/g, respectively for alumina, silica gel, and hydrous titanium oxide. In 0.5 M bicarbonate solutions, the distribution coefficient for Pu(IV) sorption on alumina was lowered to about 30 mL/g (Charyulu et al., 1991). This is likely caused by a lower pH in bicarbonate solutions which favors a carbonate complex over a hydroxide complex. Sanchez et al. (1985) measured sorption of Pu(IV) and Pu(V) on goethite over a range of pH values and carbonate concentrations. They found that Pu(V) was reduced to Pu(IV) and that the Pu(IV) was strongly sorbed above a pH of about 6. High carbonate concentrations decreased sorption at a pH of 8.6. At 1 M $NaHCO_3$, plutonium sorption was completely inhibited. Carbonate concentrations at or above 1 M and pH values at or below 8.6 in a tank supernate solution will be rare at the Hanford Site. The solubility of sodium carbonate limits the carbonate concentrations in sodium nitrate-saturated aqueous waste solutions to less than about 0.3 M (Barney, 1976).

Barney et al. (1992) measured adsorption of plutonium from waste water solutions onto commercial alumina adsorbents over a pH range of 5.5 to 9.0. Plutonium adsorption Kd values (distribution of plutonium between solid and solution phases) increased from about 10 mL/g at a pH of 5.5 to about 50,000 mL/g at a pH of 9.0. The slopes of Kd versus pH curves were close to one, which indicates that one hydrogen ion is released to the solution for each plutonium ion adsorbed on the alumina surface. This behavior is typical of adsorption reactions of multivalent hydrolyzable metal ions with oxide surfaces. Plutonium precipitation was not significant in these tests because changing the initial concentration of plutonium from about 10^{-9} to 10^{-10} M did not affect the Kd values. Also, the initial plutonium concentrations were below the measured solubility limits of plutonium hydrous oxide.

Delegard et al. (1984) performed screening tests to identify waste components that could significantly affect sorption of plutonium from simulated waste tank aqueous solutions on three typical shallow sediments of the Hanford Site. They found that sorption was decreased by the chelating agents, 0.05 M EDTA and 0.1 M HEDTA, but not by carbonate at 0.05 M. Sorption distribution coefficients of Pu(IV) on a Hanford Site shallow sediment from a synthetic groundwater containing 0.003 M carbonate ranged from about 5,000 mL/g to 30,000 mL/g (Barney, 1992). Pu(VI) sorption, however, was small under the same conditions.

Partial crystallization of the fresh, amorphous iron and aluminum hydrous oxide precipitates occurs as these precipitates age in the tanks. This has been shown recently by

Hobbs (1995) who precipitated insoluble metal hydrous oxides from synthetic acidic PUREX waste solutions containing plutonium by adding a sodium hydroxide solution. After standing for 2 months, the precipitated solids (mainly iron, aluminum, uranium, and nickel hydrous oxides) were analyzed by XRD. Most of the solids were amorphous, but crystalline goethite (α-FeOOH), gibbsite [γ-Al(OH)$_3$], bayerite [α-Al(OH)$_3$], and sodium diuranate (Na$_2$U$_2$O$_7$) also were found. This crystallization did not affect the concentration of plutonium dissolved in the supernate over a 59-day period. When iron or uranium was present in the original solutions, the plutonium concentrations were more than 100 times lower than the solubility of pure PuO$_2 \cdot$xH$_2$O. This shows that plutonium was effectively coprecipitated with iron and uranium hydrous oxides.

The effects of iron hydrous oxide crystallinity on sorption of metal ions was studied by Tochiyama et al. (1994). They measured Np(V) sorption on α-FeOOH that was prepared by adding a Fe(NO$_3$)$_3$ solution to 1 M NaOH. The crystallinity was increased by heating the α-FeOOH to 130 °C and 300 °C. At 300 °C, the diffraction pattern was similar to hematite (α-Fe$_2$O$_3$). Sorption distribution coefficients were lowered from about 10^4 mL/g iron to 10^3 mL/g iron at a pH of 7 by the 300 °C heat treatment and were essentially unchanged by the 130 °C treatment. The lower neptunium sorption also might be caused by a lower surface area after heating, but this was not measured. The capacity of the hydrous oxides for plutonium sorption probably decreases somewhat with aging, but the initial capacity is so large that the effect of crystallization is likely to be small.

Isotopic exchange studies with metal ions adsorbed on alumina gel show that some metal ions are steadily incorporated into the bulk structure as the gel is aged. For example, the percentage of zinc adsorbed onto alumina gel that is isotopically exchangeable decreases steadily until at 400 hours only 10 percent of the sorbed zinc was exchangeable (Kinniburgh and Jackson, 1981). A shift to lower pH values in the sorption - pH curves for nickel, copper, zinc, and cadmium sorption on alumina gel as it ages also is evidence for incorporation of these metal ions into the gel structures. Plutonium ions initially adsorbed onto the iron and aluminum gels formed during neutralization of the acid waste solutions will also likely be incorporated into the aging gel structure. This incorporation could account for the observed irreversibility of plutonium sorption on these materials.

Any chemical mechanism for concentrating the plutonium in a waste tank would require transport of plutonium from the solid phase to the aqueous liquid phase and then to a small-volume solid phase. Sorbed plutonium must be desorbed into the aqueous phase of the waste mixture. Plutonium sorption on hydrous oxides is known to be mostly irreversible (DOE, 1988; and Alberts and Orlandini, 1981) unless the solution composition in contact with the plutonium is dramatically altered. This would require (1) the addition of organic or inorganic plutonium complexants, (2) a change in the redox potential of the solution by adding oxidants or reductants, or (3) acidification of the solution.

Dissolution of plutonium and americium (measured as total alpha activity) from two samples of actual Hanford Site tank sludges has been tested using pH 8 bicarbonate, bicarbonate combined with chemical oxidation (permanganate and ozone), and mineral acid leaching (Lumetta et al. 1993). The sample from tank 105-AW contained high concentrations of zirconium from cladding wastes; the sample from tank 110-U, was generated from the bismuth phosphate process. Both sludge samples had been previously leached with sodium hydroxide solution. Bicarbonate leaching dissolved about 4 percent and 15 percent, respectively, of the alpha activity from these sludges. Use of oxidants improved dissolution to about 6 percent and 22 percent, respectively. Only under strong acid conditions, which also dissolved the sludge matrix, were the plutonium and americium completely dissolved. These scoping studies confirm that carbonate and oxidizing conditions favor plutonium/americium dissolution by desorption but also suggest that plutonium and americium are entrapped or sorbed irreversibly on sludge particles and are not readily removable by complexants or leachants.

The desorption of heavy metals and other chemical species adsorbed onto oxides has been found to be incomplete, suggesting that the sorption reaction is not completely reversible. This is consistent with the behavior of "specifically adsorbed" cations/anions or cations/anions that are chemisorbed to oxide surfaces (Hingston, 1981). This observation has been made for a wide range of metal ions (including plutonium), anions, and organic compounds. Barney (1984) observed incomplete reversibility for sorption of uranium, neptunium, technetium, selenium, and radium on a sandstone-claystone under oxidizing and reducing conditions. Payne et al. (1994) also found that a portion of uranium sorbed onto iron oxides was irreversibly incorporated within the mineral structure. Ho and Doern (1985) found that most of the uranyl species sorbed onto hematite was irreversibly retained on the solid during desorption experiments. This was attributed to multiple bonding sites on the hematite that vary in binding strength. Ticknor et al. (1986) found that sorption of cesium, cerium, and americium on rock surfaces (granite, gabbro, syenite, and anorthosite) was essentially irreversible. Mishra and Tiwary (1995) found that sorption of strontium from aqueous solution on hydrous magnesium oxide was essentially irreversible even when the temperature was increased and the pH lowered. They postulated that the strontium is bonded to the hydrous oxide surface by chemisorption reactions. Sorption of cerium and thorium oxide colloids or pseudocolloids on mineral surfaces (vitreous silica, mica, hematite or goethite) also was found to be partially irreversible (Dran et al., 1994). Irreversible sorption also has been observed for other metal ions including nickel and cobalt (Di Toro et al., 1986), zinc (Elrashidi and O'Connor, 1982a), and molybdenum (Karimian and Cox, 1978); for anions including iodine (Bors, 1992), fluoride (Peek and Volk, 1985), and borate (Elrashidi and O'Connor, 1982b); and for organics including 2,4,5,2′,4′,5′-hexachlorobiphenyl (Di Toro and Horzempa, 1982), 2,4,5-trichlorophenoxyacetic acid (Koskinen et al., 1979), bromacil and diquat (Corwin and Farmer, 1984), and phenolic compounds (Isaacson and Frink, 1984). Irreversible sorption of plutonium on the sludge particles in Hanford Site waste tanks will help prevent separation of plutonium from the neutron absorbers in the sludge.

Partially irreversible sorption is illustrated in Figure 6, which shows the relationships between sorption, desorption, and solubility of plutonium in the tanks. These curves are based on measured sorption and desorption isotherms for the sorption of neptunium, uranium, radium, and technetium on geologic solid materials (DOE, 1988). The sorption isotherm (heavy line) shows that as more plutonium is sorbed on the solid phase, plutonium concentrations in solution increase until the solubility limit for $PuO_2 \cdot xH_2O$ (vertical line) is reached. The desorption isotherms, measured by contacting the plutonium-bearing solids with plutonium-free solution, are nearly flat because of the partial irreversibility of the sorption reaction. The position of the desorption isotherm depends on the concentration of plutonium in the solid.

CONCLUSIONS

Sodium nitrate-rich alkaline wastes exist as crystalline salt cakes, high ionic strength solutions, and water-insoluble sludges in underground radioactive storage tanks on the Hanford Site. Plutonium concentrations are 10^{-7} to 10^{-6} moles per kilogram in salt cake, but range from 10^{-9} to 10^{-5} M in solution and from 10^{-6} to 10^{-4} moles per kilogram in sludge. Because of the significance of plutonium in nuclear criticality safety and as a transuranic element with special long-term storage requirements, understanding the chemical behavior of plutonium in these materials is important for current waste storage operations and in planning and designing future waste partitioning and processing operations.

The low chemical concentrations of plutonium in the waste, and the wastes' complexity, make direct identification of plutonium dissolved species and solid phases impossible. Therefore, studies must be conducted using model systems and the results extrapolated to real wastes. In model systems with only plutonium solid phases present, marked stability of Pu(IV) hydrous oxide solid phase is found. Plutonium(IV) hydrous oxide ultimately forms following introduction of Pu(III), Pu(IV), Pu(V), and Pu(VI) to sodium hydroxide solution; thus Pu(IV) is the expected solid phase oxidation state in tank waste. Only in the presence of strong oxidants or reductants are plutonium solid phases containing other oxidation states stable. Mixed nitrate/nitrite, and radiolysis products, are ubiquitous in tank waste and should stabilize Pu(IV) in the solid phase.

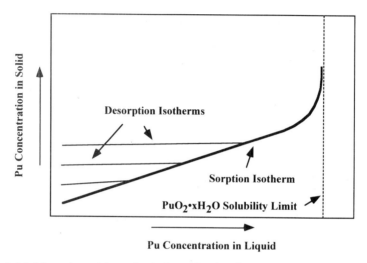

Figure 6. Model sorption and desorption isotherms for plutonium sorption on metal hydrous oxides.

Based on studies with model systems, the chemistry of plutonium in alkaline Hanford Site tank waste solutions (pH 8 to about 5 M sodium hydroxide) is expected to be controlled by the wastes' alkalinity and by the presence of key waste constituents that result in the following reactions:

Oxidation/Reduction. Plutonium oxidation state distribution in waste solutions with molar sodium hydroxide concentrations is not strongly affected by possible oxidants found in tank waste (mixed nitrate/nitrite, chromate, atmospheric oxygen). However, oxidizing chemicals added to achieve future waste processing goals (organic destruction, sludge dissolution) can oxidize plutonium from the solid and increase its solution concentration. Plutonium exists in the waste solutions as Pu(IV) or Pu(V) or a mixture of both oxidation states.

Hydrolysis. At carbonate concentrations less than about 0.1 M and in sodium hydroxide solutions between about 1 and 6 M, plutonium will hydrolyze to form hydroxide-complexed Pu(IV) species such as $[Pu(OH)_6]^{2-}$. At lower sodium hydroxide concentrations, less hydrolyzed Pu(IV) dissolved species will form. In sodium hydroxide solutions greater than 6 M (lower in the presence of added salts, which can increase hydroxide chemical activity), oxidation and hydrolysis will occur to form hydroxide-complexed Pu(V) species such as $[PuO_2(OH)_4]^{3-}$.

Carbonate Complexation. Carbonate complexation significantly enhances solubility. In solutions with pH 12 to 5 M NaOH, 1 M carbonate dissolves about 10^{-5} M plutonium by apparent formation of carbonate complexes. Increased plutonium dissolution because of high carbonate concentrations likely occurs in many Hanford Site tank waste solutions. At pH 9 to 10, 1 M (bi)carbonate can dissolve as much as $2x10^{-3}$ M plutonium to form carbonate complexes.

The trace-level concentrations of dissolved plutonium present in acidic waste solutions from the Hanford Site processing plants were coprecipitated with voluminous hydrous oxides that were formed during sodium hydroxide addition to the wastes. Hydrous oxides of aluminum, iron, chromium, zirconium, and other metals were formed in great quantities compared to the the amount of plutonium that was present. The coprecipitation mechanism was adsorption of plutonium on the large surfaces of the hydrous oxides as they were precipitating from solution. These hydrous oxide surfaces had a great excess capacity for adsorption of plutonium from the alkaline solutions. It is unlikely that significant amounts of hydrous plutonium oxide, plutonium hydrous oxide polymer, or solid solutions of plutonium in crystalline hydroxides were formed during sodium hydroxide addition to the acidic waste solutions because of the low concentrations of plutonium present and known extensive adsorption reactions of plutonium on the hydrous oxides.

As these hydrous oxide precipitates aged in the alkaline solution in the underground tanks, the plutonium was incorporated into the bulk structure of the sludge solids. Elimination of water from the initially formed hydroxyl bridges between metal ions yielded stronger oxygen bridges that hold the plutonium in the sludge structure more tightly. Isotope exchange experiments with other hydrolyzable metal ions adsorbed onto oxide surfaces show that the exchangeable fraction of the adsorbed metal ion decreases over time. This explains the difficulty experienced in extracting plutonium from the sludge solids even with strong acids and oxidizing agents. Also, this mechanism can explain the partially irreversible adsorption of plutonium on oxide surfaces. Once adsorbed onto an oxide surface, plutonium is not readily desorbed from that surface.

Plutonium-containing solids were also present in some wastes before addition of sodium hydroxide. These solids included plutonium oxide that remained undissolved and suspended in processing solutions, mainly from the PFP. Plutonium solids such as plutonium fluoride and plutonium dibutyl phosphate from chemical processing would likely be converted to the hydrous oxide, $PuO_2 \cdot xH_2O$ when made alkaline. Because $PuO_2 \cdot xH_2O$ is stable in the waste environment, it is likely to be unchanged with aging.

Numerous areas remain to be explored on the chemistry of plutonium in alkaline tank waste media. The primary fields of research are in the distribution and behavior of plutonium in sludge components. Tests to determine the affinity of plutonium for various sludge phases and elements (e.g., iron and aluminum hydrous oxides, aluminosilicates) must be performed with model systems as functions of solution compositions (pH, carbonate, redox agents), solid phases, and time. Sedimentation or other classification tests with real wastes also should be performed to determine the association of plutonium with particular phases or elements. In addition, further investigation of plutonium complexation (carbonate, aluminate, and chelating organics) in tank waste solutions should be performed as functions of hydrolysis (pH, hydroxide) and the plutonium oxidation state.

ACKNOWLEDGEMENTS

The authors acknowledge the contributions of the chemists of the Institute of Physical Chemistry of the Russian Academy of Sciences for their significant and pioneering work on the chemistry of plutonium in alkaline media and their continued work in this area being performed under the support of the US Department of Energy, Office of Science and Technology. In particular, the authors thank Professors V. F. Peretrukhin, N. N. Krot, V. P. Shilov, and A. K. Pikaev and their staff, Doctors I. G. Tananaev, A. A. Bessonov, N. A. Budantseva, I. A. Charushnikova, A. M. Fedoseev, V. I. Silin, and A. V. Gogolev for their excellent cooperation and many valuable discussions. Support, in part, for preparing this paper was provided by the Office of Science and Technology.

REFERENCES

Alberts, J. J. and K. A. Orlandini, 1981, Laboratory and Field Studies of the Relative Mobility of [239,240]Pu and [241]Am from Lake Sediments under Oxic and Anoxic Conditions, *Geochimica et Cosmochimica Acta,* 45:193-1938.

Baes, C. F. and R. E. Mesmer, 1976, *Hydrolysis of Cations,* John Wiley & Sons, New York, NY, p. 187-189.

Baldwin, J. H. and R. H. Stephens, 1996, Tank Characterization Report for Double-Shell Tank 241-AP-108, *WHC-SD-WM-ER-593, Rev. 0,* Westinghouse Hanford Company, Richland, WA.

Baldwin, J. H., L. C. Amato, and T. T. Tran, 1995, Tank Characterization Report for Double-Shell Tank 241-AW-101, *WHC-SD-WM-ER-470,* Westinghouse Hanford Company, Richland, WA.

Barney, G. S., 1976, Vapor-Liquid-Solid Phase Equilibria of Radioactive Sodium Salt Wastes at Hanford, *ARH-ST-133,* Atlantic Richfield Hanford Company, Richland, WA.

Barney, G. S., 1984, Radionuclide Sorption and Desorption Reactions with Interbed Materials from the Columbia River Basalt Formation, in *Geochemical Behavior of Disposed Radioactive Waste,* G. S. Barney, J. D. Navratil, and W. W. Schulz, Eds., American Chemical Society Symposium Series 246, Washington, D.C., p. 3-24.

Barney, G. S., 1992, Adsorption of Plutonium on Shallow Sediments at the Hanford Site, *WHC-SA-1516-FP,* Westinghouse Hanford Company, Richland, WA.

Barney, G. S., K. J. Lueck, and J. W. Green, 1992, Removal of Plutonium from Low-Level Process Wastewaters by Adsorption, in *Environmental Remediation, Removing Organic and Metal Ion Pollutants,* G. F. Vandergrift, D. T. Reed, and I. R. Tasker, Eds., American Chemical Society Symposium Series 509, Washington, D.C., p. 34-46.

Benedict, M., T. H. Pigford, and H. W. Levi, 1981, *Nuclear Chemical Engineering,* second edition, McGraw-Hill Book Company, New York, NY, p. 458-461.

Bennett, D. A., D. Hoffman, H. Nitsche, R. Russo, R. A. Torres, P. A. Baisden, J. E. Andrews, C. E. A. Palmer, and R. J. Silva, 1992, Hydrolysis and Carbonate Complexation of Dioxoplutonium(V), *Radiochimica Acta,* 56:15-19.

Bors, J., 1992, Sorption and Desorption of Radioiodine on Organo-Clays, *Radiochimica Acta,* 20:235-238.

Bourges, J., 1972, Preparation et Identification du Plutonium a l'Etat d'Oxydation-V en Milieu Basique, *Radiochem. Radioanal. Letters,* 12:111-115.

Bourges, J., 1973, Etude Electrochimique des Etats d'Oxydation du Plutonium en Milieu Hydroxyde de Sodium, *CEA-R-4406,* Commissariat a l'Energie Atomique, France.

Boyd, T. E., R. L. Kochen, and J. D. Navratil, 1983, Actinide Aqueous Waste Treatment Using Ferrites, *Radioactive Waste Management and the Nuclear Fuel Cycle,* 4:195-209.

Bratsch, S. G., 1989, Standard Electrode Potentials and Temperature Coefficients in Water at 298.15 K, *J. Phys. Chem. Ref. Data,* 18:1-21.

Bratzel, D. R., 1985, Characterization of Complexant Concentrate Supernatant, in Process Aids - A Compilation of Technical Letters, *WHC-IP-0711,* 17, Westinghouse Hanford Company, Richland, WA.

Braun, D. J., L. D. Muhlestein, T. B. Powers, and M. D. Zentner, 1994, High-Level Waste Tank Subcriticality Safety Assessment, *WHC-SD-WM-SARR-003,* Westinghouse Hanford Company, Richland, WA, p. B-9 to B-20.

Budantseva, N. A., I. G. Tananaev, A. M. Fedoseev, A. A. Bessonov, and C. H. Delegard, 1997, Investigation of the Behavior of Plutonium(V) in Alkaline Media, *PNNL-11624,* Pacific Northwest National Laboratory, Richland, WA.

Budantseva, N. A., I. G. Tananaev, A. M. Fedoseev, and C. H. Delegard, in press, Behavior of Plutonium(V) in Alkaline Media, *J. Alloys Compd.*

Capdevila, H., P. Vitorge, E. Giffaut, and L. Delmau, 1996, Spectrophotometric Study of the Dissociation of the Pu(IV) Carbonate Limiting Complex, *Radiochimica Acta,* 74:93-98.

Castaing, B. A., 1993, 101-AY, 102-AY, & 106-C Data Compendium, *WHC-SD-WM-TI-578, Rev. 0,* Westinghouse Hanford Company, Richland, WA.

Charyulu, M. M., I. C. Pius, A. Kadam, M. Ray, C. K. Sivaramakrishnan, and S. Patil, 1991, The Behaviour of Plutonium in Aqueous Basic Media, *J. Radioanal. and Nucl. Chem.,* 152:479-486.

Chisholm-Brause, C., S. D. Conradson, C. T. Buscher, P.G. Eller, and D. E. Morris, 1994, Speciation of Uranyl Sorbed at Multiple Binding Sites on Montmorillonite, *Geochimica et Cosmochimica Acta,* 58:3625-3631.

Clark, D. L., D. E. Hobart, and M. P. Neu, 1995, Actinide Carbonate Complexes and Their Importance in Actinide Environmental Chemistry, *Chem. Rev.,* 95:25-48.

Clark, D. L., D. E. Hobart, P. D. Palmer, J. C. Sullivan, and B. E. Stout, 1993, ^{13}C NMR Characterization of Actinyl(VI) Carbonate Complexes in Aqueous Solution, *J. Alloys Compd.,* 193:94-97.

Clark, D. L., S. D. Conradson, M. P. Neu, P. D. Palmer, W. Runde, and C. D. Tait, 1997, XAFS Structural Determination of Np(VII). Evidence for a Trans Dioxc Cation Under Alkaline Solution Conditions, *J. Am. Chem. Soc.,* 119:5259-5260.

Connick, R. E., W. H. McVey, and G. E. Sheline, 1949, Note on the Stability of Plutonium(IV), (VI) in Alkaline Solution, Papers 3.110 and 3.150, *The Transuranium Elements,* G. T. Seaborg, J. J. Katz, and W. M. Manning, Ed., McGraw-Hill Book Company, New York, NY.

Corwin, D. L. and W. J. Farmer, 1984, Nonsingle-Valued Adsorption-Desorption of Bromacil and Diquat by Freshwater Sediments, *Environ. Sci. Technol.,* 18:507-514.

Cunningham, B., 1954, Preparation and Properties of the Compounds of Plutonium, Chapter 10 in *The Actinide Elements,* G. T. Seaborg and J. J. Katz, Ed., McGraw-Hill Book Company, New York, NY.

Davis, J. A., R. O. James, and J. O. Leckie, 1978, Surface Ionization and Complexation at the Oxide/Water Interface, *J. Colloid Interface Sci.,* 63: 480-499.

Delegard, C. H., 1985, Solubility of $PuO_2 \cdot xH_2O$ in Alkaline Hanford High-Level Waste Solution, *RHO-RE-SA-75 P,* Rockwell Hanford Operations, Richland, WA. Also published in 1987, *Radiochimica Acta,* 41:11-21.

Delegard, C. H., 1987, Solubility of $PuO_2 \cdot xH_2O$ in PUREX Plant Metathesis Solutions, *RHO-RE-ST-53 P,* Rockwell Hanford Operations, Richland, WA.

Delegard, C. H., T. D. Elcan, and B. E. Hey, 1994, Chemistry of the Application of Calcination/Dissolution to the Hanford Tank Waste Inventory, *WHC-EP-0766,* Westinghouse Hanford Company, Richland, WA.

Delegard, C. H., 1995a, Chemistry of Proposed Calcination/Dissolution Processing of Hanford Site Tank Wastes, *WHC-EP-0832,* Westinghouse Hanford Company, Richland, WA.

Delegard, C. H., 1995b, Calcination/Dissolution Chemistry Development Fiscal Year 1995, *WHC-EP-0882,* Westinghouse Hanford Company, Richland, WA.

Delegard, C. H., 1996, Liaison Activities with the Institute of Physical Chemistry, Russian Academy of Sciences: FY 1996, *WHC-SP-1186,* Westinghouse Hanford Company, Richland, WA.

Delegard, C. H., 1997, Liaison Activities with the Institute of Physical Chemistry, Russian Academy of Sciences: FY 1997, *PNNL-11682,* Pacific Northwest National Laboratory, Richland, WA.

Delegard, C. H. and S. A. Gallagher, 1983, Effects of Hanford High-Level Waste Components on the Solubility of Cobalt, Strontium, Neptunium, Plutonium and Americium, *RHO-RE-ST-3 P,* Rockwell Hanford Operations, Richland, WA.

Delegard, C. H., G. S. Barney, and S. A. Gallagher, 1984, Effects of Hanford High-Level Waste Components on the Solubility and Sorption of Cobalt, Strontium, Neptunium, Plutonium, and Americium," in *Geochemical Behavior of Disposed Radioactive Waste,* G. S. Barney, J. D. Navratil, and W. W. Schulz, Eds., American Chemical Society Symposium Series 246, Washington, D.C., p. 95-114.

Di Toro, D. M. and L. M. Horzempa, 1982, Reversible and Resistant Components of PCB Adsorption-Desorption: Isotherms, *Environ. Sci. Technol.,* 16:594-602.

Di Toro, D. M., J. D. Mahony, P. R. Kirchgraber, A. L. O'Bryne, L. R. Pasquale, and D. C. Piccirilli, 1986, Effects of Nonreversibility, Particle Concentration, and Ionic Strength on Heavy Metal Sorption, *Environ. Sci. Technol.,* 20:55-61.

DiCenso, A. T., L. C. Amato, J. D. Franklin, G. L. Nuttall, K. W. Johnson, P. Sathyanarayana, and B. C. Simpson, 1994, Tank Characterization Report for Single-Shell Tank 241-S-104, *WHC-SD-WM-ER-370, Rev. 0,* Westinghouse Hanford Company, Richland, WA.

DOE, 1988, Site Characterization Plan, Reference Repository Location, Hanford Site, Washington, *DOE/RW-0164,* Volume 3, U.S. Department of Energy, Washington, DC, p. 4.1-106.

Dran, J.-C., G. D. Mea, V. Noulin, J.-C. Petit, and V. Rigato, 1994, Interaction of Pseudocolliods with Mineral Surfaces: The Fate of the Scavenged Cation, *Radiochimica Acta*, 66/67:221-227.

Elrashidi, M. A. and G. A. O'Connor, 1982a, Influence of Solution Composition on Sorption of Zinc by Soils, *Soil Sci. Soc. Am. J.*, 46:1153-1158.

Elrashidi, M. A. and G. A. O'Connor, 1982b, Boron Sorption and Desorption on Soils, *Soil Sci. Soc. Am. J.*, 46:27-31.

Ewart, F. T., J. L. Smith-Briggs, H. P. Thomason, and S. J. Williams, 1992, The Solubility of Actinides in a Cementitious Near-Field Environment, *Waste Mgmt.*, 12:241-252.

Felmy, A. R., D. Rai, J. A. Schramke, and J. L. Ryan, 1989, The Solubility of Plutonium Hydroxide in Dilute Solution and in High-Ionic-Strength Brines, *Radiochimica Acta*, 48:29-35.

Gerber, M. S., 1996, The Plutonium Production Story at the Hanford Site: Processes and Facilities History, *WHC-MR-0521* Rev.0, Westinghouse Hanford Company, Richland, WA.

Gephart, R. E. and R. E. Lundgren, 1996, Hanford Tank Clean up: A Guide to Understanding the Technical Issues, *PNL-10773*, Pacific Northwest Laboratory, Richland, WA.

Gevantman, L. H. and K. A. Kraus, 1949, Chemistry of Plutonium(V). Stability and Spectrophotometry, Paper 4.16 in *The Transuranium Elements*, G. T. Seaborg, J. J. Katz, and W. M. Manning, Ed., McGraw-Hill Book Company, New York, NY.

Haire, R. G., M. H. Lloyd, M. L. Beasley, and W. O. Milligan, 1971, Aging of Hydrous Plutonium Dioxide, *J. Electron Microsc.*, 20:8-16.

Herting, D. L., 1994, Characterization of Sludge Samples from Tank 241-AN-107, in Process Aids - A Compilation of Technical Letters by Process Laboratories and Technology, *WHC-IP-0711-26*, Westinghouse Hanford Company, Richland, WA.

Herting, D. L., 1995, Report of Scouting Study on Precipitation of Strontium, Plutonium, and Americium from Hanford Complexant Concentrate Waste, *WHC-SD-WM-DTR-040*, Westinghouse Hanford Company, Richland, WA.

Hindman, J. C., 1954, Ionic and Molecular Species of Plutonium in Solution, Chapter 9 in *The Actinide Elements*, G. T. Seaborg and J. J. Katz, Eds., McGraw-Hill Book Company, New York.

Hingston, F. J, 1981, A Review of Anion Adsorption, in *Adsorption of Inorganics at Solid-Liquid Interfaces*, M. A. Anderson and A. J. Rubin, Eds., Ann Arbor Science, Ann Arbor, Michigan, p. 51-90.

Ho, C. H. and D. C. Doern, 1985, The Sorption of Uranyl Species on a Hematite Sol, *Can. J. Chem.*, 63:1100.

Hobbs, D. T. and D. G. Karraker, 1996, Recent Results on the Solubility of Uranium and Plutonium in Savannah River Site Waste Supernate, *Nucl. Technology*, 114:318-324.

Hobbs, D. T. and T. B. Edwards, 1993, Solubility of Plutonium in Alkaline Salt Solutions, *WSRC-TR-93-131*, Westinghouse Savannah River Company, Aiken, SC.

Hobbs, D. T., 1995, Effects of Coprecipitation of Uranium and Plutonium in Alkaline Salt Solution, *WSRC-TR-95-0462*, Westinghouse Savannah River Company, Savannah River, South Carolina.

Hodgson, K. M., 1995, Tank Characterization Report for Double-Shell Tank 241-AZ-101, *WHC-SD-WM-ER-410, Rev. 0*, Westinghouse Hanford Company, Richland, WA.

Hsi, C-K. D., and D. Langmuir, 1985, Adsorption of Uranyl onto Ferric Oxyhydroxide: Application of the Surface Complexation Site-Binding Model, *Geochimica et Cosmochimica Acta*, 49:1931-1941.

Isaacson, P. J. and C. R. Frink, 1984, Nonreversible Sorption of Phenolic Compounds by Sediment Fractions: The Role of Sediment Organic Matter, *Environ. Sci. Technol.*, 18:43-48.

Jo, J., B. J. Morris, and T. T. Tran, 1996, Tank Characterization Report for Double-Shell Tank 241-AN-102, *WHC-SD-WM-ER-545, Rev. 1*, Westinghouse Hanford Company, Richland, WA.

Johnson, G. L. and E. B. Fowler, 1967, The Separation of Plutonium from Platinum by Coprecipitation with Ferric Iron Followed by Ion Exchange on Inorganic Exchange Materials, *LA-3149*, Los Alamos Scientific Laboratory, Los Alamos, New Mexico.

Karimian, N. and F. R. Cox, 1978, Adsorption and Extractability of Molybdenum in Relation to Some Chemical Properties of Soil, *Soil Sci. Soc. Am. J.*, 42:757-761.

Karraker, D. G., 1993, Solubility of Plutonium in Waste Evaporation, *WSRC-TR-93-578*, Westinghouse Savannah River Company, Aiken, SC.

Karraker, D. G., 1994, Radiation Effects on the Solubility of Plutonium in Alkaline High Level Waste, *WSRC-MS-94-0278X (Rev. 2)*, Westinghouse Savannah River Company, Aiken, SC.

Karraker, D. G., 1995, Plutonium(VI) Solubility Studies in Savannah River Site High-Level Waste Supernate, *WSRC-TR-95-0244*, Westinghouse Savannah River Company, Aiken, SC.

Kinniburgh, R. D. and M. L. Jackson, 1981, Cation Adsorption by hydrous Metal Oxides and Clay, in *Adsorption of Inorganics at Solid-Liquid Interfaces*, M. A. Anderson and A. J. Rubin, Eds., Ann Arbor Science, Ann Arbor, Michigan, p. 91-160.

Koskinen, W. C., G. A. O'Connor, and H. H. Cheng, 1979, Characterization of Hysteresis in the Desorption of 2,4,5-T from Soils, *Soil Sci. Soc. Am. J.*, 43:871-874.

Krot, N. N. and A. D. Gel'man, 1967, Preparation of Neptunium and Plutonium in the 7-Valent State, *Dokl. Akad. Nauk SSSR*, 177:124-126.

Krot, N. N., A. D. Gel'man, M. P. Mefod'eva, V. P. Shilov, V. F. Peretrukhin, and V. I. Spitsyn, 1977, *Semivalentnoe Sostoyanie Neptuniya, Plutoniya, Ameritsiya*, Nauka, Moscow, USSR. Available in English as Heptavalent States of Neptunium, Plutonium, and Americium, *UCRL-trans-11798*, Lawrence Livermore National Laboratory, Livermore, CA.

Krot, N., V. Shilov, A. Bessonov, N. Budantseva, I. Charushnikova, V. Perminov, and L. Astafurova, 1996, Investigation on the Coprecipitation of Transuranium Elements, from Alkaline Solutions by the Method of Appearing Reagents, *WHC-EP-0898*, Westinghouse Hanford Company, Richland, WA.

Kupfer, M. J., A. L. Boldt, B. A. Higley, K. M. Hodgson, L. W. Shelton, B. C. Simpson, R. A. Watrous, M. D. LeClair, G. L. Borsheim, R. T. Winward, R. M. Orme, N. G. Colton, S. L. Lambert, D. E. Place, and W. W. Schulz, 1997, Standard Inventories of Chemicals and Radionuclides in Hanford Site Tank Wastes, *HNF-SD-WM-TI-740, Rev. 0*, Lockheed Martin Hanford Corporation, Richland, WA.

Laitinen, H. A., 1960, *Chemical Analysis*, McGraw-Hill Book Company, Inc., New York, New York, p. 169.

Lierse, Ch., 1985, Chemisches Verhalten von Plutonium in Natürlichen, Aquatischen Systemen: Hydrolyse, Carbonatkomplexierung und Redoxreaktionen, dissertation, Institut für Radiochemie der Technischen Universität München, Garching, Germany, as reported in Puigdomènech, I. and J. Bruno, 1991, Plutonium Solubilities, *SKB TR 91-04*, Svensk Kärnbränslehantering AB, Stockholm, Sweden.

Lindsay, W. L., 1979, *Chemical Equilibria in Soils*, John Wiley & Sons, New York, NY, p. 135 and 40.

Lloyd, M. H. and R. G. Haire, 1978, The Chemistry of Plutonium in Sol-Gel Processes, *Radiochimica Acta*, 25:139-148.

Lumetta, G. J., B. M. Rapko, M. J. Wagner, J. Liu, and Y. L. Chen, 1996, Washing and Caustic Leaching of Hanford Tank Sludges: Results of FY 1996 Studies, *PNNL-11278, Rev. 1*, Pacific Northwest National Laboratory, Richland, WA.

Lumetta, G. J., M. J. Wagner, N. G. Colton, and E. O. Jones, 1993, Underground Storage Tank Integrated Demonstration, Evaluation of Pretreatment Options for Hanford Tank Wastes, *PNL-8537*, Pacific Northwest Laboratory, Richland, WA.

Maiti, T. C. and J. H. Kaye, 1992, Measurement of Total Alpha Activity of Neptunium, Plutonium, and Americium in Highly Radioactive Hanford Waste by Iron Hydroxide Precipitation and 2-Heptanone Solvent Extraction, *PNL-SA-20460*, Pacific Northwest Laboratory, Richland, WA.

Meisel, D., C. D. Jonah, S. Kapoor, M. S. Matheson, and M. C. Sauer, 1993, Radiolytic and Radiolytically Induced Generation of Gases from Synthetic Wastes, *ANL-93/43*, Argonne National Laboratory, Argonne, IL.

Mishra, S. P. and D. Tiwary, 1995, Ion Exchangers in Radioactive Waste Management. Part VIII: Radiotracer Studies on Adsorption of Strontium Ions on Hydrous Manganese Oxide, *Radiochimica Acta*, 69:121-126.

Musikas, C., 1971, Peroxyde de Plutonium Pentavalent, *Radiochem. Radioanal. Letters*, 7:375-379.

Novikov, A. I. and I. A. Starovoit, 1969, Separation, Concentration, and Determination of the Valence Forms of Plutonium (III, IV, VI) by the Method of Coprecipitation with Zirconium and Iron Hydroxides, *Radiokhimiya*, 11:339-341

Payne, T. E., J. A. Davis, and T. D. Waite, 1994, Uranium Retention by Weathered Schists - The Role of Iron Minerals, *Radiochimica Acta*, 66/67:297-303.

Peek, D. C. and V. V. Volk, 1985, Fluoride Sorption and Desorption in Soils, *Soil Sci. Soc. Am. J.*, 49:583-6.

Peretrukhin, V. F. and D. P. Alekseeva, 1974, Polarographic Properties of Higher Oxidation States of Plutonium in Aqueous Alkali Solutions, *Sov. Radiochem.*, 16:823-826.

Peretrukhin, V. F. and V. I. Spitsyn, 1982, Electrochemical Determination of the Oxidation Potentials and the Thermodynamic Stability of the Valence States of the Transuranium Elements in Aqueous Alkaline Media, *Bull. Acad. Sci. USSR, Div. Chem. Sci.*, No. 4:826-831.

Peretrukhin, V. F., F. David, and A. Maslennikov, 1994, Electrochemical Properties and Thermodynamic Stability of Pu and Neighboring Actinides in the (IV) and (V) Oxidation States in Aqueous Alkaline Media and Radwastes, *Radiochimica Acta*, 65:161-166.

Peretrukhin, V. F., S. V. Kryutchkov, V. I. Silin, and I. G. Tananaev, 1996, Determination of the Solubility of Np(IV)-(V), Pu(III)-(VI), Am(III)-(VI) and Tc(IV),(V) Hydroxo Compounds in 0.5 - 14 M NaOH Solutions, *WHC-EP-0897*, Westinghouse Hanford Company, Richland, WA.

Peretrukhin, V. F., V. P. Shilov, and A. K. Pikaev, 1995, Alkaline Chemistry of Transuranium Elements and Technetium and the Treatment of Alkaline Radioactive Wastes, *WHC-EP-0817*, Westinghouse Hanford Company, Richland, WA.

Pérez-Bustamante, J. A., 1965, Solubility Product of Tetravalent Plutonium Hydroxide and Study of the Amphoteric Character of Hexavalent Plutonium Hydroxide, *Radiochimica Acta*, 4:67-72.

Pikaev, A. K., A. V. Gogolev, S. V. Kryutchkov, V. P. Shilov, V. N. Chulkov, L. I. Belyaeva, and L. N. Astafurova, 1996, Radiolysis of Actinides and Technetium in Alkaline Media, *WHC-EP-0901*, Westinghouse Hanford Company, Richland, WA.

Pius, I. C., Charyulu, M. M., B. Venkataramani, C. K. Sivaramakrishnan, and S. K. Patil, 1995, Studies on Sorption of Plutonium on Inorganic Ion Exchangers from Sodium Carbonate Medium, *J. Radioanal. Nucl. Chem., Letters*, 199, (1), p. 1-7.

Puigdomènech, I. and J. Bruno, 1991, Plutonium Solubilities, *SKB TR 91-04*, Svensk Kärnbränslehantering AB, Stockholm, Sweden.

Rai, D. and J. L. Ryan, 1982, Crystallinity and Solubility of Pu(IV) Oxide and Hydrous Oxide in Aged Aqueous Suspensions, *Radiochimica Acta*, 30:213-216.

Rai, D., R. J. Serne, and D. A. Moore, 1980, Solubility of Plutonium Components and their Behavior in Soils, *Soil Sci. Soc. Am. J.*, 44: 490.

Robouch, P. and P. Vitorge, 1988, Solubility of $PuO_2(CO_3)$, *Inorg. Chim. Acta*, 140:239-242.

Roetman, V. E., S. P. Robyler, and H. Toffer, 1994, Estimation of Plutonium in Hanford Site Waste Tanks, Based on Historical Records, *WHC-SD-SQA-SA-20356, Rev. 0*, Westinghouse Hanford Company, Richland, WA.

Sanchez, A. L., J. W. Murray, and T. H. Sibley, 1985, The Adsorption of Plutonium IV and V on Goethite, *Geochimica et Cosmochimica Acta*, 49:2297-2307.

Shannon, R. D. and C. T. Prewitt, 1969, *Acta Cryst.*, B25:925.

Shilov, V. P., N. N. Krot, N. Budantseva, A. Yusov, A. Garnov, V. Perminov, and L. Astafurova, 1996a, Investigation of Some Redox Reactions of Neptunium, Plutonium, Americium, and Technetium in Alkaline Media, *WHC-EP-0886*, Westinghouse Hanford Company, Richland, WA.

Shilov, V. P., L. N. Astafurova, A. Yu. Garnov, and N. N. Krot, 1996b, Reaction of H_2O_2 with Suspensions of $Np(OH)_4$ and $Pu(OH)_4$ in Alkali Solution, *Radiochemistry*, 38:217-219.

Shindler, P. W., 1981, Surface Complexes at Oxide-Water Interfaces, in *Adsorption of Inorganics at Solid-Liquid Interfaces*, M. A. Anderson and A. J. Rubin, Eds., Ann Arbor Science, Ann Arbor, Michigan, p. 1-50.

Stupin, D. Yu. and M. I. Ozernoi, 1995, Coprecipitation of [152]Eu with Iron(III) Hydroxide Formed upon Reduction of Sodium Ferrate(VI) in Aqueous Medium, *Radiokhimiya*, 37:359-362.

Tananaev, I. G., 1989, Forms of Hexavalent Plutonium and Americium in Basic Aqueous Solutions, *Sov. Radiochem.*, 31:303-307.

Tananaev, I. G., 1992, Hydroxide Compounds of Pu(V), *Sov. Radiochem.*, 34:161-163.

Tananaev, I. G., S. P. Rozov, and V. S. Mironov., 1992, Speciation of Pu(VII) in Basic Aqueous Solutions, *Sov. Radiochem.*, 34:331-334.

Temer, D. J. and R. Villarreal, 1995, Sludge Washing and Alkaline Leaching Tests on Actual Hanford Tank Sludge: A Status Report, *LAUR 95-2070*, Los Alamos National Laboratory, Los Alamos, NM.

Temer, D. J. and R. Villarreal, 1996, Sludge Washing and Alkaline Leaching Tests on Actual Hanford Tank Sludge: FY 1996 Results, *LAUR 96-2839*, Los Alamos National Laboratory, Los Alamos, NM.

Thiyagarajan, P., H. Diamond, L. Soderholm, E. P. Horwitz, L. M. Toth, and L. K. Felker, 1990, Plutonium(IV) Polymers in Aqueous and Organic Media, *Inorg. Chem.*, 29:1902-1907.

Ticknor, K. V., 1993, Actinide Sorption by Fracture-Filling Minerals, *Radiochimica Acta*, 60:33-42.

Ticknor, K. V., T. T. Vandergraaf, and D. G. Juhnke, 1986, The Effect of Laboratory Time-Scale Hydrothermal Alteration on Igneous Rok Coupon Radionuclide Sorption/Desorption, *TR-376*, Atomic Energy of Canada Limited Research Company, Chalk River, Ontario, Canada.

Tochiyama, O., S. Endo, and Y. Inoue, 1994, Sorption of Neptunium(V) on Various Iron Oxides and Hydrous Iron Oxides, *Radiochimica Acta*, 66/67:105-111.

Toth, L. M., H. A. Friedman, G. M. Begun, and S. E. Doris, 1984, Raman Study of Uranyl Ion Attachment to Thorium(IV) Hydrous Polymer, *J. Phys. Chem.*, 88:5574-5577.

Toth, L. M., H. A. Friedman. and M. M. Osborne, 1981, Polymerization of Pu(IV) in Aqueous Nitric Acid Solution, *J. Inorg. Nucl. Chem*, 43, (11):2929-2934.

Valenzuela, B. D. and L. Jensen, 1994, Tank Characterization Report for Single-Shell Tank 241-T-101, *WHC-SD-WM-ER-382, Rev. 0A*, Westinghouse Hanford Company, Richland, WA.

Varlashkin, P. G., G. M. Begun, and J. R. Peterson, 1984, Electrochemical and Spectroscopic Studies of Plutonium in Concentrated Aqueous Carbonate and Carbonate-Hydroxide Solutions, *Radiochimica Acta*, 35:211-218.

Wester, D. W. and J. C. Sullivan, 1983, The Absorption Spectra of Pu(VI), -(V) and -(IV) Produced Electrochemically in Carbonate-Bicarbonate Media, *Radiochem. Radioanal. Letters*, 57:35-42.

Yamaguchi, T., Y. Sakamoto, and T. Ohnuki, 1994, Effect of the Complexation on Solubility of Pu(IV) in Aqueous Carbonate System, *Radiochimica Acta*, 66/67:9-14.

Yariv, S. and H. Cross, 1979, *Geochemistry of Colloid Systems,* Springer-Verlag, New York, NY, p. 178.

II. Actinide Complexation and Solubility

DISSOCIATION CONSTANTS OF CARBOXYLIC ACIDS AT HIGH IONIC STRENGTHS

Jiri Mizera,[1] Andrew H. Bond,[2] Gregory R. Choppin, and Robert C. Moore[3]

Department of Chemistry, The Florida State University,
Tallahassee, FL 32306-3006 USA

[1]On leave from the Department of Nuclear Chemistry, The Czech Technical University, Brehova 7, 115 19 Praha 1, Czech Republic
[2]Present address: Chemistry Division, Argonne National Laboratory, 9700 S. Cass Ave., Argonne, IL 60439 USA
[3]Sandia National Laboratory, P. O. Box 5800 MS 1341, Albuquerque, NM 87185-1320 USA

ABSTRACT

Dissociation constants (pK_a) of acetic, oxalic, citric, and ethylenediaminetetraacetic acids have been determined potentiometrically using a hydrogen ion glass electrode at 25 °C in solutions of ionic strengths ranging from 0.3 to 5.0 m NaCl. The observed ionic strength dependencies of the pK_a values have been described successfully over the entire ionic strength range using both the Specific Ion Interaction Theory and the Pitzer model. The variation of the pK_a values at high ionic strengths with the type and concentration of supporting electrolyte is discussed and compared to literature data.

INTRODUCTION

Determination of stability constants of metal complexation by a specific ligand using potentiometric titrimetry requires accurate constants characterizing proton dissociation of the acidic ligand. This requirement becomes more critical when dealing with relatively weak metal-ligand complexation, since uncertainty in the acid dissociation constants characterizing formation of the free ligand (which is responsible for metal binding) results in greater uncertainty in the related metal complex stability constant.

Critical surveys of stability constants[1-3] provide numerous references from which "recommended" values have been selected. It is often possible to find a constant obtained for similar experimental conditions at low ionic strengths and 25 °C. However, there is a shortage of reported literature constants above 1-2 m ionic strength that are suitable for establishing "recommended" pK_a values at such high ionic strengths.

Dissociation constants of some carboxylic acids at high NaCl concentrations and 25 °C have been determined in our laboratories in connection with studies on actinide chemistry in support of the Waste Isolation Pilot Plant (WIPP).[4-7] These acid dissociation constants were calculated from potentiometric titration data using a recently reported

Actinide Speciation in High Ionic Strength Media, edited by Reed *et al.*
Kluwer Academic / Plenum Publishers, New York, 1999

113

differential analysis method[4] developed in this laboratory. In some instances, substantial differences were found between these reported pK_a values and the available literature data.

As part of the WIPP Actinide Term Test Program, studies of complexation of magnesium by carboxylate ligands at high ionic strengths of NaCl are underway.[8] Prior to the determination of metal-ligand stability constants by potentiometric titrimetry, we have determined the dissociation constants of acetic, oxalic, citric, and ethylenediaminetetraacetic (EDTA) acids at 25 °C in 0.3-5 m NaCl solutions. Specific Ion Interaction Theory (SIT) and Pitzer modeling have been performed for all systems and the pertinent parameters are reported and compared to the available literature data.

EXPERIMENTAL

Reagents and Procedures

ACS reagent grade chemicals were used without further purification. All chemicals were obtained from Fisher Scientific, except for Na_2EDTA (Aldrich), oxalic acid (Fisher and Mallinckrodt), and potassium hydrogen phthalate and phenolphthalein (Baker). Deionized water (Barnstead Nanopure) was used for all solutions.

All stock solutions were prepared at the desired ionic strength (0.3, 1, 2, 3, 4, and 5 m) with NaCl as the background electrolyte. Contributions to the ionic strength from the carboxylic acids were included in the calculations. Ligand solutions were prepared by dissolving known amounts of the respective acid (oxalic and citric) or its sodium salt (acetate) in the NaCl solutions. EDTA solutions were prepared by dilution of a volumetric standard of Na_2EDTA with the appropriate NaCl solution. NaOH titrant solutions were prepared by adding a known aliquot of 50 % (CO_2 free) NaOH to an N_2 saturated NaCl solution to adjust to the desired ionic strength. The resulting NaOH solution was standardized, using dry potassium hydrogen phthalate, to a phenolphthalein endpoint. Carbon dioxide contamination of the NaOH titrant was estimated from a Gran plot[9] and solutions showing greater than 2 % CO_2 were discarded. These standardized NaOH solutions were used to standardize HCl solutions, which had been prepared by diluting concentrated HCl with NaCl solutions to the desired ionic strength. All but the NaOH and HCl containing solutions were filtered using Nalgene 0.2 μm filters.

Potentiometric Apparatus

A Fisher Accumet 950 pH/ion meter equipped with a Corning semi-micro combination glass electrode was used. The KCl reference solution was replaced with a saturated NaCl solution to minimize effects of the liquid junction potential. The titrant was delivered by a Metrohm Dosimat 665 motor-driven piston burette. A water jacketed cell (30 mL) was held at 25±0.1 °C using a Fisher Isotemp Model 910 thermostated circulator, and magnetic stirring was employed. The titrations were performed under a N_2 atmosphere using N_2 scrubbed with 4 M NaOH to remove CO_2 and passed through 1 M NaCl to humidify the gas.

Potentiometric Titration Procedure

The general procedure has been described by Martell and Motekaitis.[9] The electrode was precalibrated at 25 °C using standard 4.00±0.01 and 7.00±0.01 buffers. The pH meter readings (pHr) were converted to hydrogen ion concentrations (pcH = -log [H^+]) using calibration plots constructed by linear regression on HCl/NaOH titration data collected prior to each pK_a determination. The titrand solution, typically 15-21 mL, containing ligand ($\approx$ 5 x 10^{-3} M acetate, $\approx$ 5 x 10^{-3}-1 x 10^{-2} M oxalate, $\approx$ 5 x 10^{-3} citrate, or $\approx$ 1 x 10^{-3}-4 x 10^{-3} M EDTA), a known amount of HCl (in the acetate and EDTA systems), and NaCl was titrated to pHr > 10 with standardized NaOH ($\approx$ 0.1 M) at the same ionic strength.

Calculation of the Dissociation Constants

The computer program BEST[9] was used to calculate acid dissociation constants from pcH profiles. BEST refines estimated stability constants by minimizing the sum of the weighted squares of the deviation of the observed pcH from the calculated pcH using the mass balance equations for each component. Some results obtained by BEST were tested and compared well to those determined using a differential analysis program.[4]

The stepwise dissociation constant of an H_nL acid is defined as (charges omitted):

$$pK_{ai} = -\log \frac{[H][H_{n-i}L]}{[H_{n-i+1}L]} \tag{1}$$

For all calculations, including construction of the pcH vs. pHr calibration curves, pK_w values determined at the appropriate ionic strengths of NaCl were used.[10]

RESULTS AND DISCUSSION

The dissociation constants of the carboxylic acids measured at 25 °C in the range 0.3-5 m NaCl are listed in Table 1, and their ionic strength dependencies are shown in Figures 1-4. The errors reported in Table 1 represent estimated standard deviations of duplicate pK_a determinations, except for pK_{a1} of oxalic acid and pK_{a1} and pK_{a2} of EDTA where titration data in the presence of magnesium were used as duplicate data sets for these pK_a values. Preliminary speciation calculations and analysis of the Mg^{2+} stability constants[8] indicated that the pcH regions in which the fully protonated forms of oxalic acid and EDTA predominate are sufficiently separated from the regions in which complexation occurs that the profiles at low pcH can be used for calculation of the dissociation constants.

The reproducibility of the titration experiments corresponds to the precision typically obtained by potentiometry with a hydrogen ion glass electrode. The large errors for pK_{a1} of oxalic acid reflect the difficulty in calculating this constant when less than 5 % of the diprotonated species is present at the beginning of the forward titration. The low precision of pK_{a1} observed for EDTA is likely attributable to overlap of pK_{a1} and pK_{a2}, which differ only slightly.

The pK_a values presented in this work are in general agreement with those in the literature, however, in a few instances there are significant differences. Our acid dissociation constants for acetic acid agree well with those reported by Chen et al.,[4] but are consistently higher than the results of Mesmer et al.,[11] and Belevantsev et al.,[12] where the differences reach approximately 0.04 log units in 5 m NaCl. Our values of pK_{a2} for oxalic acid are within the range of literature values and close to those of Choppin and Chen,[5] but are somewhat higher than the values reported by Kettler et al.[13] Such differences increase as the ionic strength increases and reach approximately 0.13 log units in 5 m NaCl. The pK_a values for citric acid agree well with the critically selected values,[1,2] whereas the constants reported by Choppin et al.[6] are generally higher, reaching an average difference of approximately 0.15 log units for all three pK_a values in 5 m NaCl. The third and fourth acid dissociation constants for EDTA are in agreement with those of Chen et al.,[7] while the pK_{a1} and pK_{a2} values reported here are higher than the reference data. These deviations are similar for both pK_{a1} and pK_{a2} and are again more pronounced at higher ionic strengths, reaching about 0.2 log units in 5 m NaCl. These deviations probably are related to the difficulty in calculating values for species with similar dissociation constants.

The experimental pK_a values have been modeled using the SIT[14,15] and Pitzer methods. The SIT model is considered valid to ionic strengths of 2-3 m; it was successfully applied here to describe the variation of pK_{ai} up to 5 m (NaCl). The Pitzer model, described later, requires more interaction parameters and is valid over the entire ionic strength range of this study.

In the SIT model, the general equation describing the variation of a dissociation constant with ionic strength (in molal units) takes the form:[16]

$$pK_{ai} = pK_{ai}^0 - \Delta Z^2 \frac{A\sqrt{I}}{1 + aB\sqrt{I}} + \Delta \varepsilon I \tag{2}$$

Table 1. Dissociation constants of carboxylic acids in NaCl at 25 °C.

	Acetic Acid	Oxalic Acid	
I, molal	pK_a	pK_{a1}	pK_{a2}
0.3	4.52±0.02[1]	1.06±0.05[2]	3.67±0.02[1]
1	4.52±0.02	0.9±0.2[2]	3.52±0.02[1]
2	4.65±0.02[1]	0.9±0.1[2]	3.54±0.02[1]
3	4.78±0.01	0.86±0.07[2]	3.55±0.02[1]
4	4.96±0.02[1]	1.04±0.05[2]	3.74±0.02[1]
5	5.13±0.01	1.2±0.2[2]	3.85±0.02[1]

	Citric Acid		
I, molal	pK_{a1}	pK_{a2}	pK_{a3}
0.3	2.92±0.04	4.25±0.01	5.464±0.008
1	2.87±0.02	4.11±0.02	5.179±0.005
2	2.84±0.04	4.09±0.01	5.076±0.006
3	2.79±0.01	4.116±0.002	5.084±0.006
4	2.93±0.06	4.24±0.02	5.20±0.04
5	2.96±0.07	4.32±0.03	5.22±0.04

	EDTA			
I, molal	pK_{a1}	pK_{a2}	pK_{a3}	pK_{a4}
0.3	2.14±0.05[2]	2.56±0.02[2]	6.02±0.02[1]	9.18±0.02[1]
1	2.1±0.1[2]	2.34±0.07[2]	6.04±0.02	8.76±0.02
2	1.9±0.1[2]	2.42±0.02[2]	6.29±0.02	8.68±0.02
3	2.036±0.006[2]	2.55±0.02[2]	6.51±0.02[1]	8.68±0.02[1]
4	2.1±0.1[2]	2.50±0.01[2]	6.76±0.01	8.82±0.02
5	2.26±0.06[2]	2.625±0.009[2]	7.01±0.02[1]	8.96±0.02[1]

[1]The error has been defined as 0.02 log units and is estimated for a single experiment.
[2]These errors are derived using an experiment where Mg^{2+} was present. See the discussion in the text.

where pK_{ai}^0 is the thermodynamic dissociation constant (at zero ionic strength). ΔZ^2 is the difference between the squares of the charges of the products and reactants multiplied by stoichiometric coefficients ν_i, given by the relation:

$$\Delta Z^2 = \left| \Sigma(\nu_i Z_i^2)_{products} - \Sigma(\nu_j Z_j^2)_{reactants} \right| \qquad (3)$$

$\Delta\varepsilon$ is the difference of the coefficients characterizing specific cation-anion interactions between a given species and the background electrolyte:

$$\Delta\varepsilon = \left| \Sigma(\nu_i\varepsilon_i)_{products} - \Sigma(\nu_j\varepsilon_j)_{reactants} \right| \qquad (4)$$

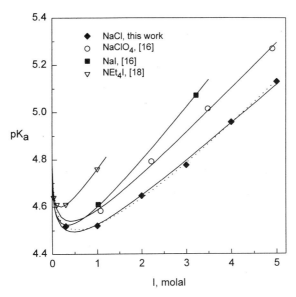

Figure 1. Ionic strength dependence of the dissociation constants of acetic acid in various electrolytes. Symbols represent experimental values, solid lines the SIT fit, dashed lines the Pitzer model.

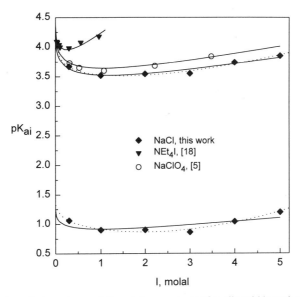

Figure 2. Ionic strength dependence of dissociation constants of oxalic acid in various electrolytes. Symbols represent experimental values, solid lines the SIT fit, dashed lines the Pitzer model.

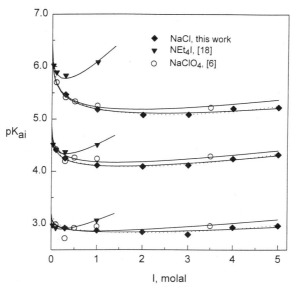

Figure 3. Ionic strength dependence of dissociation constants of citric acid in various electrolytes. Symbols represent experimental values, solid lines the SIT fit, dashed lines the Pitzer model.

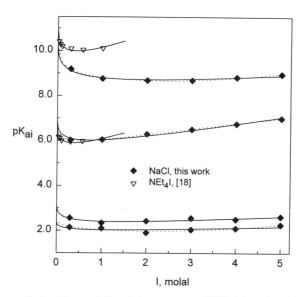

Figure 4. Ionic strength dependence of dissociation constants of EDTA in various electrolytes. Symbols represent experimental values, solid lines the SIT fit, dashed lines the Pitzer model.

The constant A in equation (2) is the Debye-Hückel coefficient (0.5091 at 25 °C), and aB is an empirical parameter involving "the mean distance of closest approach of ions" from the Debye-Hückel limiting law and a value of 1.5 was used as suggested by Scatchard.[17]

Plotting $pK_{ai} + 0.5091\Delta Z^2 \sqrt{I}/(1 + 1.5\sqrt{I})$ vs. I results in a straight line with a slope of $\Delta\varepsilon$ and an intercept of pK_{ai}^0. The results of the SIT calculations are presented in Table 2. The reported errors are those for R^2, calculated by the linear regression subroutine in the Quattro Pro 5.0 for Windows spreadsheet program.

In Table 2, the low correlation coefficients for pK_{a1} of the three polyprotic acids reflect the low precision of these dissociation constants, rather than limitations of the SIT model as no obvious curvature of the SIT plots is observed. Deviations of the experimentally determined constants from the SIT fits are generally smaller than the estimated standard deviations.

The satisfactory description of the dissociation constants of all four carboxylic acids in NaCl up to 5 m suggested that the model may be useful in describing literature data on carboxylic acid dissociation in different background electrolytes. Table 3 lists average SIT parameters for various background electrolytes calculated using data from the present work and from the literature. The critically selected values for NaCl are presented for comparison. Errors on the SIT parameters are the estimated standard deviations of mean values if more than a single reference was available, otherwise they are the calculated errors from linear regression by the spreadsheet program. Our data, selected literature data, and the SIT calculations are depicted in Figures 1-4.

As seen from Table 3 and in Figures 1-4, there are significant differences in the ionic strength dependencies of the pK_a values in different ionic media. For background electrolytes with Na^+ as cation and Cl^-, ClO_4^-, or I^- as anions, these differences are minimal at $I < 0.1$ m where the Debye-Hückel limiting law is valid, and the SIT calculations yield comparable thermodynamic constants for the different electrolytes. At higher ionic strengths the differences in the "specific ionic interactions" (behavior of the activity coefficients) become more pronounced, which results in an increase in pK_a values in the approximate order NaCl < NaClO$_4$ < NaI. Acid dissociation constants for acetic, oxalic, and citric acids in NaClO$_4$ supporting electrolyte are similar to those in NaCl.

Larger differences in pK_a values are observed between NaCl and tetraethylammonium iodide (NEt$_4$I).[18] Due to the higher value of the specific ion interaction parameter, $\Delta\varepsilon$, the SIT fits for NEt$_4$I deviate from the NaCl curves in the Debye-Hückel limiting law region, and reach significantly higher values before 1 M ionic strength. (N.B. due to the absence of density data, the NEt$_4$I molarity has not been converted to molality.) In

Table 2. SIT parameters describing the ionic strength dependence of dissociation constants of carboxylic acids.

Acid, constant	pK_{ai}^0	ΔZ^2	$\Delta\varepsilon$	R^2
acetic, pK_{a1}	4.76±0.01	2	0.177±0.004	1.00
oxalic, pK_{a1}	1.25±0.09	2	0.08±0.02	0.72
oxalic, pK_{a2}	4.21±0.04	4	0.13±0.01	0.97
citric, pK_{a1}	3.20±0.04	2	0.05±0.01	0.85
citric, pK_{a2}	4.81±0.03	4	0.108±0.006	0.99
citric, pK_{a3}	6.31±0.04	6	0.09±0.01	0.96
EDTA, pK_{a1}	2.4±0.1	2	0.07±0.02	0.65
EDTA, pK_{a2}	3.11±0.07	4	0.11±0.01	0.94
EDTA, pK_{a3}	6.91±0.06	6	0.34±0.02	0.99
EDTA, pK_{a4}	10.28±0.06	8	0.14±0.02	0.95

Table 3. Average SIT parameters describing the ionic strength dependence of dissociation constants of carboxylic acids in different supporting electrolytes.

Medium	Acid, pK_{ai}	pK_{ai}^{0}	$\Delta\varepsilon$	$pK_{ai}^{0(1)}$	Reference
NaCl	acetic, pK_a	4.74±0.01	0.175±0.002	4.757	This study, 4, 11, 12
NaClO$_4$	acetic, pK_a	4.79±0.02	0.206±0.006		12
NaI	acetic, pK_a	4.754±0.008	0.253±0.003		12
NEt$_4$I, 37 °C[2,3]	acetic, pK_a	4.761±0.003	0.298±0.004		18
NaCl	oxalic, pK_{a1}	1.23±0.02	0.076±0.001	1.252	This study, 13
NaCl	oxalic, pK_{a2}	4.24±0.02	0.12±0.01	4.266	This study, 5, 13
NaClO$_4$	oxalic, pK_{a2}	4.26±0.05	0.158±0.008		5
NEt$_4$I, 37 °C[3]	oxalic, pK_{a2}	4.38±0.03	0.69±0.03		18
NaCl	citric, pK_{a1}	3.16±0.04	0.06±0.02	3.128	This study, 2, 6
NaClO$_4$	citric, pK_{a1}	3.2±0.1	0.084±0.007		6
NEt$_4$I, 37 °C[3]	citric, pK_{a1}	3.10±0.01	0.39±0.01		18
NaCl	citric, pK_{a2}	4.80±0.02	0.116±0.006	4.761	This study, 2, 6
NaClO$_4$	citric, pK_{a2}	4.88±0.07	0.113±0.005		6
NEt$_4$I, 37 °C[3]	citric, pK_{a2}	4.79±0.03	0.56±0.03		18
NaCl	citric, pK_{a3}	6.29±0.04	0.10±0.02	6.396	This study, 2, 6
NaClO$_4$	citric, pK_{a3}	6.33±0.03	0.122±0.002		6
NEt$_4$I, 37 °C[3]	citric, pK_{a3}	6.44±0.05	0.92±0.06		18
NaCl	EDTA, pK_{a1}	2.36±0.01	0.04±0.02		This study, 7
NaCl	EDTA, pK_{a2}	3.06±0.04	0.10±0.01		This study, 7
NaCl	EDTA, pK_{a3}	6.88±0.03	0.341±0.004	6.237	This study, 7
NEt$_4$I, 37 °C[3]	EDTA, pK_{a3}	6.62±0.06	0.72±0.07		18
NaCl	EDTA, pK_{a4}	10.19±0.09	0.16±0.02	10.948	This study, 7
NEt$_4$I, 37 °C[3]	EDTA, pK_{a4}	11.01±0.06	0.82±0.07		18

[1]Recommended values from reference 2.
[2]The SIT parameters have been adjusted to $aB = 2.5$.
[3]The SIT parameters have been adjusted to $A = 0.523$ according to reference 18.

addition, the accuracy of the SIT model with $aB = 1.5$ may be decreased, and this parameter may require adjustment to obtain better correlation with the experimental data, as has been done for the acetate system (see the footnote to Table 3; the Debye-Hückel parameter A has been changed to the value of 0.523 at 37 °C[18] where appropriate).

Differences in constants from NEt$_4^+$ and Na$^+$ media can be ascribed to the noncomplexing properties of the NEt$_4^+$ cation. In contrast, alkali metal cations may be weakly bound by carboxylates resulting in lower "apparent" dissociation constants. As seen from the figures, the differences in carboxylic acid dissociation constants in NEt$_4^+$ and Na$^+$ solutions increase in the order acetic < oxalic < citric < EDTA which is consistent with increasing stability of the Na$^+$-carboxylate complexes.

The difference in the "apparent" pK_a values in NEt$_4$I and in the other electrolytes has been used to calculate stability constants of alkali and alkaline-earth metal-carboxylate complexes.[19-21] However, constants obtained from these calculations may be subject to error if the specific cation-iodide interactions are neglected or if the metal is added to the system with a different anion. Moreover, these calculations inevitably neglect differences in the activity coefficients of NEt_4^+ and alkali or alkaline-earth cations of the background electrolyte.

In this study, the SIT model accurately represents the variation of pK_{ai} with ionic strength up to 5 m NaCl. The SIT approach usually fails above approximately 3 m ionic strength, requiring an alternative modeling strategy. The Pitzer formalism[22] has been successful in representing activity coefficients to high ionic strengths. In its simplest form, the Pitzer equation is given by:

$$\ln \gamma_i = \ln \gamma_i^{DH} + \sum_j B_{ij}(I)m_{j_j} + \sum_j \sum_k C_{ijk}m_{j_j}m_{k_k} + \ldots \qquad (5)$$

where:

$$B_{mx} = \beta_{mx}^{(0)} + \beta_{mx}^{(1)}g(\alpha_{1mx}\sqrt{I}) + \beta_{mx}^{(2)}g(\alpha_{2mx}\sqrt{I}) \qquad (6)$$

$$C_{mx} = \frac{C_{mx}^{\varphi}}{2(\sqrt{|z_m||z_x|}} \qquad (7)$$

The first term in equation (5) is a modified form of the Debye-Hückel equation which predominates in dilute solutions. The parameters $\beta_{mx}^{(0)}$, $\beta_{mx}^{(1)}$, and $\beta_{mx}^{(2)}$ are most significant at high, low, and intermediate ionic strengths, respectively. Additional parameters describing ion interactions between similarly charged ions (q) and ternary groups (y) may be required to accurately model complicated data sets. A complete description of the Pitzer formalism along with parameters for many neutral and ionic species is available.[23]

Standard chemical potentials and Pitzer parameters for the acid dissociation data were calculated using the MacNONLIN 2.0* software and relevant parameters were taken from Harvie, Møller, and Weare.[26] For acetate we recommend the parameters reported by Novak.[27] The oxalate, citrate, and EDTA parameters were calculated using the MacNONLIN code and gave an excellent representation of our data. All values were given equal weight in calculations. Solubility data for the organic acids in NaCl media was not available; therefore, Pitzer parameters and standard chemical potentials for the fully protonated acids were set equal to zero. Because of the limited amount of data at lower ionic strengths, it was not possible to determine acceptable values for the $\beta^{(1)}$ parameter. It has been suggested that setting $\beta^{(1)} = 0$ is an oversimplification leading to systematic deviations in the concentration dependence of the activity coefficients.[28] However, adequate methods for estimating Pitzer parameters for multiply charged species are not available. Therefore, in this work we chose to use average values for $\beta^{(1)}$ for the same type of electrolyte pairs and calculate values for $\beta^{(0)}$ and C^{ϕ} using the high ionic strength data. Based on literature values[23] for $\beta^{(1)}$, the values of 0.29, 1.74, 5.22, and 11.6 were used for 1-1, 2-1, 3-1, and 4-1 type electrolytes.

Pitzer modeling was limited to NaCl media as this represents the major component of WIPP brines. Table 4 lists the standard chemical potentials and Table 5 the Pitzer parameters used and calculated in this work. All like ion and ternary Pitzer parameters were set to zero.

---

*MacNONLIN is the Macintosh version of NONLIN. The program was developed by A.R. Felmy and uses MINPACK nonlinear least squares programs with the GMIN chemical equilibrium program which is based on a Gibbs free energy minimization procedure.[24, 25]

Table 4. Dimensionless standard chemical potentials used in Pitzer modeling of carboxylic acids.

Species	Standard chemical potential, μ^0/RT	Reference
H_2O	-95.6635	26
Na^+	-105.651	26
Cl^-	-52.955	26
H^+	0	26
OH^-	-63.435	26
$HAc(aq)$	-158.3	27
Ac^-	-147.347	11
$H_2Ox(aq)$	0	
HOx^-	3.20935	This study
Ox^{2-}	13.0165	This study
$H_3Cit(aq)$	0	
H_2Cit^-	7.47630	This study
$HCit^{2-}$	18.6233	This study
Cit^{3-}	33.4059	This study
$H_4EDTA(aq)$	0	
H_3EDTA^-	5.76060	This study
H_2EDTA^{2-}	12.8724	This study
$HEDTA^{3-}$	28.7072	This study
$EDTA^{4-}$	53.0511	This study

The dashed lines in Figures 1-4 represent the results of the Pitzer model calculations performed for solutions with NaCl as the background electrolyte. The model for acetic acid very closely resembles the SIT fit, while the variation of pK_{a1} for oxalic acid is more accurately described by the Pitzer approach than by the SIT model, especially above I = 2 m. The second acid dissociation constant for oxalic acid and the relatively flat dependencies observed for citric acid and EDTA show excellent agreement between the Pitzer models and the experimental data from NaCl solutions.

The SIT fits for the acetic, oxalic, citric, and EDTA acids are satisfactory for all media, including the NaCl systems to 5 m ionic strength. The Pitzer models for NaCl solutions generally describe the variation in pK_a with ionic strength better than the SIT models. These acid dissociation constants and modeling parameters will be used in an upcoming manuscript describing the interactions of Mg^{2+} with carboxylate ligands in NaCl solutions.[8]

ACKNOWLEDGMENTS

This work was supported at Sandia National Laboratory by the United States Department of Energy under Contract DE-AC04-94AL85000 and at The Florida State University under Contract AH5590. The authors gratefully acknowledge the assistance of Dr. J.-F. Chen in the preparation of this manuscript.

Table 5. Pitzer parameters for dissociation of carboxylic acids in NaCl media.

Species i	Species j	$\beta_{ij}^{(0)}$	$\beta_{ij}^{(1)}$	$\beta_{ij}^{(2)}$	C_{ij}^{ϕ}	Reference
Na^+	Cl^-	0.0765	0.2664	0	0.00127	26
H^+	Cl^-	0.1775	0.2945	0	0.008	26
Na^+	OH^-	0.0864	0.253	0	0.0044	26
Na^+	Ac^-	0.1426	0.2200	0	-0.00629	27
Na^+	HOx^-	-0.2448	0.2900	0	0.068	This study
Na^+	Ox^{2-}	-0.2176	1.7400	0	0.122	This study
Na^+	H_2Cit^-	-0.1296	0.2900	0	0.013	This study
Na^+	$HCit^{2-}$	-0.0989	1.740	0	0.027	This study
Na^+	Cit^{3-}	0.0887	5.220	0	0.047	This study
Na^+	H_3EDTA^-	-0.2345	0.2900	0	0.059	This study
Na^+	H_2EDTA^{2-}	-0.1262	1.740	0	0.054	This study
Na^+	$HEDTA^{3-}$	0.5458	5.220	0	-0.048	This study
Na^+	$EDTA^{4-}$	1.016	11.60	0	0.001	This study

Species i	Species j	Species k	$\theta_{i,j}$	$\psi_{i,j,k}$	
Na^+	H^+	Cl^-	0.036	-0.004	26
Na^+	Ac^-	Cl^-	-0.090	0.01029	27

REFERENCES

1. R.M. Smith and A.E. Martell. *Critical Stability Constants, Vols. 1-6*, Plenum Press, New York (1974-1977, 1982, 1989).
2. R.M. Smith and A.E. Martell. *NIST Critical Stability Constants of Metal Complexes Database, Version 1.0*, Gaithersburg, MD (1993).
3. G. Anderegg. *Critical Survey of Stability Constants of EDTA Complexes*, IUPAC Chemical Data Series-No. 14, Pergamon Press, Oxford (1977).
4. J.-F. Chen, Y.-X. Xia, and G.R. Choppin, Differential analysis of potentiometric titration data to obtain protonation constants, *Anal. Chem.* 68:3973 (1996).
5. G.R. Choppin and J.-F. Chen, Complexation of Am(III) by oxalate in $NaClO_4$ media, *Radiochim. Acta* 74:105 (1996).
6. G.R. Choppin, H.N. Erten, and Y-X. Xia, Variation of stability constants of thorium citrate complexes with ionic strength, *Radiochim. Acta* 74:123 (1996).
7. J.-F. Chen, Y.-X. Xia, and G.R. Choppin, Ionic strength dependence of the deprotonation constants of 1,10-phenanthroline, 8-hydroxyquinoline, and some carboxylic acids, (1998) manuscript in preparation.
8. A.H. Bond, J. Mizera, G.R. Choppin, and R.C. Moore, Interactions of magnesium with carboxylate ligands at high ionic strengths, (1998) manuscript in preparation.
9. A.E. Martell and R.J. Motekaitis. *Determination and Use of Stability Constants*, VCH Publishers, New York (1988).
10. J.-F. Chen, S. Lis, M. Borkowski, Y.-X. Xia, and G.R. Choppin, *Complexation of Am(III), Th(IV), Np(V), and U(VI) by Organic Ligands*, A collection of monthly progress reports submitted to Sandia National Laboratory, Department of Chemistry, The Florida State University, Tallahassee, FL (1997).

11. R.E. Mesmer, C.S. Patterson, R.H. Busey, and H.F. Holmes, Ionization of acetic acid in NaCl media: Potentiometric study to 573 K and 130 bar, *J. Phys. Chem.* 93:7483 (1989).

12. V.I. Belevantsev, I.V. Mironov, and B.I. Peshchevitskii, Influence of changes in the ionic background on the dissociation constants of a monobasic acid, *Russ. J. Inorg. Chem.* 27:29 (1982).

13. R.M. Kettler, D.A. Palmer, and D.J. Wesolowski, Dissociation quotients of oxalic acid in aqueous sodium chloride media to 175 °C, *J. Soln. Chem.* 20:905 (1991).

14. G. Scatchard, Concentrated solutions of strong electrolytes, *Chem. Rev.* 19:309 (1936).

15. E.A. Guggenheim. *Applications of Statistical Mechanics*, Oxford University Press, Oxford (1966).

16. I. Grenthe, J. Fuger, R. Lemire, A. Muller, C. Nguyen-Trung, and H. Wanner. *Chemical Thermodynamics of Uranium*, NEA-TDB, OECD, France (1991).

17. G. Scatchard. *Equilibrium in Solution. Solution and Colloidal Chemistry*, Harvard University Press, Cambridge (1976).

18. P.G. Daniele, C. Rigano, and S. Sammartano, Ionic strength dependence of formation constants-I, Protonation constants of organic and inorganic acids, *Talanta* 30:81 (1983).

19. A. De Robertis, C. De Stefano, C. Rigano, S. Sammartano, and R. Scarcella, Studies on acetate complexes. Part 1. Formation of proton, alkali metal, and alkaline-earth metal ion complexes at several temperatures and ionic strengths, *J. Chem. Res. (M)* 42:629 (1985).

20. P.G. Daniele, C. Rigano, and S. Sammartano, The formation of proton and alkali metal complexes with organic ligands of biological interest in aqueous solution. Thermodynamics of Li^+, Na^+, and K^+-dicarboxylate complex formation, *Thermochim. Acta* 62:101 (1983).

21. P.G. Daniele, A. De Robertis, C. De Stefano, A. Gianguzza, and S. Sammartano, Studies on polyfunctional O-ligands. Formation thermodynamics of simple and mixed alkali metal complexes with citrate at different ionic strengths in aqueous solution, *J. Chem. Res. (M)* 2316 (1990).

22. K.S. Pitzer, Thermodynamics of electrolytes. I. Theoretical basis and general equations, *J. Phys. Chem.* 77:268 (1973).

23. K.S. Pitzer. *Activity Coefficients in Electrolyte Solutions*, CRC Press, Boca Raton, FL (1991).

24. C.E. Harvie, J.P. Greenberg, and J.H. Weare, A chemical equilibrium algorithm for highly non-ideal multiphase systems: Free energy minimization, *Geochim. Cosmochim. Acta* 51:1045 (1987).

25. S. Babb, C. F. Novak, and R.C. Moore, *NONLIN User's manual*, Sandia National Laboratory Report (1997) in press.

26. C.E. Harvie, N. Møller, and J.H. Weare, The prediction of mineral solubilities in natural waters: The Na-K-Mg-Ca-H-Cl-SO_4-OH-HCO_3-CO_3-CO_2-H_2O system to high ionic strength at 25 °C, *Geochim. Cosmochim. Acta* 48:723 (1984).

27. C.F. Novak, M. Borkowski, and G.R. Choppin, Thermodynamic modeling of neptunium(V)-acetate complexation in concentrated NaCl media, *Radiochim. Acta* 74:111 (1996).

28. T. Fanghänel, V. Neck, and J.I. Kim, Thermodynamics of neptunium(V) in concentrated salt solutions: II. Ion interactions (Pitzer) parameters for Np(V) hydrolysis species and carbonate complexes, *Radiochim. Acta* 69:169 (1995).

CORRELATION OF EQUILIBRIUM CONSTANTS WITH IONIC STRENGTH BY SIT, PITZER AND PARABOLIC MODELS

Miting Du* and Gregory R. Choppin

The Department of Chemistry
The Florida State University
Tallahassee, FL 32306-3006

* Present address: Isotope Technology R & D, Chemical Technology Division, ORNL, Oak Ridge, TN 37831

ABSTRACT

The p^cK_a values of acetic acid (HAc) and hydrofluoric acid (HF), as well as pK_w values of H_2O, were measured in this study at 25 °C and ionic strengths of I = 0.1 ~ 14 m ($NaClO_4$). Potentiometric titration was the experimental technique for these measurements using either a glass pH electrode or a fluoride ion selective electrode.

With these values plus literature data over the range of ionic strengths from 0.1 to 14 m, a Parabolic model was developed to correlate these equilibrium constants (p^cK_a of acids, pK_w of H_2O) with ionic strengths. This correlation was compared with that by the Specific Interaction Theory (SIT) and the Pitzer model. The calculated parameters of the SIT, the Pitzer and the Parabolic models are reported.

INTRODUCTION

A primary concern of radioactive wastes deposition in geological repositories is the possible release of radionuclides (mainly long lived actinides) into the environment. In salt-bed geologic conditions at WIPP, the underground brine solutions of different ionic strengths (0.8 ~ 8.8 m)[1] would be the main path for possible radionuclide migration. To predict the probability of such migration requires speciation calculations in brines of different ionic strengths. In turn, this demands a data bank of equilibrium constants (log β, p^cK_a, pK_w, etc.) over a wide range of ionic strengths. The complexes that must be considered include those formed by inorganic (Cl^-, NO_3^-, F^-, etc.) and organic (oxalate, citrate, EDTA, etc.) ligands in the wastes.

Technically the equilibrium constants for the speciation calculations can be measured in a laboratory at any ionic strength. But the demand for these constants at all possible ionic strengths would require a large amount of such work. A practical approach towards acquiring the necessary data base is to correlate a number of measured constants for each complex at

Actinide Speciation in High Ionic Strength Media, edited by Reed *et al.*
Kluwer Academic / Plenum Publishers, New York, 1999

various values of ionic strength to allow estimation of unmeasured values at other ionic strengths. The current models favored for such correlation are the Specific Interaction Theory[2] (SIT) and the Pitzer model[3]. In this study p^cK_a values of acetic acid (HAc) and hydrofluoric acid (HF), as well as pK_w values of H_2O were measured or collected from the literature for T = 25°C and I = 0.1 to 14 m ($NaClO_4$) and correlated as function of ionic strength by the SIT and the Pitzer model and by a new approach we term the Parabolic Model.

For a protonation reaction of a monoprotic acid in solution, HL $\Longleftrightarrow$ $H^+ + L^-$, the concentration equilibrium constant, ^cK_a, can be related to the thermodynamic equilibrium constant, ^0K_a, and the activity coefficients of the three equilibrated species.

$$^cK_a = \frac{m_H \cdot m_{L^-}}{m_{HL}} = \frac{a_H \cdot a_{L^-}}{a_{HL}} \cdot \frac{\gamma_{HL}}{\gamma_H \cdot \gamma_{L^-}} = {}^0K_a \frac{\gamma_{HL}}{\gamma_H \cdot \gamma_{L^-}} \tag{1}$$

where m_i is the concentration of species i in molality, γ_i the activity coefficient of species i and $^0K_a = (a_{H^+} \cdot a_{L^-})/a_{HL}$ where a = activity of i. The logarithmic expression of eq. (1) is:

$$p\,^cK_a = p\,^0K_a + \log\gamma_{H^+} + \log\gamma_{L^-} - \log\gamma_{HL} \tag{2}$$

Theoretical

The SIT and the Pitzer models give different descriptions of the activity coefficient as a function of ionic strength, which results in different modeling equations for the variation of ^cpK_a with I_m.

The Specific Interaction Theory (SIT). The SIT equation uses a two term expression for the activity coefficient of species i as a function of I_m:

$$\log \gamma_i = -z_i^2 D(I_m) + \epsilon_i I_m \tag{3}$$

where $D(I_m)$ is the Debye-Hückel term equal to $(0.509\sqrt{I_m}/(1+B\sqrt{I_m})$; z_i is the valency of ion i; B=1.5; ϵ_i is the SIT parameter for species i and accounts for the short range non-electrostatic interactions between i and other species in the system. The assumption that ϵ values are zero for ions of same sign and for uncharged species results in a general SIT equation for p^cK_a of a monoprotic acid as a function of I_m:

$$pK_a = p\,^0K_a - \Delta z^2 D(I_m) + \Delta\epsilon I_m \tag{4}$$

where

$$\Delta z^2 = (-1+1)^2 - z_H^2 - z_L^2 = -2 \tag{5}$$

$$\Delta\epsilon = \epsilon_{HL} - \epsilon_H - \epsilon_L = 0 - \epsilon_H - \epsilon_L \tag{6}$$

126

Equation (4) indicates a linear relationship when $(p^cK_a + \Delta z^2D)$ is plotted against I_m with $+\Delta\epsilon$ as the slope and p^0K_a as the intercept. Such linearity is observed within $I_m = 0$ to $2\sim3$ m, but not at higher ionic strengths indicating the limit of validity of the SIT approach.

The Pitzer Model. The polynomial expression of the Imperfect Gas Model (PV = RT + BP + C'P^2 + D'P^3 + ...) describes the relationship between the pressure of a gas and interparticle potentials. Pitzer proposed a similar relation for the excess Gibbs energy using a polynomial relating the osmotic pressure of a solution and the potentials of mean force of solute species in a mixed electrolyte system[3]. The equation has a term $f(I_m)$ which includes the D-H limiting law in an extended form.

Differentiation of the Pitzer expression for the Gibbs energy equation leads to expressions which for $\ln \gamma_i$ of tracer level H$^+$, L$^-$, HL in a background electrolyte NaClO$_4$ can be written as specific equations (with some simplifications and footnote changes):

$$\ln \gamma_H = f^{\gamma} + I_m(2B_{HClO_4} + EC_{HClO_4}) + I_m^2(C_{NaClO_4} + \Psi_{H,HClO_4}) + I_m[2\phi_{H,Na} + I_mB'_{NaClO_4}] \tag{7}$$

$$\ln \gamma_L = f^{\gamma} + I_m(2B_{NaL} + EC_{NaL}) + I_m^2(C_{NaClO_4} + \Psi_{L,HClO_4}) + I_m[2\phi_{L,ClO_4} + I_mB'_{NaClO_4}] \tag{8}$$

$$\ln \gamma_{HL} = 2(\lambda_{Na,HL} + \lambda_{ClO_4,HCl})I_m \tag{9}$$

where the coefficients B, B', and C, etc, are calculated with the Pitzer equations for the parameters of $\beta^{(x)}$ and C^{ϕ}, while ϕ and ψ are parameters for two- and three-particle-interactions.

$$B_{HClO_4} = \beta^{(0)}_{HClO_4} + \beta^{(1)}_{HClO_4}g(2\sqrt{I_m}) \tag{10}$$

$$B_{NaL} = \beta^{(0)}_{NaL} + \beta^{(1)}_{NaL}g(2\sqrt{I_m}) \tag{11}$$

where

$$g = \frac{2[1 - (1 + 2\sqrt{I_m})e^{-2\sqrt{I_m}}]}{(2\sqrt{I_m})^2} \tag{12}$$

$$B'_{NaClO_4} = \left(\frac{\beta^{(1)}_{NaClO_4}}{2I_m^2}\right)[-1 + (1 + 2\sqrt{I_m} + 2\sqrt{I_m})e^{-2\sqrt{I_m}}] \tag{13}$$

127

$$C_{NaClO_4} = (\tfrac{1}{2})C^{\Phi}_{NaClO_4} \qquad (14)$$

$$C_{NaL} = (\tfrac{1}{2})C^{\Phi}_{NaL} \qquad (15)$$

$$f^{\gamma} = -0.392\left[\frac{\sqrt{I_m}}{(1+1.2\sqrt{I_m})}+\left(\frac{2}{1.2}\right)\ln(1+1.2\sqrt{I_m})\right] , \quad E = \Sigma m_i|z_i| \qquad (16)$$

In this study, Eq. (7), (8) and (9) for $\ln \gamma_i$ expressions of H^+, L^- and HL were used with Eq. (2). In the case of fitting by the Pitzer model of $p^c K_a$ data in $NaClO_4$ as the background electrolyte, 15 parameters are involved in the computation (most obtained from the literature for the present analysis).

The Parabolic Model. We propose a new approach to reduce the number of parameters for modeling constants at $I_m > 2\text{-}3$ m. The approach is empirical and is an extension of the SIT model in which a term $(\Delta\delta I_m^2)$ is added:

$$p^c K_a = p^0 K_a - \Delta z^2 D + \Delta\epsilon I_m + \Delta\delta I_m^2 \qquad (17)$$

For the simple HL case, as for the SIT term $\Delta\epsilon$, we propose that $\Delta\delta$ can be separated into δ_i components for each i species: $\Delta\delta = \delta_{HL} - \delta_H+ - \delta_L-$.

Curve Fitting Programs

Correlation of $p^c K_a$ values vs. I_m for the three models was performed using the Automatic Optimizer function of the Quattro Pro for Windows (Borland, Version 5.0). For the SIT model, linear regression of $(p^c K_a - \Delta z^2 D)$ vs. I_m, with sufficient values of $p^c K_a$ at low I_m range yields the values of $p^0 K_a$ and the SIT parameter $\Delta\epsilon$. For the Parabolic model, a non-linear regression analysis of Eq. (17) on $p^c K_a$ values from 0 - 14 m was used to obtain $p^0 K_a$ and the parameters $\Delta\epsilon$ and $\Delta\delta$. The Pitzer equations for log γ of H^+, L^- and HL were used with known values of $\beta^{(0)}$, $\beta^{(1)}$ and C^{Φ} for $NaClO_4$, $HClO_4$ and NaL; the values were used with equation (2) and by non-linear regression, $p^0 K_a$ and the parameters ϕ and Ψ calculated.

EXPERIMENTAL

Chemicals

Analytical grade $NaClO_4$, NaF and NaOH, purchased from Fisher Chemical Co., were used with no additional purification. Distilled, deionized water (E-Pure, Barnstead) was used in all experiments for aqueous solutions. Solutions of $HClO_4$, HAc (acetic acid), HF and HCl of desired concentrations were diluted from the concentrated acids (Fisher).

Procedures

Titration with pH Electrode for p^cK_a of HAc. The pH titration for p^cKa of HAc was conducted with a pre-calibrated pH electrode (Corning Glass Combination Electrode and Fisher Scientific Accumet 925 or 950 meters) in a 25 ml water-jacket glass cup controlled to $25.0 \pm 0.1°C$ with an Isotemp refrigerated circulator. Each 15.0 ml of titrate (0.001 M HAc + 0.01 M $HClO_4$ + $NaClO_4$ for desired I_m) in the cup was stirred by a magnetic stirrer while the cup was sealed with a rubber stopper in which only a small hole was open to air. The titrate solution in cup was flushed with nitrogen gas to remove the dissolved CO_2 and to minimize its influence on the pH reading. The NaOH titrant (0.01-0.03 M) at the same ionic strength as that of titrate was delivered into the cup by a Schott Gerate automatic buret via a Teflon tubing. Variable volumes (as small as 0.002 ml) of titrant (adjusted to give approximatedly a constant pH increment) was added and the pH readings were recorded accordingly. The first derivative of the titration curve as dV/dpH_m (pH_m is the pH meter reading) was analyzed by the method of ref. 4.

The pH_m values are related to pcH (= - log concentration of H^+) by the equation:

$$pcH = s\, pH_m + b \qquad (18)$$

where s and b are the slope (= 1.00 in all calibrations) and intercept (listed in Table 1) of the linear calibration line.

Titrations for p^cK_a of HF. For determination of the p^cK_a of HF data, an F^- ion selective electrode (Orion 96-09 ISE) and a Fisher Scientific Accumet 925 meter were employed at room temperature in an 80 ml titration vessel (sealed by a plastic lid with holes for electrode, N_2 gas and titrant tubings). Two types of titration operations were used in experiments. In the "normal" titration, 50.0 ml of $NaClO_4$ solution of desired ionic strength was titrated with known concentration of NaF at the same ionic strength ($NaClO_4$); the calibration curve was made by plotting -log $[F]_T$, added vs. the ISE meter reading. In other titrations, a concentration of $[F^-] \geq 5 \times 10^{-4}$ M was placed in the vessel and titrated with 0.01 M $HClO_4$ at the same ionic strength. Defining $\bar{n}_H = [HF]/C_H = (C_F-[F^-])/C_H$, the p^cK_a value was calculated by:

$$p^cKa + \log[F^-] = \log\frac{[HF]}{[H^+]} = \log\frac{\bar{n}_H}{1-\bar{n}_H} \qquad (19)$$

At high ionic strengths, the calibration curve for the electrode was not linear for $[F^-]$ < 10^{-5}M due to interference by H^+. Higher concentration of F^- overcome the influence of H^+ ions which was the purpose of the second type of titration. In this alternate approach, the mV reading of F^- ISE was kept constant by adding known concentrations of NaF and $HClO_4$ in turn; this changes the $\bar{n}_H$ values while keeping the $[F^-]$ constant. As more NaF and $HClO_4$ were added to the titration solution, pH_m was decreased further while the F^- ISE readings were constant. The p^cK_a of HF was calculated from the data for changing C_F, C_H and constant $[F^-]$.

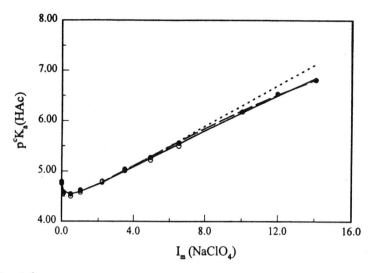

Figure 1. Plot of p^cK_a values of acetic acid compared with models: -----, SIT; – – –, parabolic; ——, Pitzer. Solid points are from this study, open points from literature.

RESULTS

Experimental Measurements of p^cK_a Values

The p^cK_a of HAc between $I = 0.1$ and 14 m and $T = 25°$ C were measured by pH titrations at least in duplicate with constant titrate and titrant conditions (i.e., 0.001 m HAc, 0.01 m HClO$_4$ and 0.03 m NaOH) at $I = 0.1 \sim 14$ m. Table 1 summarizes the computed measurements of p^cK_a, pK_w and b values at different ionic strengths. The measured p^cK_a values agree very well with literature data[5] at $I_m \le 6$ m; for the p^cK_a; values at $I_m > 6$ m, there are no published literature data.

The published literature data[6,7] for HF covers an ionic strength range up to 12 m. In this study, the p^cK_a of hydrofluoric acid at ionic strengths at 1.0 m, 10.0 m and 12.0 m measured experimentally by titration methods with pH and F⁻ ion selective electrodes are listed in Table 1. The data are in good agreement with the literature data.

Correlation of p^cK_a & pK_w with I_m by Models of SIT, Parabolic and Pitzer up to $I = 14.0$ m

The literature data plus those measured in this study were used in the model fitting. The fit of the model calculations to the experimental values of p^cK_a (HAc and HF) and pK_w of H$_2$O based on the three models are shown in Figures 1- 3. The values from this study are shown in the graphs as solid circles, while those from the literature are the open circles. The solid point at $I_m = 0$ m is the extrapolated value from this study and the open one at $I_m = 0$ m is taken from the literature for each system. Table 2 lists the p^oK_a, $\Delta\epsilon$ and $\Delta\delta$ values of HAc, HF and H$_2$O from the three different models. Table 3 is a summary of the Pitzer parameters used in the curve fittings of the three HAc, HF and H$_2$O systems where the numbers in *normal* are the parameters taken from literature, the numbers in *italics* are those taken directly or calculated from the literature, those in *bold italics* are the fitting results.

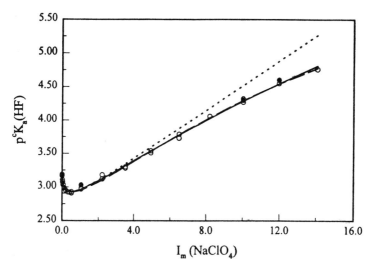

Figure 2. Plot of p^cK_a values of hydrofluoric acid compared with models: -----, SIT; – – –, parabolic; ——, Pitzer. Solid points are from this study, open points from literature.

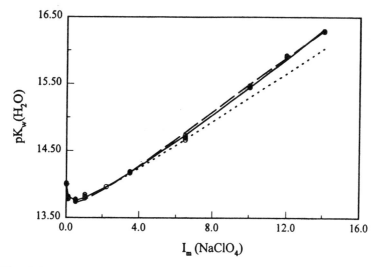

Figure 3. Plot of p^cK_a values of water compared with models: -----, SIT; – ·· –, parabolic; ——, Pitzer. Solid points are from this study, open points from literature.

Table 1. Summary of p^cK_a, pK_w and b values Measured by pH Titration at Different Ionic Strengths (NaClO$_4$)

I (M)	I_m(m)	p^cK_a	pK_w	$b*$
		a. Acetic Acid		
0.10	0.104	4.585±0.016	13.80±0.02	-0.015±0.010
0.50	0.508	4.545±0.010	13.77±0.01	0.162±0.011
1.00	1.042	4.623±0.017	13.82±0.02	0.298±0.010
3.00	3.509	5.041±0.005	14.18±0.01	0.753±0.001
5.00	6.501	5.561±0.005	14.70±0.01	1.286±0.004
7.00	10.018	6.179±0.005	15.46±0.01	1.920±0.006
8.00	11.974	6.535±0.030	15.91±0.01	2.283±0.017
9.00	14.061	6.812±0.010	16.29±0.01	2.485±0.025
		b. Hydrofluoric Acid		
1.00	1.04	3.03±0.02		F⁻ ISE
7.00	10.02	4.30±0.04	15.61±0.02	1.800
8.00	11.97	4.60±0.10		F⁻ ISE

*b values from pcH/pH conversion: $pcH = pH_m + b$ at different ionic strengths

Table 2. Regression Results of log $\beta°$, $\Delta\epsilon$, $\Delta\delta$ Values by Three Models

	SIT		Parabolic			Pitzer
	$p°K_a$	$\Delta\epsilon$	$p°K_a$	$\Delta\epsilon$	$\Delta\delta$	$p°K_a$
HAc	4.782	-0.2123	4.793	-0.20692	0.001403	4.713
H$_2$O	14.013	-0.1910	14.014	-0.1841	-0.00148	14*
HF	3.189	-0.1785	3.184	-0.18905	0.00232	3.078

132

Table 3. Pitzer Parameters

	HAc	H_2O	HF
$\beta^{(0)}_{NY}$	0.0554	0.0554	0.0554
$\beta^{(1)}_{NY}$	0.2755	0.2755	0.2755
C^{ϕ}_{NY}	-0.0012	-0.0012	-0.0012
$\beta^{(0)}_{MY}$	0.1747	0.1747	0.1747
$\beta^{(1)}_{MY}$	0.2931	0.2931	0.2931
C^{ϕ}_{MY}	0.00819	0.00819	0.00819
$\beta^{(0)}_{NL}$	0.1426	0.0864	0.0215
$\beta^{(1)}_{NL}$	0.3237	0.253	0.2107
C^{ϕ}_{NL}	-0.00629	0.0044	0.0
$\phi_{M,N}$	0.036	0.036	0.036
$\phi_{L,Y}$	*-0.05065*	*-0.05***	*0.0633*
$\Psi_{M,NY}$	*-0.01058*	*-0.011*	*-0.0157*
$\Psi_{N,YL}$	*0.00034*	N/A	*-0.0046*
$\lambda_{N,ML}$	0	0	0
$\lambda_{Y,ML}$	0	0	0
p^0K_a	*4.713*	14	*3.078*

"NY": the background electrolyte $NaClO_4$; "M": H, Eu, UO_2, respectively; "L": Ac, OH, F, Cl, respectively.

As seen in Table 3, the $\beta^{(x)}$, C^{ϕ} parameters for $NaClO_4$ and $HClO_4$ are common to all the three equilibrium systems due to the same background electrolyte. The $\phi_{Na,H}$ values for the three systems are the same for the same reason, while $\phi_{OH,ClO4}$ is taken from $\phi_{OH,Cl}$. The Pitzer parameter calculations for p^cK_a of HAc, HF and pK_w of H_2O use an assumption of $\lambda_{n,x} = 0$ for interactions of neutral species. The fitted results are the three-particle-interaction parameter Ψ and the two-particle-interaction parameter ϕ for L^-, ClO_4^-, in *bold italics* in Table 3.

DISCUSSION

General Trend of p^cK_a Variation with Ionic Strength

The general trend of p^cK_a variation with increasing ionic strength is an initial decrease to a minimum at $I \approx 0.5$ m followed by an increase with further increases in I_m. This trend reflects log γ_{H^+} and log γ_{A^-}. Two factors can account for the variation in log γ: the ionic interparticle potential and the difference between the calculated and "real" concentrations of

Table 4. Summary of Parabolic Fittings of pK_a

Acids	p^0K_a	$\Delta\epsilon$	$\Delta\delta$	Dev.*	I_m**
Citric (1)	3.095	-0.1186	0.00226	2.3×10^{-3}	14
(2)	4.778	-0.1553	0.00281	1.25×10^{-3}	14
(3)	6.330	-0.1163	-0.00043	1.63×10^{-3}	14
Glycolic	3.804	-0.20780	0.0101	9.6×10^{-5}	3.5
Cl-acetic	2.875	-0.1244	-0.0167	3.7×10^{-4}	3.5
Glycine (1)	2.521	-0.3567	0.0436	3.56×10^{-3}	3.5
(2)	9.984	-0.4283	0.0289	7.65×10^{-3}	3.5
H-azide	4.648	-0.1925	0.00353	2.4×10^{-5}	4.5
Lactic	3.860	-0.1887	0.0	7.5×10^{-6}	2.1
Aspartic (1)	4.111	-0.3183	0.010	5.39×10^{-3}	3.5
(2)	10.254	-0.409	0.010	8.14×10^{-3}	3.5
Alanine (1)	2.418	-0.4000	0.0462	2.48×10^{-3}	3.5
(2)	10.15	-0.3224	0.00325	8.9×10^{-4}	3.5
Propanoic	4.887	-0.1763	-0.0122	4.4×10^{-5}	3.5
Oxalic (2)	4.150	-0.2480	0.00636	1.81×10^{-2}	14

* Dev. = $\Sigma(a_i - a'_i)^2$ in the curve fitting based on the Parabolic model
** Ionic strength to which Parabolic model fits.

the chemical species as the concentration increases. At very low ionic strengths, when the distance between the ions are large, the decrease in log γ can be described by Debye-Hückel theory. But with increasing ionic strength, the H^+ and L^- are not insulated from the effects of adjacent ions. Moreover, at higher concentrations, a growing fraction of the water is associated with the solvation spheres of the ions so calculations of concentrations assuming all the water acts as "free" solvent leads to incorrect (low) ionic concentrations. This results in increasing values of log γ.

Mathematical Comparison of Three Models

The results of correlation of the p^cK_a values with I_m using the three models (Fig. 1 - 3) show that the SIT model fits the data points well for $I_m \leq 3$ m, while the Parabolic and the Pitzer models provide a satisfactory correlation to 14 m. The three models can be compared mathematically when the Pitzer equation is in revised form of Millero[8]:

SIT: $\qquad p^cK_a = p^0K_a - \Delta z^2 D(I_m) + \Delta\epsilon\, I_m$
Parabolic: $\quad p^cK_a = p^0K_a - \Delta z^2 D(I_m) + \Delta\epsilon\, I_m - \Delta\delta\, I_m^2$
Pitzer: $\qquad p^cK_a = p^0K_a - \Delta z^2 f^\gamma + \Delta B^{(0)} I_m + \Delta B^{(1)} f^1 + \Delta C I_m^2$

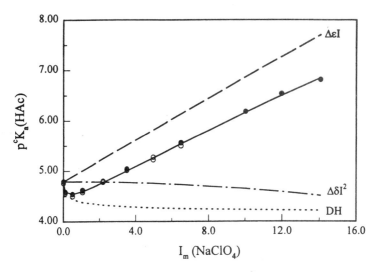

Figure 4. Variation of the Debye-Huckel (----), $\Delta\varepsilon I$ (– – –) and $\Delta\delta I^2$ (–.–.–) terms in the parabolic model for HAc. The solid line is the fit of the model to the experimental data.

All three models use the Debye-Hückel equation in their second term; although the value of B (see eq. 3) is 1.5 in the SIT and Parabolic modeling while it is 1.2 in f^γ of Pitzer. Moreover, an additional term, $(2/1.2)\ln(1+1.2\sqrt{I_m})$, is included in f^γ to account for the "Hard Core Effect".

The main difference among the three models is in the additional terms for short range interactions. The SIT model uses a single term with a first power dependence on I_m and defines the coefficient of the term as a constant. The Parabolic model also uses this term but adds an additional one with a second power dependence of I_m. This third term in the Parabolic model reflects the final term in the Pitzer model and is significant for $I_m \geq 3m$. The later terms in all three models are specific to the electrolyte system which is related to the short range interactions and the influence at short distances of the structure of the solvent and the ion-solvent interactions.

Development of these models is based on the summation of the excess Gibbs free energy of the electrolyte solution coming from the contributions of i) long range interactions or electrostatic interactions, ii) short range interactions, iii) concentration dependence of dielectric constants. For long range interactions where the interparticle potential varies as r^{-1} at low ionic strength, the Debye-Hückel equation describes satisfactorily the decrease of $\log \gamma$ with I_m. For short range interactions at intermediate strengths, an empirical repulsive term describes the increase of $\log \gamma$ with I_m.

Figure 4 shows the geometric significance of each term of the Parabolic model in the variation of the p^cK_a of acetic acid with I_m, showing the effect of the D-H term over $0 < I_m < 0.5$ m, the first power term for the range 0.5 m $> I_m < 2\sim3$ m and the second power term for the high concentrations, ($I_m > 2\sim3$ m).

Analysis on Correlation Results by Three Models

The results of curve fitting with the three models are compared in Table 3. The p^oK_a values by the Parabolic or SIT models are always a bit higher for the same acid than those from the Pitzer model. This may result from the correction term for the "hard-core effect" in f^γ, which intensifies the curvature of the fitting curves at $I_m = 0.5\sim4$ m, resulting in a lower intercept value. The same trend was found in other studies, e.g. the solubility study of yttrium carbonate by Grenthe and Spahiu[9]. In that study, data points at relatively higher ionic strengths (e.g. at $I_m = 4m$) were included in the SIT fitting to minimize the difference in calculations by the two models.

The curve fittings of p^cK_a values vs. ionic strength based on the Pitzer model in the Fig. 1 - 3 show good correlation of p^cK_a vs I_m. Fifteen parameters are involved in the curve fitting for the case of p^cK_a values of which at least 5 are unknowns. In solving for unknown $\beta^{(x)}$ and C^{ϕ} values for a species, the selection of known parameters are decisive in the Pitzer modeling. Calculations, assuming $\lambda_{xx} = 0$ can result in two sets of $\beta^{(x)}$ and C^{ϕ} values although the resulting curves are the same indicating a mathematical compensation in the Pitzer parameters. The Millero revision of the Pitzer equation has a polynomial form similar to the Parabolic equation with I_m as the variable. This results in the Parabolic model with only two coefficients as variables and with similar curve fitting power as the Pitzer model.

The Parabolic model has been applied to correlate literature p^cK_a of 10 acids and $\log \gamma_{\pm}$ of 34 pure electrolytes vs the ionic strength. The Parabolic parameters, p^0K_a, $\log \gamma_{\pm}^0$, $\Delta\epsilon$ and $\Delta\delta$, have been obtained and listed in Table 4 and 5.

Relating $\Delta\epsilon$, $\Delta\delta$ with ϵ_i & δ_i by a Revision of Pitzer's Subequations

In the parabolic model, $\Delta\epsilon$ and $\Delta\delta$ values, together with p^0K_a are sufficient parameters to describe an equilibrium system. As these parameters are specific to the equilibrium system, they may be defined as "system-specific", while the "species-specific" or "ion-specific" parameters are most desired by chemists.

Application of "ion-specific" parameters is based on a hypothesis: the interactive behavior of the same ion in different equilibrium systems is similar (if not the same). Thus the parameters (e.g. ϵ_i in SIT model, $\beta^{(x)}$ and C^{ϕ} in the Pitzer model) can be obtained from one system and used in another system as known values to calculate the unknown parameters for other species in the new system. For examples, $\beta^{(x)}$ and C^{ϕ} values for the pure electrolyte $NaClO_4$, and for trace $HClO_4$ and $NaAc$ in electrolyte mixtures can be used to calculate unknown ϕ_{xx} and ψ_{xxx} values for ion interactions of like charge and three-body interactions in the equilibrium system $HL \Longleftrightarrow H^+ + L^-$. The calculation of p^cK_a at different ionic strengths involves $\log \gamma_{H^+}$, $\log \gamma_{Ac^-}$ and $\log \gamma_{HAc}$, as well as $\beta^{(x)}$, C^{ϕ}, ϕ and ψ values of the involved electrolytes. The Pitzer model is the most successful in this approach although it requires a large number of parameters even in the calculation for a simple equilibrium system. When the parameters can be mutually compensating (one fitted curve gives more than one set of parameters with different selection of known parameters), the curve fitting is actually based on a polynomial form of the Pitzer formulism, which is comparable with the Parabolic model.

It is possible for the Parabolic model to relate $\Delta\epsilon$ & $\Delta\delta$ with ϵ_i & δ_i by adopting subequations in the Pitzer equations for B, B' and C. In the Parabolic fitting for $\log \gamma_{\pm}$ of a 1:1 pure single electrolyte ML, $\Delta\epsilon$ and $\Delta\delta$ values can be compared with the Pitzer expression:

$$\log \gamma_{\pm} = f^{\gamma} + (B_{ML} + B^{\phi}_{ML})m + 1.5 C^{\phi}_{ML}m^2$$

then,

$$\Delta\epsilon = (B_{ML} + B^{\phi}_{ML}) = 2\beta^{(0)}_{ML} + \beta^{(1)}_{ML}e^{-\alpha\sqrt{Im}} + 2\beta^{(1)}_{ML}/(\alpha^2\sqrt{Im}) \quad (....)$$

$$\Delta\delta = 1.5 C^{\phi}_{ML}$$

if the difference between $D(I_m)$ and f^{γ} is ignored. The Pitzer equations for $\log \gamma_i$ in an electrolyte mixture give the following expressions for $\Delta\epsilon$ and $\Delta\delta$:

$$\Delta\epsilon = 2(B_{MClO4} + mz^2_{\pm}B'_{NaClO4} + \phi_{M,Na})$$

Table 5. Summary of Parabolic Fittings of log γ±

Electrolyte	log $\gamma_{\pm}^{0}$ *	$\Delta\epsilon$	$\Delta\delta$	Dev.**	$I_m^{\ddagger}$
HCl	-0.0274	-0.1412	0.0011	2.4×10^{-4}	16
HBr	-0.0035	-0.1451	-0.005	2.4×10^{-5}	8
HI	0.0108	-0.1741	-0.002	6.4×10^{-5}	10
HNO$_3$	0.007	-0.0631	0.0014	3.0×10^{-5}	28
HClO$_4$	-0.0158	-0.1334	-0.0042	2.2×10^{-5}	14
HF	-0.8501	0.5116	-0.0702	3.6×10^{-2}	5
NaCl	-0.0047	-0.023	-0.0035	2.8×10^{-6}	6
NaBr	-0.0048	-0.045	-0.0027	$1.4e \times 10^{-5}$	7
NaI	-0.0060	-0.075	-0.0015	1.6×10^{-5}	10
NaNO$_3$	-0.0126	0.0452	-0.0034	9.7×10^{-6}	6
NaClO$_4$	-0.0087	-0.012	-0.0008	8.2×10^{-6}	10
NaOH	-0.0199	-0.051	-0.0028	3.6×10^{-4}	10
NaSCN	-0.00001	-0.053	-0.0007	2.5×10^{-6}	10
NaAc	-0.0012	-0.085	0.00130	1.5×10^{-6}	3.5
KCl	-0.0085	0.0086	-0.0035	4.0×10^{-6}	5
KBr	-0.0188	0.1274	-0.0031	1.1×10^{-4}	14
KI	-0.0036	-0.017	-0.0012	1.5×10^{-6}	4.5
KNO$_3$	-0.0101	0.154	-0.0157	3.4×10^{-5}	2.5
KClO$_4$	0.00173	0.339	-0.4498	8.1×10^{-6}	0.3
KOH	-0.0306	-0.102	-0.0006	1.1×10^{-4}	15
KF	-0.0216	-0.036	-0.0016	7.0×10^{-5}	12
LiCl	-0.0104	-0.105	-0.0019	1.4×10^{-5}	12
LiBr	0.01699	-0.132	-0.0025	3.7×10^{-4}	14
LiI	0.00407	-0.164	0.0050	1.6×10^{-5}	2
LiNO$_3$	-0.0065	-0.086	0.0018	1.8×10^{-5}	20
LiClO$_4$	0.00473	-0.147	0.0003	1.8×10^{-5}	4.5
LiOH	-0.0062	0.133	-0.055-	9.2×10^{-5}	1.5
NH$_4$Cl	-0.0037	0.127	0.0032	1.6×10^{-5}	4
NH$_4$NO$_3$	-0.0201	0.065	-0.0031	1.1×10^{-4}	8
NH$_4$ClO$_4$	-0.0065	0.146	-0.0335	2.4×10^{-5}	2
NH$_4$SCN	0.04685	0.009	-0.0001	1.0×10^{-5}	15

* Theoretically log $\gamma_{\pm}^{0}$ should be equal to zero but this is a fitted intercept.
** Dev. = $\Sigma(a_i - a'_i)^2$ in the curve fitting based on the Parabolic model.
$\ddagger$ Ionic strength to which the Parabolic model fits.

$$\Delta\delta = 2C_{MClO4} + EC_{NaClO4} + \psi_{M,NaClO4} + \psi_{L,NaClO4}$$

If $\Delta\epsilon$ and $\Delta\delta$ are considered as constants and not a function of I_m, only $\beta^{(0)}_{ML}$ and C^ϕ are obtained from B and C (simplifying the expressions).

Such a manner of selecting parameters is somewhat arbitory. If this hypothesis is correct, the validity of the parameters or subequations used can be tested by using the parameters from one equilibrium system to describe other systems. But unknown parameters always exist in such curve fittings. Errors (if any) in those known parameters could be tolerated and compensated in the new parameters. Even in the simple case for the fitting pK_w of H_2O, the unknowns $\phi_{OH,ClO4}$, $\psi_{H,NaClO4}$ and $\psi_{OH,NaClO4}$ exist and the new parameters obtained by fitting the curve may not be unique.

The confirmation of parameters or subequations requires a large data bank for $\log\gamma$ values of metal ions, ligands, and charged or uncharged ion pairs. For example, the determination of solubilities of uncharged neutral species at different ionic strengths to obtain the variation of $\log\gamma$ variation with I_m allows evaluation of $\lambda_{X,HL}$ values instead of using the normal assumption of $\lambda_{X,HL} = 0$. As the experimentally determined number of parameters increases, the calculated parameters in equilibrium constant fittings would become less ambiguous and concern over a fit as due to mutual compensation diminishes.

SUMMARY

This study shows that the proposed Parabolic Model with two coefficients correlates satisfactorily equilibrium constants (p^cK_a, pK_w) as a function of ionic strength to $I = 14$ m. It represents an empirical extension of the SIT model and uses fewer unknown parameters than the theoretically based Pitzer model. As a result, the fit can be obtained with less experimental data.

Examination of the relationship between the $\Delta\epsilon$ (SIT) and $\Delta\delta$ (Parabolic) coefficients and parameters of the Pitzer equations indicates the basis for success for the Parabolic model.

Acknowledgement

This research was supported by a grant from the USDOE-OBES Office of Chemical Sciences.

REFERENCES

1.   a) L. H. Brush, SAND 90-0266 (**1990**); b) C. F. Novak, SAND 91-1299 (**1991**); Sandia National Laboratory.

2.   I. Grenthe, et al; *Chemical Thermodynamics of Uranium*; NEA-TDB , OECD, France, **1991**.

3.   K. S. Pitzer; *Activity Coefficients in Electrolyte Solutions*; CRC Press, Boca Raton, **1991**.

4.   J. F. Chen, Y. X. Xia, G. R. Choppin; *Analytical Chemistry*; **1996**, <u>68</u>, 3973.

5.   A. M. Martell, R. M. Smith; *Critically Selected Stability Constants of Metal Complexes Database*; Vers. 2.0, NIST, Gaithersburg, **1995**.

6.   G. T. Hefter; *J. Solution Chem.*, **1982**, <u>11</u>, 45.

7.   G. T. Hefter; *Polyhedron*, **1984**, <u>3</u>, 75.

8.   F. J. Millero, D. J. Hawke; *Marine Chemistry*, **1992**, <u>40</u>, 19.

9.   I. Grenthe, K. Spahiu; *J. Chem. Soc. Faraday Trans.*, **1992**, <u>88(9)</u>, 1267.

ACTINYL(VI) CARBONATES IN CONCENTRATED SODIUM CHLORIDE SOLUTIONS: CHARACTERIZATION, SOLUBILITY, AND STABILITY

W. Runde, M.P. Neu, and S.D. Reilly

Chemical Science and Technology Division, Los Alamos National Laboratory, Los Alamos, NM 87545, USA

INTRODUCTION

Salt formations, like the Waste Isolation Pilot Plant (WIPP) site near Carlsbad New Mexico and the Gorleben site in Niedersachsen Germany, are proposed to serve as permanent repositories for transuranic waste. Safety assessment required for the licensing of these sites necessitates the modeling of potential radionuclide releases from proposed nuclear waste repositories and the estimation of actinide solubilities in surrounding groundwaters. An understanding of actinide speciation and complex stability in the natural aquifer systems is the basis for developing reliable release models and predicting solubility. The predominant geochemical complexation reactions of actinides, hydrolysis and carbonate complex equilibria, have been studied in dilute solutions. Until recently, however, analogous studies in concentrated electrolyte solution, which are critical for the prediction of actinide behavior in complex brine systems, had not been performed.

We are investigating the formation and stability of U(VI) and Pu(VI) complexes in concentrated NaCl solutions relevant for WIPP and Gorleben. We are performing these studies using a multimethod (X-ray diffraction, ^{13}C-NMR, Diffuse reflectance, conventional UV-Vis-NIR absorption, and X-ray absorption spectroscopies) approach. Considering the interactions between actinyl and chloride ions is necessary to predict actinyl solubilities in brine solution. Neglecting actinyl chloride complexation results in a significant underestimation of the actinyl solubility.[1] Previously hexavalent actinides have been reported to form both weak outer-sphere[2] and inner-sphere complexes[3] with chloride. The mono- and bischloro chloride species, AnO_2Cl^+ and AnO_2Cl_2, and their formation constants have been reported for both U(VI) and Pu(VI).[4,5] Formation of the trischloro complex has been reported only for U(VI).[4] The complexation mechanism of U(VI) and Pu(VI) with chloride in NaCl and synthetic WIPP brine solutions was studied spectrophotometrically.

Building on the actinyl chloride interactions, we are studying the formation and stability of carbonate solid and solution compounds in the presence of sodium chloride. The monocarbonate, AnO_2CO_3, has been identified as the predominant solid phase formed in carbonate media within the pH range of natural brines. We determined the solubility of AnO_2CO_3 (An = U, Pu) at different NaCl concentrations and calculated the standard chemical

Actinide Speciation in High Ionic Strength Media, edited by Reed *et al.*
Kluwer Academic / Plenum Publishers, New York, 1999

potential using the Pitzer approach.[6]   Monomeric and trimeric actinyl(VI) species, $AnO_2(CO_3)_n^{2-2n}$ (n = 1 - 3) and $(UO_2)_3(CO_3)_6^{6-}$, have been proposed as solution species.[4,7] The formation of the polymeric species hasn't been reported in previous solubility studies and its relevance under repository conditions is questionable.

We studied the equilibrium:

$$3\ UO_2(CO_3)_3^{4-} + 3\ H^+ \rightleftharpoons (UO_2)_3(CO_3)_6^{6-} + 3\ HCO_3^- \qquad (1)$$

and determined the formation constant of the trimeric U(VI) species as a function of the NaCl concentration.

EXPERIMENTAL

Stock Solutions. Plutonium(VI) stock solutions were prepared by dissolving ^{239}Pu metal in 7 M $HClO_4$ and fuming aliquots of this solution to near dryness with concentrated $HClO_4$, diluting with H_2O, then determining the total Pu concentration by liquid scintillation counting (LSC) and verifying the oxidation state purity using conventional absorbance spectrophotometry.   The isotopic composition of the material was determined radioanalytically. Uranium(VI) stock solutions were prepared by dissolving uranyl nitrate in 1 M $HClO_4$.

Preparation and Characterization of Actinyl Carbonates. Actinyl carbonates, AnO_2CO_3, (An = U, Pu) were prepared by bubbling CO_2 through a stirred acidic stock solution (pH = 4) for 3 to 5 days, washing the resulting precipitate with distilled deionized water, redissolving, and repeating precipitation. For plutonium, ozone was also bubbled through the suspension for the final 2 days to re-oxidize any reduced plutonium.   The resulting pale pink-tan Pu and pale yellow U solids were characterized using powder X-ray diffraction (Inel, CPS-120) and extended X-ray absorbance fine structure (EXAFS) spectroscopy. EXAFS data were recorded at Stanford Synchrotron Radiation Laboratory (SSRL): unfocused beamline 4-2, Si-(220), double-crystal monochromator, 3.0 GeV, 60-100 mA.

Solubility of Actinyl Carbonates. Solubility experiments were performed using approximately 50 mg of AnO_2CO_3 (An = U, Pu) and 5–20 mL of NaCl solution contained in closed vessels with 100% CO_2 bubbling through. Solution pHs were adjusted by addition of 0.01M Na_2CO_3/NaCl or 0.1M HCl/NaCl. Combination pH glass electrodes (Orion, Ross) used in solubility experiments were filled with 3 M NaCl. The electrode was calibrated in pH units against certified buffer solutions (Fisher). The solution p[H] was calculated using p[H] = pH_{obs} - ΔpH, where ΔpH values are experimentally determined using analytical solutions prepared and measured for each ionic strength. Solution $[CO_3^{2-}]$ was calculated using log $[CO_3^{2-}]$ = log K^* + log P_{CO_2} + 2pH, where K^* is the equilibrium constant for the formation of CO_3^{2-} from CO_2 gas in equilibrium in NaCl solutions which was calculated for each ionic strength using the program NONLIN.[8] Samples removed from the solid-liquid mixtures for analysis were filtered using 250 nm pore size filters (Millipore). Total plutonium solution concentrations were determined using LSC (Packard, Model Tricarb 2500). Soluble uranium concentrations were determined using ICP-OES (Varian). Uranium and plutonium speciation in the chloride solutions was determined using optical absorbance spectroscopy (Perkin-Elmer, Lambda 9 and Varian, Cary 5).

Determination of Thermodynamic Constant for Uranyl Bis- and Triscarbonate Equilibrium. Sodium carbonate solutions were prepared by dissolving ^{13}C-enriched Na_2CO_3 (99.9% ^{13}C, Cambridge Isotopes) in a known masses of distilled, deionized water.   Uranyl nitrate solutions were prepared by dissolving $UO_2(NO_3)_2$, (recrystallized from nitric acid) in a known masses of distilled, deionized water. Aliquots of uranyl stock solutions were added dropwise to individual 30 mM sodium carbonate solutions to yield a final uranium concentration of 10 mM. A small, known amount of D_2O (Cambridge Isotopes, 99.9% D) and NaCl (Baker, reagent) appropriate for the desired ionic strength were added to individual solutions. The p[H] of the resulting solutions was determined using a combination pH electrode and the calibration and ionic strength corrections described above. All NMR sample solutions were loaded into Wilmad 5 mm o.d. 507-PP pyrex glass NMR tubes. FT ^{13}C NMR spectra were recorded on a Bruker AMX500 spectrometer with a 5 mm braodband probe operating at 125.76 MHz with ^2H field-frequency lock. The spectral reference was set for all ^{13}C NMR spectra relative to the carbonyl carbon of external acetone $-d_6$ set at $\delta = 206.0$.

RESULTS AND DISCUSSION

Stabilization of Actinyl(VI) in Chloride Media. Modeling actinyl solubility in brine is based upon accurate experimental quantification of the ion interaction between actinyl and chloride. The replacement of water molecules by chloride ions in the inner coordination sphere of the actinyl ion is expected to change the absorption spectra significantly (as is known for the strong hydroxide, carbonate, or humate complexes). Figure 1 shows the absorption spectra of uranyl in 0.1 m $HClO_4$ and sodium chloride solutions to 6 m. The analytical $HClO_4$ concentration was maintained at 0.1 M to avoid U(VI) hydrolysis. Addition of NaCl results in a significant change in the absorption spectra, although the shift of the absorption maxima is far smaller than those of strong U(VI) complexes, such as the ~30 nm shift upon carbonate complexation.[9] At [NaCl] ≤ 1.5m an isosbestic point at 392.4 nm indicates the existence of only two uranyl species in solution, UO_2^{2+} and UO_2Cl^+. At higher NaCl concentrations the absence of an isosbestic point indicates further complexation of uranyl with chloride and the formation of at least one more species, such as UO_2Cl_2 and/or $UO_2Cl_3^-$. The tetrachloro compound is the limiting chloro species and is only formed at very high chloride concentrations (above 10 M). To confirm the stoichiometry of the limiting U(VI) solution species in chloride solutions, we determined and compared the coordination of the corresponding solid. We crystallized the uranyl chloride species from a

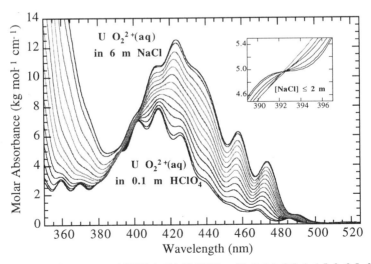

Figure 1. UV-Vis absorption spectra of U(VI) in 0.1 M $HClO_4$ with 0, 0.1, 0.5, 1, 1.5, 2, 2.5, 3, 3.5, 4, 4.5, 5, 5.5, 6m [NaCl].

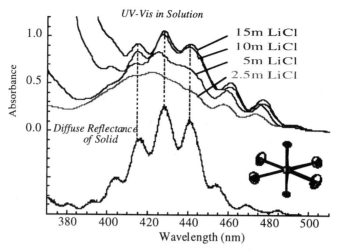

Figure 2. Diffuse reflectance spectra of [K(18-crown-6)]$_2$UO$_2$Cl$_4$ compared with UV-Vis absorbance spectra of U(VI) in acidic concentrated NaCl and LiCl solutions.

KCl solution by addition of excess of 18-crown-6 and isolated single crystals of (18-crown-6)UO$_2$Cl$_2$ and [K(18-crown-6)]$_2$UO$_2$Cl$_4$. Figure 2 shows the diffuse reflectance spectra of [K(18-crown-6)]$_2$UO$_2$Cl$_4$ and for comparison the absorption spectra of U(VI) in highly concentrated NaCl and LiCl solution. The diffuse reflectance spectra contains the same features as observed in the U(VI) absorption spectra at high concentrations of LiCl concentration (>10m), supporting the existence of UO$_2$Cl$_4^{2-}$ in solution.

Characterization and Solubility of AnO$_2$CO$_3$. The actinyl carbonates, UO$_2$CO$_3$ and PuO$_2$CO$_3$, were precipitated from actinide solutions (pH 4) by saturating the solution with 100% CO$_2$. The XRD powder patterns of both solids (Figure 3) agree well with eachother and with those previously reported.[10-12] To determine the local coordination environments of the actinides we measured the EXAFS spectra of each solid (Figure 4). The Fourier transform moduli show the An---An distance of 4.2 Å, consistent with the Pmmn structure reported.[12] Actinyl An=O distances were determined to be 1.76 for U and 1.74 for Pu. The equatorial An-carbonate bond distances are very similar to those for the uranyl tris carbonate and trimeric bis carbonate species, i.e. 2.45 Å for An--O$_{(bidentate)}$.[7]

The stability of AnO$_2$CO$_3$ is governed mainly by the H$^+$ concentration, or pH, and the CO$_2$ partial pressure which are coupled experimental parameters according to the CO$_2$ dissociation equilibrium:

$$CO_2(g) \rightleftharpoons CO_2(aq) \rightleftharpoons H_2CO_3(aq) \rightleftharpoons HCO_3^- + H^+ \rightleftharpoons CO_3^{2-} + 2H^+ \quad (2)$$

The U(VI) and Pu(VI) solubility experiments are performed under a CO$_2$ atmosphere to ensure the stability of AnO$_2$CO$_3$ and to avoid hydrolysis. The pale pink-tan plutonyl solid is found to be stable for only weeks to months under these conditions, showing evidence of increasing fractions of polymeric Pu(IV) hydroxide; and in solution we observe the formation of PuO$_2^+$. At low ionic strength Pu(VI) is found to be reduced within days, while concentrated NaCl appears to stabilize the higher oxidation state presumably via chloro

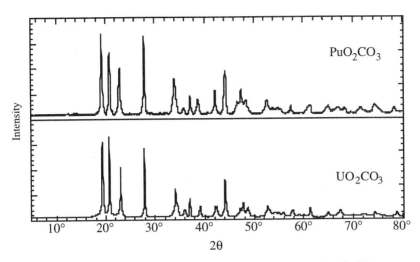

Figure 3. Powder X-ray diffraction patterns of UO_2CO_3 and PuO_2CO_3.

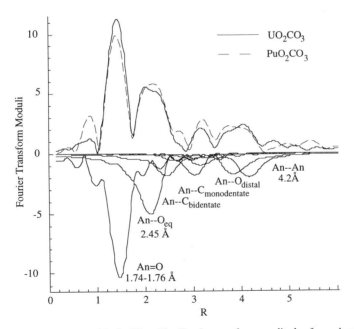

Figure 4. EXAFS of UO_2CO_3 and PuO_2CO_3. The Fourier transform amplitudes for each solid is show. The shells of neighboring scatterers which comprise the fit to the data are based upon idealized Pmmn structure and are plotted here with negative intensity.

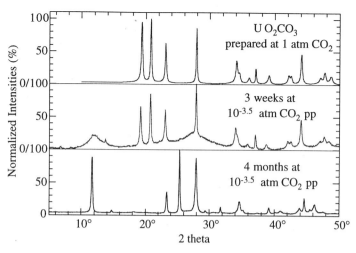

Figure 5. Transformation of UO_2CO_3 by changing the CO_2 partial pressure from 1 atm to $10^{-3.5}$ atm followed by X-ray diffraction.

complexation. Radiolytically produced chlorine and hypochlorite have been reported to be responsible for creating an oxidizing medium in acidic to near neutral solutions, and maintaining the hexavalent oxidation state of Pu.[13] Absorbance spectra of the solutions phase show that these oxidizing species were absent in our experiments (pH 3-4), likely removed from the system by continuos bubbling with CO_2. Because of the higher redox stability of U(VI), we do not observe a similar reduction of UO_2CO_3 with change of the uranium oxidation state However, when the CO_2 partial pressure is lowered from 1 atm (pH 4) to $10^{-3.5}$ atm, we observe a complete phase transformation within 3 weeks. The amount of UO_2CO_3 decreases and an amorphous phase forms (Figure 5). After 4 months UO_2CO_3 transformed into a crystalline phase with significantly different Bragg reflections. The powder pattern of the new phase compares well with that reported for the sodium uranate, $Na_2U_7O_{22}$.[14]

The primary parameter needed to estimate the actinide concentration in solution is the solubility product of the solubility limiting solid phase. We determined the solubility of uranyl(VI) and plutonyl(VI) carbonate in solutions with different NaCl concentration. The speciation was monitored using conventional UV-Vis-NIR absorbance spectroscopy to verify the presence of the free actinyl ion or its chloro complexes and the absence of hydroxo- or carbonato complexes. The apparent solubility products of UO_2CO_3 and PuO_2CO_3 as a function of the NaCl concentration are shown in Figure 6. The apparent solubility product, $\log K'_{sp} = [AnO_2^{2+}][CO_3^{2-}]$, includes the formation of actinyl chloro complexes at higher chloride concentration summed in the term $[AnO_2^{2+}]$. The solubility product increases by about one order of magnitude with the addition of NaCl, and remains nearly constant at $[NaCl] \geq 0.5m$. We parameterized the experimental values by using the Pitzer approach.[6] The interaction of actinyl with chloride was interpreted as ion-associations and modeled using Pitzer ion-interaction parameters calculated from osmotic data for UO_2Cl_2 recommended by Goldberg.[15] The solubility product of PuO_2CO_3 is found to be similar to that of UO_2CO_3. While the solubility product for uranyl carbonate in 0.1 M NaCl, $\log K_{sp}(UO_2CO_3) = -13.08 \pm 0.07$, agrees well with the previously determined value by Pashalidis et al.[12] $(\log K_{sp}(UO_2CO_3) = -13.35 \pm 0.14$ in 0.1 M NaClO$_4$), the solubility product of plutonyl carbonate, $\log K_{sp}(PuO_2CO_3) = -12.9$, differs by about one order of magnitude $(\log K_{sp}(PuO_2CO_3) = -13.98 \pm 0.12$[12]). We note that Pashalidis *et al.* used ^{242}Pu, which is much less radioactive than the ^{239}Pu used in the present work, and the radiation damage of the solid phase is expected to be far less. However, the cause for the previously reported solubility product of PuO_2CO_3, which is lower than that of uranyl carbonate, remains unclear. The solubility product of PuO_2CO_3 in 5.6 m NaCl was found to be log K_{sp} = -12.5. This value may be compared with the previously reported value of log K_{sp} = -13.5 ± 0.3 in 3.5 m NaClO$_4$.[17] The solubility difference of about one order of magnitude at high NaCl/NaClO$_4$ concentrations agrees well with that found for pentavalent neptunyl.[1] To precisely determine the solubility product of plutonyl carbonate we are ongoing to investigate the Pu(VI) solubility at oxidizing conditions, i. e. by addition of hypochlorite or electrochemical control.

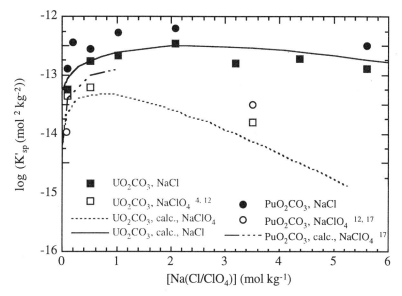

Figure 6. Apparent solubility product of AnO$_2$CO$_3$ (An = U, Pu) as a function of the NaCl concentration. The experimental data were parameterized using the Pitzer approach.

Stability of the Trimeric Species, $(UO_2)_3(CO_3)_6{}^{6-}$, as a Function of NaCl. The uranyl carbonate system was studied using ^{13}C NMR spectroscopy of 10 mM ^{13}C-enriched $UO_2(CO_3)_3{}^{4-}$ at NaCl concentrations ranging from 0.05 to 5 m. Initial solutions were prepared with a carbonate-to-uranium ratios of 3:1 to favor the formation of the monomeric triscarbonato complex. Careful titration with HClO$_4$ leads to the protonation of the carbonate ligand resulting in a decrease in the carbonate:uranyl ratio to 2:1. The species formed has been shown to be the trimeric biscarbonato, $(UO_2)_3(CO_3)_6{}^{6-}$ based on NMR, EXAFS and single X-ray crystal diffraction studies.[7] Since the trimeric species is so highly charged, we anticipated a large effect of the electrolyte concentration on the equilibrium between these two species. We titrated the pH of individual uranyl carbonate solutions of varying NaCl concentrations and, using NMR data collection parameters and chemical shifts previously reported,[7,18] measured NMR spectra at each pH. Indeed, the equilibrium constant is dependent on NaCl concentration; an example of the changes observed at a given pH as a function of ionic strength is shown in Figure 7. By integrating the resonances corresponding to each species at each pH and using known uranyl concentrations, measured p[H], and calculated carbonate concentrations, we calculated the equilibrium constant relating the two species (Eq. 1) at each NaCl concentration. The values determined for the equilibrium constant range from log K_{eq} = 20.85±0.23 with no NaCl in

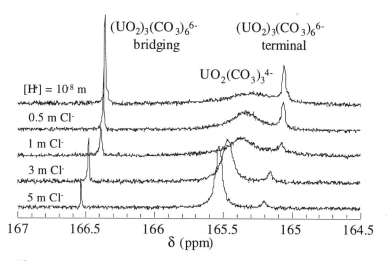

Figure 7. ^{13}C NMR spectra of 10 mM uranyl carbonate at $[H^+] = 10^{-8}$ M as a function of sodium chloride.

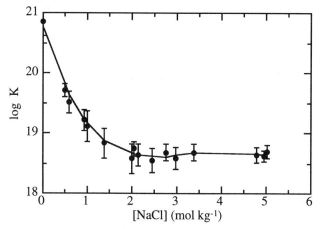

Figure 8. Formation constant of the trimeric species $(UO_2)_3(CO_3)_6^{6-}$ according to Eq. (1) as a function of the NaCl concentration.

solution to 18.69 ± 0.11 in 5.0 m NaCl (Table 1, Figure 8) The greatest change occurs for NaCl concentrations up to 1.4 m; the equilibrium varies little when NaCl concentrations are between 2.0 and 5 m.

Solubility data of uranyl in carbonate media have been interpreted with the formation of only monomeric species, $UO_2(CO_3)_n^{2-2n}$ where $n = 1 - 3$.[4] The trimeric species, $(UO_2)_3(CO_3)_6^{6-}$, has been unequivocally proven to exist in solution by NMR and other spectroscopic studies when the uranyl and carbonate concentrations are fairly high (in the millimolar range).[7] In these studies the solutions were prepared by addition of acid to the uranyl carbonate solutions to form the trimeric species without precipitation. In the solubility experiments at $[CO_3^{2-}]$ higher than 10^{-8} M, the U(VI) solution concentration increases due to the formation of the bis- and triscarbonato complexes. However, we observe a large scattering in the U(VI) solubility data depending on whether or not the experiment is

Table 1. Equilibrium Constants for the First Protonation Step of the Triscarbonato Uranium(VI) Complex as a Function of the Sodium Chloride Concentration.

[NaCl], m	log K'_{eq}	[NaCl], m	log K'_{eq}
0.000	20.85 ± 0.23	2.148	18.64 ± 0.18
0.496	19.71 ± 0.11	2.454	18.54 ± 0.20
0.597	19.51 ± 0.18	2.754	18.68 ± 0.13
0.938	19.21 ± 0.18	2.975	18.58 ± 0.18
0.998	19.11 ± 0.25	3.373	18.68 ± 0.14
1.380	18.83 ± 0.25	4.784	18.64 ± 0.12
2.000	18.58 ± 0.25	4.945	18.61 ± 0.09
2.036	18.74 ± 0.12	5.015	18.69 ± 0.11

performed from undersaturation or oversaturation. The undersaturation experiments result in a lower solubility, presumably due to the formation of only monomeric U(VI) carbonates. Starting the solubility experiment from the alkaline side by addition of acid, similar to the sample preparation for NMR studies, allows for the stabilization of uranyl in solution via formation of the more soluble trimeric compound. Figure 9 shows the UV-Vis absorbance of U(VI) in carbonate solution taken from an oversaturation experiment compared with those of the pure trimeric species and a solution with mostly $UO_2(CO_3)_3^{4-}$. The solution from the solubility experiment clearly matches the absorbance features of $(UO_2)_3(CO_3)_6^{6-}$ and differs significantly from those corresponding to $UO_2(CO_3)_3^{4-}$. These solutions have also been studied using ^{13}C NMR with the same conclusion. Thus far we found only the monomeric species in solution when solubility experiments are performed from undersaturation.

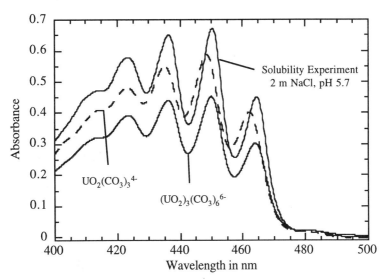

Figure 9. Absorbance spectrum of a U(VI) solubility solution ([NaCl] = 2m) at pH 5.7 and 1 atm CO_2 partial pressure.

Concluding Remarks. We have shown that actinyl(VI) ions form mono-, bis-, tris-, and tetrachloro complexes. In natural aquifer systems the chloride concentration in solution is such that only the mono-, bis- and perhaps the trischloro complexes are important. We found that Pu(VI) is unstable towards reduction in highly concentrated chloride in the absence of an oxidizing agent. Under these conditions the solubility product of plutonyl carbonate can only be determined after relatively short times of solid-liquid phase equilibration. We determined that plutonyl carbonate is about as soluble as uranyl carbonate in NaCl. We learned that increasing ionic strength favors the triscarbonato uranyl complex with respect to formation of the trimeric biscarbonato uranyl complex, consistent with the trend observed in sodium perchlorate.[4,7] Data from the NEA review of uranium thermodynamics yield an equilibrium constant for Eq. 1 of log K = 17.82 and Allen, *et al.*[7] reported 18.1 ± 0.5 in 2.5 m NaClO$_4$; for comparison, we have determined log K = 18.6 in 2.5 m NaCl. We observed that the formation of the $(UO_2)_3(CO_3)_6^{6-}$ clearly depends on the details of sample preparation and its formation has to be considered in solubility experiments performed from oversaturation at $[CO_3^{2-}] \geq 10^{-8}$ M.

ACKNOWLEDGEMENTS

EXAFS experiments were performed at the Stanford Synchrotron Radiation Laboratory (SSRL) in collaboration with Steve D. Conradson, Materials Science and Technology Division, Los Alamos National Laboratory. SSRL is supported by the Office of Basic Energy Sciences, Division of Chemical Sciences, U.S. Department of Energy. Full details of the EXAFS data and analysis will be reported elsewhere. Uranium studies were supported by the Waste Isolation Pilot Plant, under Contract No. AN-1756 with Sandia National Laboratory. Plutonium studies were supported by Los Alamos National Laboratory Directed Research Development and the Nuclear Materials Stabilization Task Group, EM-66 of the U.S. D.O.E.

REFERENCES

1. W.H. Runde, M.P. Neu, D.L. Clark. Neptunium(V) hydrolysis and carbonate complexation: experimental and predicted neptunyl solublity in concentrated NaCl using the Pitzer approach. *Geochim. Cosmochim. Acta* 60(12):2065 (1996).

2. S.P. Awasthi, M. Sundaresan. Spectrophotometric & calorimetric study of uranyl cation/chloride anion system in aqueous solution. *Indian J. Chem.* 20A:378 (1981).

3. T.W. Newton, F.B. Baker. *J. Phys. Chem.* 61:934 (1957).

4. I. Genthe, J. Fuger, R. J.M. Konings, R.J. Lemire, A.B. Muller, Ch. Nguyen-Trung, H. Wanner. *Chemical Thermodynamics 1. Chemical Thermodynamics of Uranium.* OECD, Elsevier Science Publishers B.V., North Holland, 1992.

5. J. Fuger, I.L. Khodakovsky, E.I. Sergeyeva, V.A. Medvedev, J.D. Navratil. The Chemical Thermodynamics of Actinide Elements and Compounds. Part 12. The Actinide Aqueous Inorganic Complexes. IAEA, Vienna, 1992.

6. K.S. Pitzer. *Activity Coefficients in Electrolyte Solutions.* CRC Press, 1991.

7. P.G. Allen, J.J. Bucher, D.L. Clark, N.M. Edelstein, S.A. Ekberg, J.W. Gohdes, E.A. Hudson, N. Kaltsoyannis, W.W. Lukens, M.P. Neu, P.D. Palmer, T. Reich, D.K. Shuh, C.D. Tait, B.D. Zwick. Multinuclear NMR, Raman, EXAFS, and X-ray diffraction studies of uranyl carbonate complexes in near-neutral aqueous solution. X-ray structure of $[C(NH_2)_3]_6[(UO_2)_3(CO_3)_2] \cdot 6.5 H_2O$, *Inorg. Chem.* 34:4797 (1995).

8. NONLIN 94.05.13b developed by A.R. Felmy (Pacific Northwest National Laboratory). A chemical equilibrium program based on the Gibbs free energy minimization procedure by Harvie, *et al. Geochim. Cosmochim. Acta* 48:723 (1984).

9. S.O. Cinneide, J.P. Scanlan, M.J. Hynes. Equilibria in uranyl carbonate systems I. The overall stability constant of $UO_2(CO_3)_3^{4-}$. *J. Inorg. Nucl. Chem.* 37:1013 (1975).

10. J.R. Clark, C.L. Christ. Some observations on rutherfordine. *Am. Min.* 41: 844 (1957).

11. J.D. Navratil, H.L. Bramlet. Preparation and characterization of plutonyl(VI) carbonate. *J. Inorg. Nucl. Chem.* 35:157 (1973).

12. I. Pashalidis, W. Runde, J.I. Kim. A study of solid-liquid phase equilibria of Pu(VI) and U(VI) in aqueous carbonate systems. *Radiochim. Acta* 61:141 (1993).

13. K. Bueppelmann, J.I Kim, Ch. Lierse. The redox behavior of plutonium in saline solutions under radiolysis effects. *Radiochim. Acta* 44-45:65 (1988).

14. JCPDS 5-132 and Wamser et al. J. Am. Chem. Soc. 74:1022 (1952).

15. R.N. Goldberg. Evaluated activity and osmotic coefficients for aqueous solutions: bi-univalent compounds of lead, copper, manganese, and uranium. *J. Phys. Chem. Ref. Data* 8(4):1005 (1979).

16. W. Runde, G. Meinrath, J.I. Kim. A study of solid-liquid phase equilibria of trivalent lanthanide and actinide ions in carbonate systems. *Radiochim. Acta* 58/59:93 (1992).

17. P. Robouch, P. Vitorge. Solubility of $PuO_2(CO_3)$. *Inorg. Chim. Acta* 140:239 (1987).

18. L. Ciavatta, D. Ferri, I. Grenthe, F. Salvatore. The first acidification step of the tris(carbonato)dioxo-uranate(VI) ion, $UO_2(CO_3)_3^{4-}$. *Inorg. Chem.* 20:463 (1981).

SOLUBILITY OF NaNd(CO_3)$_2$.6H$_2$O(c) IN MIXED ELECTROLYTE (Na-Cl-CO$_3$-HCO$_3$) AND SYNTHETIC BRINE SOLUTIONS

Linfeng Rao[*], Dhanpat Rai[*], Andrew R. Felmy[*] and Craig F. Novak[#]

[*]Pacific Northwest National Laboratory[a]
Richland, Washington 99352
[#]Sandia National Laboratories[b]
Albuquerque, New Mexico 87185

ABSTRACT

Experiments were conducted to evaluate the solubility of neodymium(III) in mixed electrolyte solutions (Nd^{3+}-Na^+-Cl^--CO_3^{2-}-HCO_3^--H_2O) and in two synthetic brines under different partial pressures of CO_2. The stable solid phase in solutions containing concentrated NaCl (2 m and 4 m)/Na_2CO_3 (0.1 to 2.0 m) and NaCl (2 m)/$NaHCO_3$ (0.1 to 1.0 m) was identified by X-ray diffraction to be NaNd(CO_3)$_2$.6H$_2$O(c), which is the same equilibrium solid phase previously observed in solutions of single electrolytes (Na_2CO_3 or $NaHCO_3$ in the absence of NaCl). Using Pitzer's specific ion-interaction approach, ternary ion interaction parameters were developed for Na^+-Cl^--$Nd(CO_3)_3^{3-}$ and incorporated into the previous data base for the single electrolyte, non-chloride systems. The resulting model developed from these experiments provided satisfactory interpretation of the concentration of neodymium in the mixed electrolyte solutions of this study. Model predictions were tested against the experimental measurements of the solubility of Nd(III) in two synthetic

[a]Pacific Northwest National Laboratory is operated for the U.S. Department of Energy by Battelle Memorial Institute under Contract DE-AC06-76RLO 1830.

[b]Sandia National Laboratories are operated by Sandia Corporation, a Lockheed Martin Company, for the U.S. Department of Energy.

brines equilibrated with carbon dioxide at partial pressures of $P_{CO2} = 10^{-3.5}$ atm and 1 atm. The model predictions of the total neodymium concentration in these two brines agree with the experimental results showing deviations less than one order of magnitude. In addition, the model provides a good prediction of the solubility of $NaAm(CO_3)_2(c)$ in 5.6 m NaCl solution equilibrated with carbon dioxide over a wide range of carbonate concentrations.

INTRODUCTION

Nuclear waste repositories situated in salt formations, such as the Waste Isolation Pilot Plant (WIPP) near Carlsbad, New Mexico USA, may contain highly concentrated NaCl and $MgCl_2$ brines.[1] Information on the chemical behavior of actinide elements in concentrated electrolytes is required for evaluating the potential performance of such repositories. In particular, the stable chemical forms of actinides, including solid compounds and aqueous species, and their concentrations in brines must be determined. To obtain such information, Pacific Northwest National Laboratory (PNNL) has been conducting systematic studies of the solubilities of trivalent actinides in various electrolyte solutions covering wide concentration ranges.[2,3] Our goal is to establish a thermodynamic model that can reliably predict the stable solid phases and the dissolved concentrations of trivalent actinides in brines. Taking Pitzer's ion interaction approach for evaluation of activity coefficients,[4,5] we have conducted studies in the following phases to build and test the model: 1) solubility studies of actinides in single electrolyte solutions (Na_2CO_3 or $NaHCO_3$) to identify the stable solid compounds, determine their solubility products, and develop necessary binary ion interaction parameters; 2) solubility studies of actinides in mixed electrolyte solutions (Na^+-Cl^--CO_3^{2-}-HCO_3^-) to develop necessary ternary ion interaction parameters; and 3) solubility studies of actinides in synthetic brines with compositions relevant to WIPP to test the model developed in Phases 1 and 2. The results of the Nd(III) solubility studies in the single electrolyte systems (Phase 1) are described in previous publications from this laboratory.[2,3] This paper presents the results from the studies of mixed electrolyte solutions (Phase 2) and the model testing with synthetic brines (Phase 3).

Several sparingly-soluble carbonate solids containing trivalent actinides (or lanthanides as chemical analogs) have been observed in carbonate solutions,[6-16] including $M_2(CO_3)_3$,[12,13] $MOHCO_3$[14,15] and $NaM(CO_3)_2$,[3,16] where M represents trivalent actinides or lanthanides. The stability field of these solid carbonate compounds depends on solution conditions such as pH, partial pressure of carbon dioxide over the solution, and concentrations of components such as carbonate and sodium ion. For example, previous studies[2] of the solubility of Am(III) and Nd(III) in single carbonate or bicarbonate solutions showed that $AmOHCO_3$ was the stable solid phase in dilute sodium carbonate and sodium bicarbonate solutions, while $NaNd(CO_3)_2 \cdot 6H_2O$ was found to be the stable solid phase in concentrated sodium carbonate and sodium bicarbonate solutions.[3] The solubility products of these two compounds were determined to be $10^{-22.5}$ for $AmOHCO_3$ and $10^{-21.39}$ for $NaNd(CO_3)_2 \cdot 6H_2O$, respectively. Based on these experimental results, a set of binary ion interaction parameters for Na^+-$Nd(CO_3)_3^{3-}$ were developed (see Table 1). The established model provided satisfactory prediction of the solubility of trivalent actinides in sodium carbonate or sodium bicarbonate solutions in the absence of sodium chloride (Figure 1).

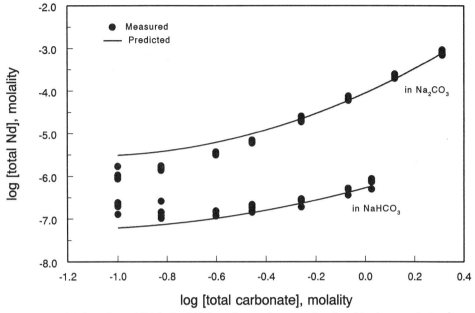

Fig.1 Solubility of $NaNd(CO_3)_2 \cdot 6H_2O(c)$ in sodium carbonate and sodium bicarbonate solutions.

Natural brines are highly concentrated multi-component electrolyte solutions. For example, brines from the WIPP Site[1] contain major ions including Na^+, K^+, Mg^{2+}, Ca^{2+}, Cl^-, and SO_4^{2-}, in addition to CO_3^{2-}/HCO_3^-. Sodium chloride is usually the major electrolyte component of these brines, with concentrations as high as 4-6 M.[17] The presence of additional ions in high concentrations, especially sodium and chloride, may affect the solubilities of actinides in brine solutions by changing the activities of actinide species. As a result, the previous data base, developed solely from measurements in sodium carbonate and sodium bicarbonate systems, may not include all of the necessary parameters to describe the solubilities of actinides in complex brines and thus, may need to be modified based on studies of mixed electrolyte systems.

The objectives of the present study are: 1) to identify the stable neodymium carbonate solid compound in concentrated sodium chloride/sodium carbonate and sodium chloride/sodium bicarbonate solutions and determine its solubility in these mixed electrolyte systems; 2) to develop the ternary ion interaction parameters necessary to allow the previously developed data base to represent mixed electrolytes; and 3) to determine the neodymium solubilities in two synthetic brines equilibrated with carbon dioxide with partial pressures of $10^{-3.5}$ atm (the partial pressure of carbon dioxide in air) and 1 atm, and to test the resulting model by comparison with these results. In addition, we discuss the application of this model to the solubility data obtained for the Am^{3+}-Na^+-Cl^--CO_3^{2-}-HCO_3^- systems.

Table 1 Pitzer Ion-Interaction Parameters Used in the Model
for Single Electrolyte Systems (Na_2CO_3 or $NaHCO_3$)[3]

Species	Ion-Interaction Parameters		
	Binary System Interactions		
	$\beta^{(0)}$	$\beta^{(1)}$	C^ϕ
Na^+-HCO_3^-	0.0277	0.0411	0
Na^+-CO_3^{2-}	0.0399	1.389	0.0044
Na^+-OH^-	0.0864	0.253	0.0044
Na^+-$Nd(CO_3)_3^{3-}$	-0.256	5.0	0.0443
	Common Ion Ternary System Interactions		
H^+-Na^+	0.036		
OH^--CO_3^{2-}	0.10		
OH^--CO_3^{2-}-Na^+	-0.017		
CO_3^{2-}-HCO_3^-	-0.04		
CO_3^{2-}-HCO_3^--Na^+	0.002		

MATERIALS AND PROCEDURES

Solutions and Reagents

All chemical reagents used in this work were A.R. grade or higher. Double-distilled
acids (HNO_3 and HCl) were used when needed.   All solutions were prepared with distilled
and deionized water.

Table 2 Compositions of the Two Synthetic Brines

Components	Concentration (molality)	
	ERDA6B	GSeep
KCl	0.109	0.40
$CaCl_2$	-	0.0088
$MgCl_2$	-	0.72
Na_2SO_4	0.192	0.346
NaBr	0.0124	0.0195
NaCl	5.31	3.99
H_3BO_3	0.0711	0.164

NaNd(CO$_3$)$_2$.6H$_2$O(c) was used as the initial solid in the solubility experiments in mixed electrolyte (Na$^+$-Cl$^-$-CO$_3^{2-}$-HCO$_3^-$). Nd(OH)$_3$(c) was used as the initial solid in the solubility experiments in synthetic brines. These two compounds were prepared using the procedures described in previous publications from this laboratory.[3,18]

Two synthetic brines, named as ERDA6B and GSeep, were prepared with dry chemicals using the compositions reported by Brush[17] and shown in Table 2. These compositions were selected because they represent the compositions of some natural brines collected from the WIPP site. ERDA6B is basically a sodium brine while GSeep has significant amounts of magnesium and sulfate.

Solubility Experiments

The solubility experiments with the mixed Na$^+$-Cl$^-$-CO$_3^{2-}$-HCO$_3^-$ systems and with the synthetic brines were conducted in different manners. In the experiments with the mixed Na$^+$-Cl$^-$-CO$_3^{2-}$-HCO$_3^-$ systems, the carbonate concentrations were controlled by dissolving calculated amounts of Na$_2$CO$_3$ or NaHCO$_3$ and the centrifuge tubes were tightly sealed during the equilibration to avoid contact with air. After the solubility experiments, the concentrations of carbonate or bicarbonate were verified by carbon analysis. On the contrary, in the experiments with synthetic brines, the centrifuge tubes were lightly capped so that the suspensions were in contact with the atmosphere (either air or 100% CO$_2$ in a controlled-atmosphere chamber) and the carbonate concentrations were controlled by the pH of the suspension and the partial pressure of CO$_2$ in the atmosphere. More detailed descriptions of these experiments are provided in the subsequent sections.

Table 3 Experimental Conditions for the Mixed Electrolyte Systems

Experiment #	Concentration (molality)		
	Na$_2$CO$_3$	NaHCO$_3$	NaCl
I	0.1 - 2	-	2
II	0.1 - 2	-	4
III	-	0.1 - 0.5	2

Several sets of solubility experiments were performed in the mixed Na$^+$-Cl$^-$-CO$_3^{2-}$-HCO$_3^-$ systems, with NaNd(CO$_3$)$_2$.6H$_2$O(c) as the initial solids. The concentrations of the electrolytes in the solutions are shown in Table 3. These solubility experiments were performed in 50 mL polypropylene centrifuge tubes, each containing enough initial solid to guarantee excess solid phase materials remaining after the system reached equilibrium. After the initial pH was measured, the centrifuge tubes were sealed and continuously shaken on an orbital shaker until the scheduled sampling and analysis. The equilibration periods for these experiments were 19, 39, 80, and 145 days. No particular measures were taken to control the pH or the atmosphere under which these suspensions were equilibrated. However, the pH was measured prior to the experiments and before each sampling. The pH was found to remain stable throughout the experiments: in the range of 10.4 to 10.8 for the Na_2CO_3-NaCl system and 7.9 to 8.3 for the NaHCO$_3$-NaCl system. The total carbonate concentrations were measured after each sampling to ascertain that no significant changes in the carbonate concentration had occurred.

For the solubility study in the synthetic brines, each centrifuge tube initially contained 40 ml of either ERDA6B brine or GSeep brine and 20 mg of $Nd(OH)_3(c)$ as the starting solid. The "pH" (or pH_{obs}) of these brines was measured with a pH electrode and adjusted with NaOH/HCl to cover an appropriate range. This electrode was previously calibrated using a Gran-type titration method so that the hydrogen ion concentration in the brines could be calculated from the observed pH. This method is described in the next section. The centrifuge tubes were kept lightly capped to allow equilibration of $CO_2(g)$ with the brines throughout the solubility experiments. Two sets of parallel experiments were conducted for each brine: one on the lab bench and the other in a controlled-atmosphere chamber filled with 100% CO_2, providing CO_2 partial pressures of $10^{-3.5}$ atm and 1 atm, respectively. The equilibration periods for these experiments were 15 or 17 days, 63 or 64 days, and 103 days. The experimental conditions for these solubility experiments are shown in Table 4.

Table 4 Experimental Conditions for the Synthetic Brine Systems

Experiment #	Brine	pC_H	P_{CO2}(atm)
I	ERDA6B	7.9 - 10.6	$10^{-3.5}$
II	ERDA6B	6.2 - 8.2	1
III	GSeep	7.4 - 7.8	$10^{-3.5}$
IV	GSeep	5.6 - 6.0	1

Calibration of pH Electrodes in Brine Solutions

Hydrogen ion concentration is a crucial parameter for determining thermodynamic equilibrium and speciation of carbonate species in electrolyte solutions. It is therefore necessary to determine the hydrogen ion concentration accurately. To calibrate the pH electrodes and determine the factor for the pH_{obs} - pC_H correction (pC_H = -log m_H, where m_H denotes the concentration of hydrogen ion in molality), Gran-type pH titrations with the synthetic brines were performed using the procedures described in a previous publication.[19] The relationship between pH_{obs} and pC_H is expressed as: pC_H = pH_{obs}+ A, where the correction factor A was determined to be 0.98 for ERDA6B brine and 1.20 for GSeep brine, respectively.

Analyses

At the end of each equilibration period, the pH of the solutions was measured and the solutions were then centrifuged at about 2,000 g for 7 to 10 minutes. Samples of the supernatants were filtered through membranes (approximate pore size = 0.0018 μm, effective molecular-weight cutoff = 25,000) using the procedure developed by Rai et al.[20,21] The procedure includes the following three steps: 1) the filters were washed with deionized water; 2) 2.0 ml of the supernatant was passed through the filters to saturate any possible adsorption sites on the filters and filtration containers (this filtrate was discarded); and 3) 6.0 ml of the supernatant was filtered. Of the 6.0 ml of final filtrate, 1.0 ml was used for analysis of total inorganic carbon with a Dohrmann DC180 Carbon Analyzer. The remaining 5.0 ml of filtrate was acidified with concentrated nitric acid and analyzed for sodium and neodymium concentrations. Sodium was determined by inductively coupled plasma spectroscopy (ICP), while neodymium was analyzed by the inductively coupled plasma spectroscopy - mass spectrometry (ICP-MS) technique using indium-115 as the internal standard. A Plasma-Spec ICP 2.5 from Leeman Labs Inc. and a Plasma Quad ICP-MS (PQ2-451) from VG Elemental were used for the analysis. The analytical errors in Nd concentrations (at the 95% confidence level) are estimated to be less than 10% in all cases, based on the standard deviations of the tests using NBS certified Nd standard solutions.

At the end of the sampling periods, wet samples of the solid phases were taken and analyzed by X-ray diffraction using a Philips APD 3520 X-ray diffractometer.

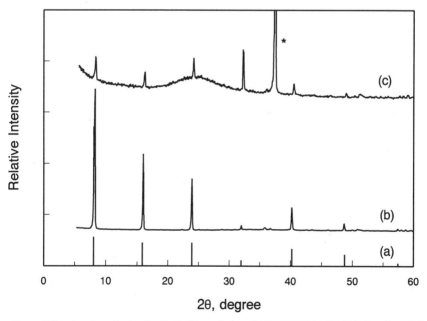

Fig.2   X-ray diffraction data obtained with Co-Kα radiation. (a) NaNd(CO$_3$)$_2$.6H$_2$O(c) from Ref [22]. (b) The neodymium solid in equilibrium with 1 M Na$_2$CO$_3$. (c) The neodymium solid in equilibrium with ERDA6B brine. The intense peak marked with "*" in (c) results from the NaCl in the brine.

RESULTS AND DISCUSSION

Solubility Studies in the Mixed NaCl-Na$_2$CO$_3$ and NaCl-NaHCO$_3$ Systems

The changes in aqueous neodymium concentrations as a function of equilibration time indicate that the steady state neodymium concentrations were reached within 19 days in the NaCl-Na$_2$CO$_3$ systems and within 39 days in the NaCl-NaHCO$_3$ systems.

The solids present at the conclusions of the experiments were analyzed by X-ray diffraction (XRD). The XRD patterns of all the final solid samples are in good agreement with those for NaNd(CO$_3$)$_2$.6H$_2$O(c) reported by Mochizuki et al.,[22] which is the same neodymium compound used as the initial solid (Figure 2). As a result, the observed aqueous concentrations of neodymium at the steady state represent the solubility of NaNd(CO$_3$)$_2$.6H$_2$O(c) in the mixed Na$^+$-Cl$^-$-CO$_3^{2-}$-HCO$_3^-$ systems.

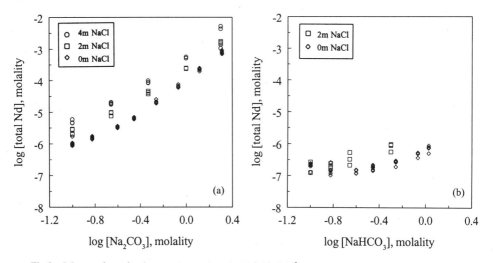

Fig.3   Measured neodymium concentrations in Na$^+$-Cl$^-$-CO$_3^{2-}$-HCO$_3^-$ mixed electrolyte solutions.
(a) in NaCl/Na$_2$CO$_3$ solutions. (b) in NaCl/NaHCO$_3$ solutions.

Figure 3 shows the solubility of NaNd(CO$_3$)$_2$.6H$_2$O(c) as a function of the concentration of carbonate or bicarbonate in the mixed electrolyte solutions from this study (with 2 m or 4 m NaCl) and in the single electrolyte systems from a previous study[3] (0 m NaCl). Two trends are observed from these results: 1) At constant concentrations of sodium chloride (2 m or 4 m), the solubility of NaNd(CO$_3$)$_2$.6H$_2$O(c) increases as the concentration of carbonate or bicarbonate is increased, in a similar pattern as in the single Na$_2$CO$_3$ or NaHCO$_3$ systems (0 m NaCl). This behavior was described in the previous model and attributed to the formation of carbonato complexes. The stability constants for the neodymium carbonato complexes and binary ion interaction parameters involving these species are included in that model.[3] Neodymium hydroxy species (Nd(OH)$_i^{(3-i)+}$, i = 1, 2, and 3) were not included in that model because calculations demonstrated that these species were not significant in solutions containing carbonate or bicarbonate. 2) At constant carbonate or bicarbonate concentrations, the solubility of $NaNd(CO_3)_2.6H_2O(c)$ increases as the concentration of sodium chloride is increased. The significant increase in the solubility of $NaNd(CO_3)_2.6H_2O(c)$ as a function of NaCl concentration is the focus of the thermodynamic analysis presented in the following section.

Thermodynamic Analysis of the Mixed NaCl-Na$_2$CO$_3$ and NaCl-NaHCO$_3$ Systems

As in the thermodynamic analysis of the single electrolyte systems,[3] the nonlinear least-squares program NONLIN was used to analyze the results of the mixed electrolyte systems. Details of NONLIN and Pitzer's expressions for activity coefficients have been outlined in previous publications[23-26] and extensive discussions on the analysis are not repeated in this paper. The NONLIN program was used in combination with the chemical equilibrium program GMIN[2] to minimize the free energy difference between the aqueous and the solid phases, based on the dissolution of $NaNd(CO_3)_2.6H_2O(c)$:

$$NaNd(CO_3)_2.6H_2O(c) \rightleftharpoons Na^+ + Nd^{3+} + 2CO_3^{2-} + 6H_2O \qquad (1)$$

The influence of chloride on the solubility of $NaNd(CO_3)_2.6H_2O(c)$ in the mixed electrolyte systems could be interpreted with two alternate approaches: one assuming that chloride forms complexes with Nd(III); the other attributing the effect to a change in activities caused by specific ion interactions. The approaches could provide equivalent predictions, so either may be used. Our thermodynamic data base already contains internally-consistent representations using both approaches, typically invoking complex formation for strong complexes and ion interaction for weak complexes. We elected to use the specific ion interaction approach, to account for the effect of NaCl on the solubility of $NaNd(CO_3)_2.6H_2O(c)$ because of the following reasons: 1) the evidence for the formation of chloride complexes in mixed electrolyte solutions is insufficient and ambiguous, and 2) in the presence of strongly complexing ligands such as carbonate, complexation of Nd(III) with chloride is probably insignificant and carbonato complexes are expected to be dominant when the carbonate concentrations are high. In addition, previous studies[27,28] indicate that chloride complexes are not required to describe the solubility behavior of Pu(III) and Cm(III) in concentrated NaCl solutions in the absence of carbonate.

Using binary ion interaction parameters for $Na^+-Nd(CO_3)_3^{3-}$, the previous model successfully interprets the solubility of $NaNd(CO_3)_2.6H_2O(c)$ in the single electrolyte systems.[3] However, when applied to the mixed electrolyte systems without modification, the model did not provide satisfactory predictions, indicating that the effects of NaCl were not represented adequately. Therefore, the data in the mixed electrolyte systems were used to calculate ion interaction parameters for $Cl^--Nd(CO_3)_3^{3-}$ and $Na^+-Cl^--Nd(CO_3)_3^{3-}$ to better reproduce these data. A comparison between the augmented thermodynamic parameter set and the experimental results is shown in Figure 4. The new interaction parameters are Θ = 0.168 for $Cl^--Nd(CO_3)_3^{3-}$ and ψ = 0.0273 for $Na^+-Cl^--Nd(CO_3)_3^{3-}$. Values of μ^o/RT for the chemical species included in this model are listed in Table 5.

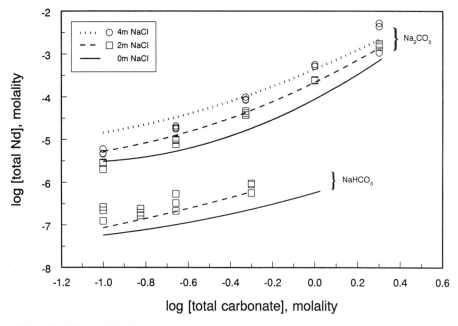

Fig.4 Solubility of $NaNd(CO_3)_2 \cdot 6H_2O(c)$ in mixed $NaCl/Na_2CO_3$ and $NaCl/NaHCO_3$ solutions.
Symbols - experimental points; lines - model predictions.

Table 5 Chemical Species and Their $\mu°/RT$ Values Used in This Study

Species	$\mu°/RT$	Reference
H_2O	-95.663	[23]
Na^+	-105.651	[23]
Cl^-	-52.955	[23]
OH^-	-63.435	[23]
CO_3^{2-}	-212.944	[23]
HCO_3^-	-236.751	[23]
$CO_2(aq)$	-155.680	[23]
Nd^{3+}	-270.926	[29]
$NdCO_3^+$	-501.292	[25]
$Nd(CO_3)_2^-$	-725.112	[25]
$Nd(CO_3)_3^{3-}$	-944.692	[25]
$NaNd(CO_3)_2 \cdot 6H_2O(c)$	-1425.70	[3]

Solubility of $NaNd(CO_3)_2 \cdot 6H_2O(c)$ in Synthetic Brines

In these experiments, $Nd(OH)_3(c)$ was used as the initial solid to be suspended in two synthetic brines equilibrated with either air or 100% CO_2 at 1 atm. Results show that steady state was reached within 64 days (Figure 5). X-ray diffraction analysis was performed on the solid phases after the last sampling (103 days). As Figure 2 shows, an intense peak for NaCl appears in the XRD pattern (spectrum c of Figure 2), which is not surprising because the brines contain highly concentrated NaCl. Nevertheless, peaks corresponding to those of $NaNd(CO_3)_2 \cdot 6H_2O(c)$ are observed. This observation suggests that the initial solids, $Nd(OH)_3(c)$, were converted to $NaNd(CO_3)_2 \cdot 6H_2O(c)$ and the aqueous concentration of Nd was controlled by the dissolution of $NaNd(CO_3)_2 \cdot 6H_2O(c)$.

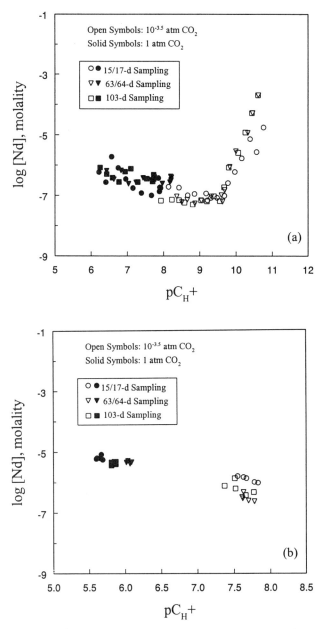

Fig.5 Measured concentrations of neodymium in synthetic brine solutions equilibrated with CO_2 at different partial pressures. (a) ERDA6B brine; (b) GSeep brine.

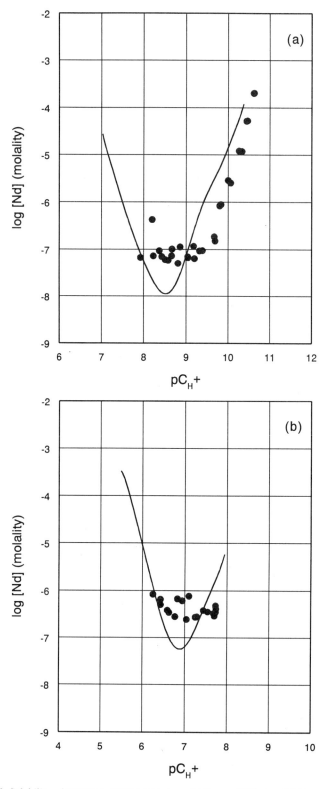

Figure 6 Solubility of Nd(III) in ERDA6B brine. (a) $P_{CO2} = 10^{-3.5}$ atm; (b) $P_{CO2} = 1$ atm.
Symbols - experimental points; solid lines - model predictions for $NaNd(CO_3)_2 \cdot 6H_2O(c)$.

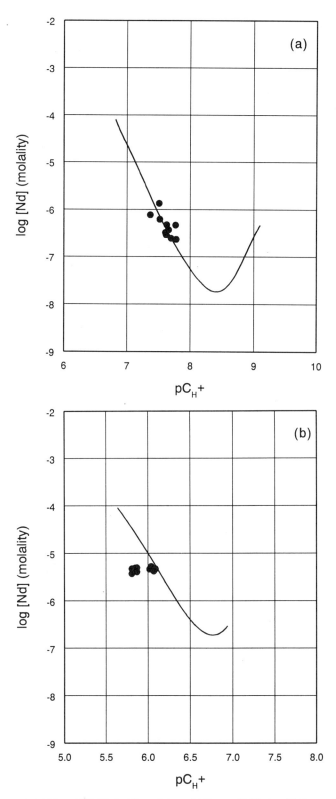

Figure 7 Solubility of Nd(III) in GSeep brine. (a) $P_{CO2} = 10^{-3.5}$ atm; (b) $P_{CO2} = 1$ atm.
Symbols - experimental points; solid lines - model predictions for $NaNd(CO_3)_2.6H_2O$(c).

The measured and predicted solubilities of Nd(III) in ERDA6B (the predominantly NaCl brine) at difference partial pressures of $CO_2(g)$ are shown in Figure 6. The predictions follow the trends in the measured data fairly well and agree with the measured data to within one order of magnitude.

The measured and predicted solubilities of Nd(III) in Gseep (the predominantly Na-K-Mg-Cl-SO_4 Gseep brine) at different partial pressures of $CO_2(g)$ are shown in Figure 7. The experimental conditions covered a narrow hydrogen ion concentration range to minimize the formation of solids other than neodymium carbonate compounds. As a result, there isn't sufficient information to establish conclusions regarding the trends of solubility in a wide pC_H range. However, at $P_{CO2(g)} = 10^{-3.5}$ or 1 atm, the predicted solubilities agree quite well with the measured data to within 0.5 log units.

Application of the Model to the Am^{3+}-Na^+-Cl^--CO_3^{2-}-HCO_3^- System

The solubility of $NaAm(CO_3)_2$ in 5.6 m NaCl solutions in contact with carbon dioxide at a partial pressure of 0.01 atm was measured by Runde and Kim.[16] The hydrogen ion concentrations varied from $pC_H = 6.4$ to $pC_H = 9.3$, giving carbonate concentrations ranging from 10^{-7} to 0.03 m. Because Nd(III) and Am(III) are considered to have analogous aqueous phase chemical behavior and because the Nd(III) data base was based in part upon data from Am(III), it is reasonable to expect that the Nd(III) model will reproduce the chemical behavior of the Am(III) system. A comparison between the model prediction and these independent data provides a test of the model under simpler chemical conditions.

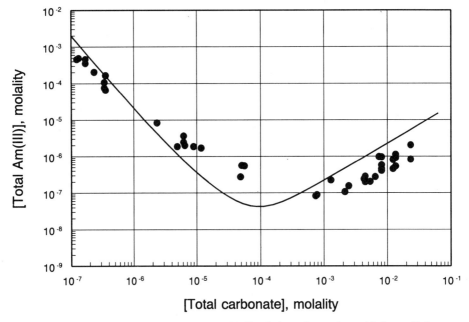

Fig.8 Solubility of $NaAm(CO_3)_2$ in 5.6 m NaCl solutions: Comparison with the prediction by the model developed in this study. Symbols - experimental; line - model prediction.

A graph comparing the data base predictions for this system with the solubility measurements is given in Figure 8. The predictions agree very well with the data at the low carbonate concentration where Am^{3+} is the dominant Am solution species. The thermodynamic parameters for the interaction between Am^{3+} and Cl^- (based on measurements with Nd^{3+}) are well established, and the comparison in the figure serves to demonstrate this once again. In addition, this comparison shows that the solubility product for nonradioactive $NaNd(CO_3)_2.6H_2O(c)$ reproduces the measured solubility of radioactive $NaAm(CO_3)_2.xH_2O(s)$, suggesting that the disorder in crystal structure of $NaAm(CO_3)_2.xH_2O(s)$ caused by the α-radiation of ^{241}Am is not sufficient to significantly alter the solubility. The predictions reproduce the data to within an order of magnitude over the rest of the concentration range for which data exist. The model overpredicts concentrations at the high carbonate end of the experimental conditions, where $Am(CO_3)_3^{3-}$ dominates, consistent with observations in the ERDA6B brine (Figure 6a). The intermediate carbonate range, where $AmCO_3^+$ and $Am(CO_3)_2^-$ dominate, shows an underprediction of the measured solubilities.

SUMMARY

$NaNd(CO_3)_2.6H_2O(c)$ was identified to be the thermodynamically stable solid phase in mixed $Na_2CO_3/NaCl$ and $NaHCO_3/NaCl$ solutions and its solubility was found to increase when the concentration of sodium chloride was increased from 0 m through 2 m to 4 m. This effect was interpreted with Pitzer's ion interaction approach and necessary ternary interaction parameters for Na^+-Cl^--$Nd(CO_3)_3^{3-}$ were accordingly developed and incorporated into the previously established data base. The resulting model provides a good representation of the solubility data from the mixed electrolyte solutions containing high concentrations of sodium chloride.

Solubilities of Nd(III) in synthetic brines were also studied to test the model developed from the studies of mixed $Na_2CO_3/NaCl$ and $NaHCO_3/NaCl$ solutions. The model predictions of the total neodymium concentration in the brines agreed with the experimental results showing deviations less than one order of magnitude.

In addition, comparison of the model prediction with experimental data of solubilities of $NaAm(CO_3)_2.xH_2O(c)$ indicates that the model provides a good prediction of the solubility of $NaAm(CO_3)_2.xH_2O(c)$ in 5.6 m NaCl solutions over a wide range of carbonate concentrations.

ACKNOWLEDGEMENTS

The work described in this report was performed for Sandia National Laboratories under Contract No. AF-3339, supported by the United States Department of Energy under Contract DE-DC04-94DL85000. All work was conducted under a quality assurance program equivalent to NQA-1 requirements.

REFERENCES

1. C.F. Novak. An Evaluation of Radionuclide Batch Sorption Data on Culebra Dolomite for Aqueous Compositions Relevant to the Human Intrusion Scenario for the Waste Isolation Pilot Plant, Sandia Report SAND91-1299, Sandia National Laboratories, Albuquerque, New Mexico (1992).

2. A.R. Felmy, D. Rai and R.W. Fulton. The Solubility of $AmOHCO_3$(c) and the Aqueous Thermodynamics of the System Na^+-Am^{3+}-HCO_3^--CO_3^{2-}-OH^--H_2O, *Radiochimica Acta*, 50, 193-204 (1990).

3. L.F. Rao, D. Rai, A.R. Felmy, R.W. Fulton and C.F. Novak. Solubility of $NaNd(CO_3)_2.6H_2O$(c) in Concentrated Na_2CO_3 and $NaHCO_3$ Solutions, *Radiochimica Acta*, 75, 141-147 (1996).

4. K.S. Pitzer. In "*Activity Coefficients in Electrolyte Solutions*", 2nd edition, Chapter 3, pp.75-153, "Ion Interaction Approach: Theory and Data Correlation", CRC Press, Boca Raton, Florida (1991).

5. K.S. Pitzer and G. Mayorga. Thermodynamics of Electrolytes II. Activity and Osmotic Coefficients for Strong Electrolytes with One or Both Ions Univalent, *J. Phys. Chem.*, 77, 2300-2308 (1973).

6. P. Vitorge. $Am(OH)_3$(s), $AmOHCO_3$(c), $Am_2(CO_3)_3$(s) Stabilities in Environmental Conditions, *Radiochimica Acta*, 58/59, 105-107 (1992).

7. J.I. Kim, R. Klenze, H. Wimmer and W. Hauser. The Carbonate Complexation of Cm(III) and Eu(III) Studied by Time-Resolved Laser-Induced Fluorescence Spectroscopy, in *Abstracts of Actinides-93 International Conference*, September 19-24, 1993, Santa Fe, New Mexico (1993).

8. G. Meinrath and J.I. Kim. Solubility Products of Different Am(III) and Nd(III) Carbonates, *Eur. J. Solid State Inorg. Chem.*, t.28, 383-388 (1991).

9. G. Meinrath and J.I. Kim. The Carbonate Complexation of the Am(III) Ion, *Radiochimica Acta*, 52/53, 29-34 (1991).

10. G. Meinrath and H. Takeishi. Solid-Liquid Equilibria of Nd^{3+} in Carbonate Solutions, *J.Alloys & Compounds*, 194, 93-99 (1993).

11. W. Runde, G. Meinrath and J.I. Kim. A Study of Solid-Liquid Phase Equilibrium of Trivalent Lanthanide and Actinide Ions in Carbonate Systems, *Radiochimica Acta*, 58/59, 93-100 (1992).

12. M. Shiloh, M. Givon and Y. Marcus. A Spectrophotometric Study of Trivalent Actinide Complexes in Solutions - III, Americium with Bromide, Iodide, Nitrate and Carbonate Ligands, *J. Inorg. Nucl. Chem.*, 31, 1807-1814 (1969).

13. F.H. Firsching and J. Mohammadzadei. Solubility Products of the Rare-Earth Carbonates, *J. Chem. Eng. Data*, 31, 40-42 (1986).

14. M.F. Bernkopf and J.I. Kim. Hydrolyse-Reaktionen und Karbonatkomplexierung von dreiwertigem Americium im natürlichen aquatischen System, *RCM-02884*, Institut fur Radiochemie, Technische Universitat, Munchen (1984).

15. R.J. Silva and H. Nitsche. Thermodynamic Properties of Chemical Species of Waste Radionuclides, in NRC Nuclear Waste Geochemistry '83, D.H.Alexander and G.F.Birchard, (eds.) NUREG/CP-0052, US Nuclear Regulatory Commission, Washington, D.C., pp.70-93 (1984).

16. W. Runde and J.I. Kim. *Chemical behavior of Trivalent and Pentavalent Americium in Saline NaCl Solutions.* RCM 01094. Munich, Germany: Institute for Radiochemistry, Technical University of Munich.

17. L.H. Brush. Test Plan for Laboratory and Modeling Studies of Repository and Radionuclide Chemistry for the Waste Isolation Pilot Plant, SAND90-0266, Sandia National Laboratories, Albuquerque, New Mexico (1990).

18. L.F. Rao, D.Rai and A.R.Felmy. Solubility of $Nd(OH)_3$(c) in 0.1 M NaCl Aqueous Solution at 25°C and 90°C, *Radiochimica Acta*, 72, 151-155 (1996).

19. D. Rai, A.R. Felmy, S.P. Juracich and L.F. Rao. Estimating the Hydrogen Ion Concentration in Concentrated NaCl and Na_2SO_4 Electrolytes, SAND94-1949 (1995).

20. D. Rai, R.G. Strickert, D.A. Moore and J.L. Ryan. Am(III) Hydrolysis Constants and Solubility of Am(III) Hydroxide, *Radiochimica Acta*, 33, 201-206 (1983).

21. D. Rai. Solubility Product of Pu(IV) Hydrous Oxide and Equilibrium Constants of Pu(IV)/Pu(V), Pu(IV)/Pu(VI), and Pu(V)/Pu(VI) Couples, *Radiochimica Acta*, 35, 97-106 (1984).

22. A. Mochizuki, K. Nagashima and H. Wakita. The Synthesis of Crystalline Hydrated Double Carbonates of Rare Earth Elements and Sodium, *Bull. Chem. Soc. Japan*, 47 (3), 755-756 (1974).

23. C.E. Harvie, N. Moeller and J.H. Weare. The Prediction of Mineral Solubilities in Natural Waters: The Na-K-Mg-Ca-H-Cl-SO$_4$-OH-HCO$_3$-CO$_3$-H$_2$O System in High Ionic Strengths at 25°C, *Geochim. Cosmochim. Acta* 48, 723-751 (1984).

24. C.E. Harvie, J.P. Greenberg and J.H. Weare. A Chemical Equilibrium Algorithm for Highly Non-Ideal Multiphase Systems: Free Energy Minimization, *Geochim. Cosmochim. Acta*, 51, 1045-1057 (1987).

25. D. Rai, A.R. Felmy, R.W. Fulton and J.L. Ryan. Aqueous Chemistry of Nd in Borosilicate-Glass/Water Systems, *Radiochimica Acta*, 58/59, 9-16 (1992).

26. D. Rai, A.R. Felmy and R.W. Fulton. Solubility and Ion Activity Product of AmPO$_4$.xH$_2$O(am), *Radiochimica Acta*, 56, 7-14 (1992).

27. A.R. Felmy, D. Rai, J.A. Schramke and J.L. Ryan. The Solubility of Plutonium Hydroxide in Dilute Solution and in High-Ionic-Strength Chloride Brines, *Radiochimica Acta*, 48, 29-35 (1989).

28. Th. Fanghanel, J.I. Kim, P. Paviet, R. Klenze and W. Hauser. Thermodynamics of Radioactive Trace Elements in Concentrated Electrolyte Solutions: Hydrolysis of Cm^{3+} in NaCl-Solutions, *Radiochimica Acta*, 66/67, 81-87 (1994).

29. D.D. Wagman, W.H. Evans, V.B. Parker, R.H. Schumn, I. Halow, S,M. Bailey, K.L. Churney and R.L. Nuttal. The NBS Tables of Chemical Thermodynamic Properties. Selected Values for Inorganic and C$_1$ and C$_2$ Organic Substances in SI Units. *J. Phys. Chem.*, Ref. Data, Vol.11, Suppl.2 (1982).

PLUTONIUM (VI) SOLUBILITY STUDIES IN SAVANNAH RIVER SITE HIGH-LEVEL WASTE

D. G . Karraker
Chemical and Hydrogen Technology Section
Savannah River Technology Center
Aiken, SC 29801

INTRODUCTION

The high-level waste tanks at Savannah River and Hanford contain a layer of insoluble solids ("sludge") and a supernatant liquid; the latter a strongly basic solution of sodium nitrate, nitrite, and aluminate at mole/L levels and sodium carbonate, sulfate, chloride, etc. at 0.1 M to 10^{-2} M or less. Plutonium is present at a 10^3 to 10^6 α d/min-mL level; the radiation level of the solution is about 4 x 10^4 R/hr from ^{137}Cs γ-rays. To conserve space in the waste tanks, the supernate is evaporated to a specific gravity of 1.35-1.40 and returned to the waste tank, where the least soluble salts crystallize. This process is repeated as necessary. The most soluble species (sodium hydroxide, nitrite, aluminate, chloride) increase in concentration during evaporation and other species (sodium carbonate, sulfate, phosphate) are precipitated with sodium nitrate.

The behavior of plutonium through this processing is both of practical and scientific interest. A serious concern is that Pu would precipitate in the high-level waste evaporator, and thus lead to nuclear safety problems. Previous investigations[1-3] have found that the solubility of Pu is enhanced by high NaOH concentrations; and . ^{60}Co irradiation has been found [4] to increase Pu solubility in some solutions. The presence of high concentrations of nitrate and nitrite could prevent an increase in Pu solubility by scavenging radiation-produced species that might oxidize Pu(IV) species to a more soluble Pu(V) or Pu(VI) species.

The behavior of Pu in this processing has been studied in synthetic solutions initially spiked with Pu(IV) tracer[3]. The work reported here investigates the behavior of Pu(VI) in under the same conditions.

EXPERIMENTAL SECTION

Solutions were prepared by dissolving reagent-grade chemicals in distilled water and mixing an equal volume of ^{239}Pu in 0.25 M HNO_3. Pu (IV) solution was purified by anion exchange and diluted directly to ca. 200 mg/mL for Pu(IV) tracer. Pu(VI) tracer was prepared from a solution of the same concentration by drop-wise addition of a dilute solution of $KMnO_4$ until a pink color persisted. Attempts to oxidize Pu(IV) with Ce(IV)

Actinide Speciation in High Ionic Strength Media, edited by Reed *et al.*
Kluwer Academic / Plenum Publishers, New York, 1999

were unsuccessful and tests found that MnO_4^- is reduced to MnO_4^{-2} in less than 5 minutes in strong base or instantly when NO_2^- is present.

Solutions were normally agitated for at least one hour before sampling;. Samples were filtered through 0.2 micron syringe filters, thus defining "solubility " as Pu species smaller than 0.2 micron in this study. After filtering, an aliquot of the solution was diluted into 9 M HNO_3, Pu separated from the bulk of salts by anion exchange and the Pu concentration determined by alpha counting and alpha pulse-height analysis.

Some samples were irradiated in a ^{60}Co source to simulate the effect of radiation on Pu solubility. These samples were made up in pairs, one sample for irradiation and a control sample to compensate for any changes in the solution not due to radiation. Initially, the ^{60}Co source had a dose rate of 4×10^4 R/hr, approximately the field in the SRS waste tanks, and irradiation 5×10^6 R required about 4 days. This source was retired during the course of this work for a new source of ca. 10^6 R. No attempt was made to investigate dose rate effects.

Electromigration experiments were performed ina glass cell where the cathode, anode and center compartment were separated by glass fritted disks. The electrode were 1 cm. x 8 cm. platinum strips. Electromigration conditions were 50 mA, 5v. at room temperature.

Experiments to simulate evaporation were performed by boiling a synthetic solution in an open beaker and sampling the solution as its volume decreased. Solid samples for x-ray diffraction were filtered from the solution, and washed with water to remove crystallized salts.

RESULTS

Previous results[4] on the solubility of NaOH solutions initially spiked with Pu(IV) tracer and irradiated with ^{60}Co gamma rays are summarized in Figure 1. Pu solubility increases both with increasing NaOH concentration and with irradiation. Other experiments[4] found that the presence of molar amounts of $NaNO_3$ and $NaNO_2$, as in SRS and Hanford waste tank supernatant liquid, suppressed the effect of radiation. Electromigration experiments found that Pu in these solutions, initially Pu(IV), migrated to the anode in 3 M NaOH .

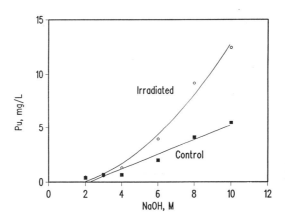

Figure 1. Pu Soluibility in NaOH . Top: solutions irradiated to 5×10^6 R. Bottom: solutions not irradiated Pu initially added as Pu(IV)

Pu(VI) Stability

Initial experiments with Pu(VI) tracer found Pu concentrations of the order of 100 mg/L to be stable; solutions sampled over a 25-day period did not change concentration within the estimated error of the experiment. The concentrations in these solutions may not represent the maximum solubility; Bourges [5] reported preparing Pu(VI) solutions of 220 mg/L . NaNO$_3$ and NaNO$_2$ additions had no effect on the concentration of the solutions. The data are shown in Table 1.

Table 1 Stability of Pu in NaOH; Initial Pu(VI) Spike

Concentration, M			Pu, mg/L		
NaOH	NaNO$_3$	NaNO$_2$	0 days	4 days	25 days
2			73	58	66
4			57	71	81
6			69	64	71
8			68	67	78
10			72	69	81
4	1	0.5	56	76	75
4	1	0.5	61	76	66
4	1	1	80	75	68
4	1	1.8	68	73	66

An electromigration experiment in 3 M NaOH found that Pu from an initial Pu(VI) spike migrated to the anode. Figure 2 shows the increase in Pu concentration in the anode compartment with time. Attempts to determine the solid phase in equilibrium with the solution by x-ray diffraction found only amorphous material.

Irradiation Effects

Experiments where irradiation of Pu(VI) solutions was performed showed that radiation reduces the concentration of Pu at lower hydroxide concentration, but had no effect at

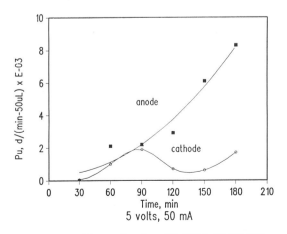

Figure 2. Electromigration of Pu(VI) in 3 M NaOH

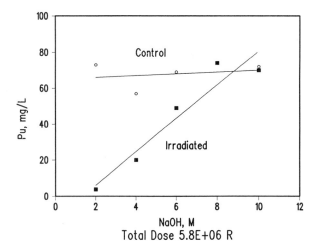

Total Dose 5.8E+06 R

Figure 3. Pu Solubility in NaOH, Pu added initially as Pu(VI) Top: unirradiated solution Bottom: solution irradiated to 5 x 10^6 R.

high NaOH (Figure 3). The presence of $NaNO_3$ and $NaNO_2$ reduced Pu solubility under irradiation in 1-2 M NaOH, but had no effect in 4 M NaOH; data are shown in Table 2. The concentrations shown in Table 2 are approximately those that would be encountered in evaporating a normal high-level waste supernatant solution.

Table 2. Pu Solubility under Irradiation in $NaOH-NaNO_3-NaNO_2$ Solutions: Initial Pu(VI) Spike

Concentration, M			Pu, mg/L	
NaOH	$NaNO_3$	$NaNO_2$	Control	Irradiated (a)
1	2.8	0.7	79	1.90
2	2.0	0.45	126	63
2	4.0	1.5	111	79
4	2.5	2.5	114	143

(a) Total dose 6 x 10^6 R

Table 3. Pu Solubility During Evaporation: Initial Pu(VI) Spike

Concentration, M			Pu, mg/L in Supernate
NaOH	$NaAlO_2$	Na_2CO_3	
1.68	0.64	0.09	81
2.05	0.84	0.03	62
2.58	1.23	--	83
2.79	1.0	0.37 (a)	160
6.79	0.68	0.17(a)	114

(a) These values are considered to be spurious. The solubility of carbonate in this system is very low[1].

Evaporation Tests

To simulate the operation of a waste farm evaporator, a synthetic waste supernatant solution was spiked with Pu(VI) tracer, evaporated by boiling in a an open beaker and sampled as the volume reduced. The evaporation was accompanied by precipitation of some species, principally $NaNO_3$, but also including $NaAlO_2$, Na_2SO_4 and Na_2CO_3. The results are shown in Table 3.

DISCUSSION

Given the manner that these experiments were performed, it is considered that the Pu concentrations found in the control samples represent concentrations of the Pu(VI) spike valence. The species in solution might be a hydroxide complex, and although the solid phase present could not be determined, a plausible composition, based on electromigration experiments, is a sodium-Pu(VI)-hydroxide. While Pu(V) may be equally stable[1,5], attainment of equilibrium in these systems without a stimulus has been shown to be slow[1]. The experiments where the solutions were exposed to radiation, heat and an accompanying composition change are more applicable to the waste tank supernatant solutions. However, in the absence of long-term studies it cannot be said that these experiments yield solutions equivalent to actual waste tank supernates, only that these solutions are similar. In these solutions, heat and concentrations changes may not readily affect Pu valence, but in the irradiated samples, the Pu valences should respond rapidly to the reducing and oxidizing radicals produced by radiation.

The Pu(VI)-spiked solutions had concentrations approximately 10 times greater the solutions prepared with Pu(IV) spikes, demonstrating the higher solubility of Pu(VI). The effects of radiation on Pu concentration are generally consistent with NaOH concentration, both in NaOH and $NaOH-NaNO_3-NaNO_2$ solutions. Among the possible causes for the observed effects, high NaOH concentrations may alter the ratio of oxidizing to reducing species produced by radiation[6]. The stability of oxidized Pu species [Pu(V), Pu(VI)] are probably increased at high NaOH due to hydroxide complexing. Both factors probably contribute to the observed effects.

Evaporation tests serve to demonstrate that the solubility of Pu(VI) or possibly Pu(V) is increased by the concentration changes involved during evaporation, and thus there is little, if any, risk of Pu precipitation in an evaporator. The higher solubility of Pu in the evaporated supernatant solution may lead to decontamination problems during In-Tank processing, so strong oxidizing solutions should be reduced before discharge to the waste tanks. It is preferable to have the traces of Pu in the waste tanks mixed with the precipitated hydroxides ("sludge") rather than the supernatant liquid.

ACKNOWLEDGEMENTS

The author is indebted to Jacqueline Middleton for much of the experimental work, which she performed with energy and enthusiasm and to Charles Parkman of the Analytical Development Section for counting an estimated 1000+ plates for the analyses.

REFERENCES

1. C. H. Delegard, Radiochim. Acta **41**, 11 (1987)
2. D. T. Hobbs and D. G. Karraker, "Recent Results on the Solubility of Uranium and Plutonium in Savannah River Site Waste Supernate" Nucl. Tech. **114**, 318 (1996)

3. D. G. Karraker, WSRC-TR-93-0578, "Solubility of Plutonium in Waste Evaporation, October 1993.
4. D. G. Karraker, WSRC-TR-94-0278X(Rev.2), "Radiation Effects on the Solubility of Plutonium in Alkaline High-Level Waste", August 1994.
5. J. Bourges, Radiochem, Radioanal.Letters, **12,** 111 (1972).
6. I. G. Dragonic and Z. D. Dragonic, "The Radiation Chemistry of Water", Academic Press, New York. (1971) p140 et.seq.

UO_2^{2+} AND NpO_2^+ COMPLEXATION WITH CITRATE IN BRINE SOLUTIONS

M. Bronikowski[1], O. S. Pokrovsky[1], M. Borkowski[2], and G.R.Choppin[1].

[1]Department of Chemistry, The Florida State University, Tallahassee Fl 32306
[2]Institute of Nuclear Chemistry and Technology, Department of Radiochemistry
03-195 Warsaw, Poland.

ABSTRACT

Complexation of uranyl (UO_2^{2+}) and neptunyl (NpO_2^+) with citrate at high ionic strengths I(m) has been investigated by solvent extraction with HDEHP. β_1^{app} values were obtained for I(m) between 0.3 m and 5.0 m NaCl. For NpO_2^+, at pHm = 4.7-6.4 only the 1:0:1 complex was formed while for UO_2^{2+} at pHm 3.0 the 1:1:1 and 1:0:1 complexes were detected with the latter species more dominant. Values obtained for β_1^{app} ranged from 2.62 ± 0.05 to 2.56 ± 0.03 for neptunyl and from 7.30 ± 0.04 to 7.03 ± 0.08 for uranyl as I increased from 0.3m to 5.0m NaCl. The β_1^{app} values were used to estimate the concentration of citrate required in a neutral brine to provide a 10% competition with hydrolysis.

INTRODUCTION

The Waste Isolation Pilot Plant (WIPP), which is being developed for the disposal of transuranic nuclear waste, is situated in bedded halite. Actinide migration from the nuclear waste repository can be modeled using reliable thermodynamic values of actinide species expected to be present in the environment of the repository. Modeling actinide behavior in WIPP requires reliable complexation constants in high ionic strength brine with organic ligands used earlier in decontamination and separation processes and which are to be included in the deposited waste.[1] One such component, citrate, is a sufficiently strong complexant to have significant potential to increase actinide migration.

Earlier investigations [2-17] of UO_2^{2+} and NpO_2^+ citrate complexation have consisted mainly of potentiometric and spectrophotometric titrations at low ionic strengths. Studies at higher ionic strength are rare and, except for an earlier study in 5 m NaCl from this laboratory,[15] only include values to 2 m $NaClO_4$.[5-9] The millimolar concentrations of citrate and metal necessary for detection with titration methods can produce a number of multiple species including dimers and polymers. For example, earlier studies of millimolar

Actinide Speciation in High Ionic Strength Media, edited by Reed *et al.*
Kluwer Academic / Plenum Publishers, New York, 1999

177

concentrations of UO_2^{2+} in 0.01 M to 0.1 M citrate[7-13] report a dimeric species, $(UO_2)_2(Cit)_2^{2-}$. Similar species have also been reported[4] for NpO_2^+. To avoid formation of dimeric and polymeric species, studies with ion exchange resins were carried out at relatively low actinyl concentrations (10^{-5} M - 10^{-6} M).[2,4,14] Complexation with inorganic ligands, OH^- and CO_3^{2-}, was also avoided in most of these studies by keeping the experimental pH below that expected for hydrolysis formation. To overcome the lack of reliable β_{101} stability constants over a sufficient range of higher ionic strengths for performance assessment in WIPP, a detailed investigation of NpO_2^+ and UO_2^{2+} with citrate in 0.1 m - 5.0 m NaCl has been performed.

EXPERIMENTAL

Reagents and Solutions

All reagents were analytical grade and all solutions were prepared in deionized water. A.C.S. certified (Fischer) sodium chloride was used without further purification to adjust ionic strength. Stock solutions of citrate (0.02 M for neptunyl and 2.0×10^{-4} M for uranyl) from sodium citrate (Mallinkrodt) were prepared immediately before use to minimize bacterial decomposition of citrate. The solution pH was adjusted with HCl or NaOH. Aqueous solutions were filtered through a Nalgene Disposable Filter (0.2 μm pore size). Di(2-ethylhexyl)-phosphoric acid, HDEHP (Sigma), was purified by the method of Peppard et al.[18] Organic extractant solutions of 0.01 M HDEHP (for NpO_2^+) and 2.0×10^{-4} M HDEHP (for UO_2^{2+}) in heptane were preequilibrated with NaCl solutions at pHm 4.5-6.0 for neptunyl and pHm 2.0-4.0 for uranyl.

^{237}Np and ^{233}U tracers (from Oak Ridge National Laboratory) were prepared as NpO_2^+ and UO_2^{2+} after purification by passage through a column of Dowex 1(X4) anion exchange resin. The resulting tracer solutions were checked for purity by α- and γ-ray spectrometry (ORTEC). The ^{237}Np tracer was three times evaporated to dryness with 6 M HNO_3. The residue, was dissolved in 0.001 M HCl with 0.01 M $NaNO_2$ which reduced the NpO_2^{2+} to NpO_2^+. Stock solutions of 10^{-3} M neptunium and 10^{-4} M uranium were prepared in HCl media at pHm 2 and 3 respectively. The oxidation state of neptunium was checked on a CARY-14 (OLIS upgraded) spectrophotometer. The absorbance band for NpO_2^+ at 980.4 nm was observed while bands for Np^{4+} at 960.0 nm and for NpO_2^{2+} at 1222.0 nm were absent. Additionally, $^{233}Pa^{5+}$, the active daughter of ^{237}Np, and any residual Np^{4+}, and NpO_2^{2+} were removed from the ^{237}Np tracer by extraction with HDEHP prior to the extraction experiments.

Instrumentation

Liquid scintillation counters, a Tri-Carb 4000 (Hewlett Packard Instruments) and a Beckman LS6500, were used to count the radioactivity in EcoLume cocktail (ICN Biomedical Co.). For pH measurements an Accumet 950, pH meter (Fisher Scientific) was used with a combination glass electrode (Corning Semi-Micro Combination). The electrode was calibrated with 4.00±0.01 and 7.00±0.01 pH buffer standard solutions (Fisher ACS certified). The pH meter readings, pHm, of the aqueous phase in the extraction experiments were converted to pcH (-log free hydrogen ion concentration) using the equation,

$$pcH = pHm + 0.255 \cdot I(m) \qquad (1)$$

where I(m) is the ionic strength in molality. This relationship was derived from calibration curves obtained by a series of HCl and NaOH solutions of known H^+ concentration in 0.3 to 5.0 m NaCl.

Solvent Extraction Procedure

For each extraction experiment, 5.00 ± 0.01 ml of the HDEHP in heptane was added to vials containing 5.00 ± 0.01 ml of the aqueous phase (NaCl and citrate). A ten microliter aliquot of the actinyl stock solution was added to each of the vials to give a final concentration of 10^{-7} M UO_2^{2+} or 10^{-6} M NpO_2^+. The mixture was agitated on a mechanical shaker at $25\pm0.5°C$ for at least 6 hours (previously determined to be sufficient to achieve equilibrium). The vials were centrifuged for two minutes after which 1.00 mL each of the aqueous and the organic phase was transferred into plastic scintillation vials containing 10 ml EcoLume coctail for α - counting. An additional aliquot of the aqueous phase was used to determine the pHm after extraction.

RESULTS

The data were analyzed as described previously[19], where the distribution coefficient, D, for an extraction system is defined as:

$$D = \Sigma[M]_o/\Sigma[M]_a \qquad (2)$$

where o and a denote the organic and aqueous phases, respectively. Assuming only a single organic species in the organic phase, equation 2 can be rewritten as:

$$D_o/D - 1 = \beta_1^{app}[L] + \beta_2^{app}[L]^2 \qquad (3)$$

Where D_o is the distribution coefficient in the absence of the ligand and L is the free ligand concentration in the aqueous phase. We define the apparent stability constants as:

$$\beta_n^{app} \equiv \beta_{10n} + \beta_{11n}[H] + \beta_{12n}[H]^2 \qquad (4)$$

with n denoting the number of ligands complexed with the metal.

The free ligand concentrations were calculated from the pcH, the total ligand concentration, and pKa values of citrate in NaCl.[20] If needed, D values for individual samples were corrected to constant pcH using the experimentally determined variation of the distribution ratio of free neptunyl and uranyl cations at each ionic strength. The relationship used was:

$$logD_{corr} = logD - n \cdot (pcH_{ave}-pcH) \qquad (5)$$

where n was obtained experimentally from the slope of the log D vs pcH plot in the absence of the ligand. The experimental value of n for UO_2^{2+} in both I= 0.1 m between pcH = 3.0-4.0 and 3.0 m NaCl between pcH = 2.25-4.75 was 1.4 so this value was used at the other ionic strengths. For NpO_2^+, the experimental values were n = 1.0 ± 0.1 in 0.3 m to 1 m NaCl and n = 0 in 2 m to 5 m NaCl over a pHm range of 3 to 6. In most cases, this correction did not exceed 5% of the D value for each sample.

Linear curves for 1/D vs. [L] plots were observed over the whole range of citrate concentrations studied for both uranyl and neptunyl systems. Some typical curves for NpO_2^+ citrate are shown in Figure 1. This linearity indicates the formation of only 1:1 metal to ligand complexes for both systems, allowing us to ignore complexes having n > 1 as defined in equation 5. From the extraction data, values of β_1^{app} for both systems were obtained from the slope of line in $(D_0/D-1)$ vs. [L] plots. These values are listed in Tables 1 and 2.

Citrate is present as three species H_2Cit^-, $HCit^{2-}$, and Cit^{3-} over the pcH range of our experiments.[7,20] Several experiments at different pcH values between 4.6-6.6 (Table 1) to determine the pcH dependence of the neptunyl citrate system showed no dependence of log

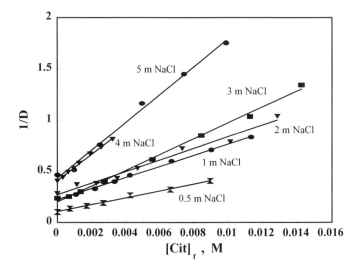

Figure 1. NpO_2^+ distribution ratio between HDEHP in heptane and aqueous NaCl solutions as a function of free ligand concentration.

β_1^{app} on pcH in 0.5 m NaCl. This indicates that for our experimental conditions only deprotonated 1:0:1 complexes were formed and $\beta_1^{app} = \beta_{101}$. The pcH dependence of the uranyl system in 3.0 m NaCl measured between pcH 3.2 and 4.5 (Table 2) showed a possible minor dependence above pcH 3.5 suggesting that minor amounts of protonated species such as 1:1:1 or 1:2:1 may have formed; however, 1:0:1 is dominant. The levels of 1:m:1 species formed relative to the 1:0:1 species did not allow resolution of β_1^{app} so we set $\beta_{101} \approx \beta_1^{app}$.

We have estimated UO_2^{2+} hydrolysis in 4.0 m and 5.0 m NaCl to be 5 to 10% of the total UO_2^{2+} concentration. At lower ionic strengths we estimated the contribution due to hydrolysis to be negligible. The hydrolysis constants, log β_{OH}, used (-5.24 and -4.97 at 4.0 m and 5.0 m (NaCl), respectively) were estimated from the value of $\beta^\circ_{OH} = -5.2\pm0.3$ using specific ion interaction theory.[22,23] For I = 3.0 m, the hydrolysis estimates remove most of the small dependence of log β_1^{app} on pcH suggesting that at I > 3.0, β_1^{app} values of UO_2^{2+} do not contain a significant contribution from β_{111}.

DISCUSSION

The variation of the stability constants for the UO_2^{2+} and NpO_2^+ systems with ionic strength are shown in Figure 2. In both systems, log β_{101} values decrease, with a broad

Table 1. Stability constants of NpO_2^+ citrate complexation (T = 25°C).

I(m)	pHm (pcH)	method[a]	log β_{101}	reference[a]
0.05 NaClO$_4$	4.3 - 5.4	iex,spc	3.67±0.08	[2]
0.05 NaClO$_4$	2 - 12	spc	2.87±0.04	[3]
0.1 NaClO$_4$	7.5	iex	4.84±0.72	[4]
1.0 NaClO$_4$	5.5, 8.4	dis	3.94±0.02	[5]
2.0 NaClO$_4$	5.0	spc	2.49±0.01	[6]
5.0 NaCl	5.8 (7.05)	dis	2.40±0.06	[15]
0.1 NaCl	5.0 (5.00)	dis	2.97±0.01	p.w.
0.3 NaCl	5.0 (5.08)	dis	2.62±0.05	p.w.
0.5 NaCl	5.1 (5.28)	dis	2.54±0.05	p.w.
0.5 NaCl	4.9 (5.05)	dis	2.53±0.05	p.w.
0.5 NaCl	6.0 (6.12)	dis	2.63±0.02	p.w.
0.5 NaCl	6.1 (6.25)	dis	2.57±0.03	p.w.
1.0 NaCl	5.0 (5.28)	dis	2.39±0.01	p.w.
2.0 NaCl	4.9 (5.43)	dis	2.50±0.07	p.w.
3.0 NaCl	5.0 (5.75)	dis	2.52±0.01	p.w.
4.0 NaCl	5.0 (6.00)	dis	2.56±0.05	p.w.
5.0 NaCl	5.5 (6.78)	dis	2.56±0.03	p.w.

a)iex: ion exchange, spc: spectrophotometry, dis: distribution between two phases, p.w.:present work

minimum at around 1.0 m (NaCl), increasing slightly to 3.0 m (NaCl) above which they become constant. This parallel behavior is not surprising since both UO_2^{2+} and NpO_2^+ have linear dioxo structures and can be expected to have parallel chelation behavior. The larger β_{101} values for UO_2^{2+} are due mainly to its higher effective charge of +3.3 compared to +2.2 for NpO_2^+.[21]

Our β_{101} values for NpO_2^+ are in reasonable agreement with spectrophotometry measurements of Rizkalla et al.[6] and Sevost'yanova[3], but they are significantly lower than that of Inoue et al.[5] obtained by the solvent extraction technique with TTA. As expected dimeric complexes of neptunyl with citrate, as well as with monoprotonated ligand (1:0:1, 1:1:1, 1:0:2, and 1:1:2 types of complexes), were reported at the high concentrations used in the spectrophotometric studies. Inoue et al.[5] performed their experiments at pH 5.5 and 8.4 using acetate and borate buffers at concentrations higher than 0.05 M. The effect of neptunyl complexation with TTA dissolved in the aqueous phase at these high pH values was not taken into account.

Moskvin et al.[2] obtained spectroscopic evidence of NpO_2^+ complexation with $HCit^{2-}$ and Cit^{3-} in the pH range from 0.9 to 6.1. From cation-exchange resin studies, they reported log β_{101} = 3.67±0.08 and log K (NpO_2HCit^{2-}) = 2.69±0.10 in 0.05 M NaClO$_4$ and pH = 4.3-5.2.

Table 2. Stability constants of UO_2^{2+} citrate complexation (T = 25°C).

I(m)	pHm (pcH)	method[a]	log β_1^{app}	log β_{101}	reference[a]
0.15 NaCl	1.9 - 2.5	pot	----	8.50±0.15	[17]
0.1 KNO$_3$	~2.5 - 4.5	pot	----	7.40±0.21	[7]
1.0 KNO$_3$	~2.5 - 4.5	pot	----	6.87±0.11	[7]
1.0 KNO$_3$	~2.5 - 4.5	pot,est	----	6.30	[13]
0.1 NaH$_2$Cit	2.2 - 2.9	iex	----	7.28	[14]
0.1 NaClO$_4$	3.91	iex	----	7.22	[14]
1.0 (H,Na)$_3$Cit	1.4 - 4.0	spc,pot	----	7.17±0.16	[8]
5.0 NaCl	5.8 (7.05)	dis	6.04±0.01	----	[15]
0.3 NaCl	3.0 (3.08)	dis	7.30±0.04	< 7.30±0.30[b]	p.w.
1.0 NaCl	3.0 (3.25)	dis	7.08±0.01	< 7.08±0.30[b]	p.w.
2.0 NaCl	3.0 (3.50)	dis	7.22±0.02	< 7.22±0.30[b]	p.w.
3.0 NaCl	3.0 (3.75)	dis	7.10±0.01	< 7.11±0.30[b]	p.w.
4.0 NaCl	3.0 (4.00)	dis	7.02±0.01	≈ 7.05±0.30[b]	p.w.
5.0 NaCl	3.0 (4.25)	dis	7.03±0.02	≈ 7.10±0.30[b]	p.w.
3.0 NaCl	2.5 (3.23)	dis	6.61±0.02	~ 6.61±0.30[b]	p.w.
3.0 NaCl	2.8 (3.50)	dis	6.99±0.02	~ 7.00±0.30[b]	p.w.
3.0 NaCl	3.0 (3.75)	dis	7.37±0.15	~ 7.38±0.33[b]	p.w.
3.0 NaCl	3.0 (3.80)	dis	7.37±0.02	~ 7.38±0.30[b]	p.w.
3.0 NaCl	3.2 (3.94)	dis	7.20±0.02	~ 7.22±0.30[b]	p.w.
3.0 NaCl	3.5 (4.23)	dis	6.94±0.20	~ 6.98±0.36[b]	p.w.
3.0 NaCl	3.5 (4.27)	dis	6.96±0.02	~ 7.00±0.30[b]	p.w.
3.0 NaCl	3.7 (4.47)	dis	6.92±0.01	~ 6.99±0.30[b]	p.w.
3.0 NaCl	3.7 (4.49)	dis	6.69±0.02	~ 6.76±0.30[b]	p.w.

a) pot: potentiometry, iex: ion exchange, spc: spectrophotometry, est.: estimation, dis: distribution between two phases, p.w.: present work
b) corrected for hydrolysis

The first value is in reasonable agreement with our data at low ionic strength. A further ion exchange determination by Rees and Daniel[4] in 0.02 M phosphate buffer should be taken only as a rough estimate of NpO_2^+ - citrate complexation due to the good possibility that the neptunyl cation studied was NpO_2^{2+}.

The literature data for the UO_2^{2+}-citrate system[7,8,13-15,17] is less reliable than that for the NpO_2^+ complexation due principally to the problems of interference by hydrolysis and

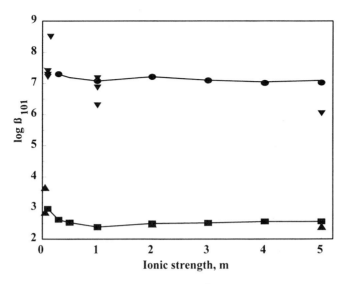

Figure 2. Variation of β_{101} with ionic strength for NpO_2^+ and UO_2^{2+} citrate systems in NaCl media. Experimental values for (■) UO_2^{2+} and (●) NpO_2^+. Literature values of [7,8,13-15](▼) UO_2^{2+} and [2,3,5,6](▲) NpO_2^+.

polymerization. Previous studies that sought to avoid hydrolysis effects used pH ranges where hydrolysis would be expected to be negligible. These pH ranges chosen varied considerably among the studies. Further, in some cases, the variation of the ionic strength of the solution is not taken into account. Potentiometric titration studies have reported β_{101} values for UO_2^{2+}-citrate. Rajan and Martell[7] reported values of 7.40±0.21 and 6.87±0.11 at I=0.1 m and 1.0 m, respectively. The 1.0 m value is 0.4 log units lower than our value at this ionic strength which is probably due to their use of KNO_3 as electrolyte. An even lower, less accurate, value of 6.30 was estimated with previous 1m KNO_3 values[7] and an averaged log K value for the mixed metal complex $UO_2InCit_2^{3-}$. Li et al.,[17] reported a log β_{101} value for UO_2^{2+}- citrate determined in NaCl media[17] of 8.5±0.15. This value was determined from potentiometric data at I=0.15 at pH between 1.9-2.5 where H_2Cit^- is the dominant species. UO_2HCit was not observed. A spectrophotometric study by Vănura and Kuča[8] used both spectrophotometric and potentiometric data in determining β_{101}. The study was done in 1.0 M Na_3Cit where H_3Cit was used to adjust pH. The log β_{101}, log β_{111}, and log β_{202} were reported to be 7.17±0.16, 9.81±0.07, and 17.00±0.14, respectively.

Two studies utilizing tracer methods in order to measure only monomeric uranyl citrate complexes have reported β_{101} values. Ohyoshi et al.,[14] used cation exchange resin to determine log β_{101} = 7.28 in 0.1 M NaH_2Cit between a pH of 2.2 and 2.9 At a higher pH value of 3.9, they measured a smaller log β_{101} value of 7.22. We observed a similar trend of decreasing log β_1^{app} values with increasing pcH. In a previous study from our laboratory,[15] log β_1^{app} values were determined by solvent extraction at a pcH of 7.0 in 5.0 m NaCl. The reported log β_1^{app} is lower than our present value which suggests increased hydrolysis and carbonate complexation at the higher pcH. Because of possible carbonate complexation, the NpO_2^+-citrate β_1^{app} is also slightly lower.

The complexation constants have been used to estimate the effect of citrate on the solubility of UO_2^{2+} and NpO_2^+ in the WIPP. The brines in the repository can be approximated as a 5 molal NaCl solution at neutral pcH if we assume the main reaction in competition with citrate complexation is actinide hydrolysis. For the hydrolysis reaction,

$$M^{n+} + iH_2O \rightarrow M(OH)_i^{n-i} + iH^+ \tag{6}$$

The hydrolysis constant can be written as:

$$\beta_i^{OH} = \frac{[M(OH)_i^{n-i}][H^+]^i}{[M^{n+}]} \tag{7}$$

We pose the question: what concentration of citrate in the repository is necessary to increase the total soluble uranyl concentration by 10% due to 1:1 uranyl citrate formation? This can be expressed as,

$$\frac{[MCit]}{[MCit]+\sum[M(OH)_i^{n-i}]} = 0.1 = \frac{\beta_{101}[Cit]}{\beta_{101}[Cit]+\sum\beta_i^{OH}[H^+]^{-i}} \tag{8}$$

The citrate is fully deprotonated at the neutral pcH of the repository as found from the relationship,

$$[Cit]_f = [Cit]_t / (10^{\,pka-pcH}-1) \tag{9}$$

where f and t denote free and total concentrations.

The total citrate concentration that produces a 10% citrate metal complexation relative to the hydrolyzed species was estimated to be 1.1×10^{-6} M for UO_2^{2+} and 3.1×10^{-7} M for NpO_2^+ using the β_{101} and pKa data for 5.0 m NaCl. The hydrolysis constants used in these calculations were[15,23]: log β_1^{OH} = -10 for NpO_2^+ and -4.97 for UO_2^{2+}; log β_2^{OH} = -12.77 for UO_2^{2+}. These estimates agree well with a previous estimate.[15] UO_2^{2+} requires a higher concentration of citrate to produce the same effect because of the more extensive hydrolysis. The estimated citrate concentrations to cause a significant (10%) increase in the solubility are quite small, indicating that the citrate in WIPP wastes could have some potential to increase neptunyl and uranyl mobility. However, WIPP brine and the waste to be buried contains other cations (e.g., Ca^{2+}, Mg^{2+}, Fe^{2+}, Ni^{2+}) in high concentrations which strongly complex citrate. The presence of these cations will raise the citrate concentration required to increase actinyl solubilities. Similarly, the higher expected pH of the repository should raise the required citrate concentration due to the increase in actinyl hydrolysis and carbonate complexation at the expense of the weaker citrate complexation. A more thorough estimate including the effect of carbonate complexation of the actinyl cations and the competition for citrate due to its significant complexation by other cations expected to be in the brine in high concentrations is required once the necessary data is available.

ACKNOWLEDGEMENTS

This work was performed as part of the Waste Isolation Pilot Plant (WIPP) Actinide Source Term Program, supported at Sandia National Laboratories by the U.S. Department of Energy under contract DE-AC04-94AL85000, and at The Florida State University under contract AH5590.

REFERENCES

1. Lappin, A.R., Garber, D.P., Hunter, R.L., Davies, P.B., System analysis, long-term radionuclide transport, and dose assessment, waste isolation pilot plant (WIPP), southeastern New Mexico. Report SAND89-0462,Sandia National Laboratories (1989).
2. Moskvin, A.I., Marov, I.N., Zolotov, Y.A., Neptunium (V) complexes with citric and tartaric acids. *Russ. J. Inorg. Chem.*, 6:926-929 (1961).
3. Sevost'yanova, E.P., Complex formation by neptunium(V) with citric acid. *Radiokhimiya*,27:24-27 (1985).
4. Rees, T.F., Daniel, S.R., Complexation of neptunium(V) by salicylate, pthalate and citrate ligands in a pH 7.5 phosphate buffered system. *Polyhedron*, 3:667-673 (1984).
5. Inoue, Y., Tochiyama, O., Takahashi, T., Study of the carboxylate complexing of Np(V) by solvent extraction with TTA and capriquat. *Radiochimica Acta*, 31:197-199 (1982).
6. Rizkalla, F.N., Nectoux, F., Dabos-Seignon, S., Pages, M., Complexation of neptunium(V)by halo- and hydroxycarboxylate ligands. *Radiochimica Acta,* 51:113-118 (1990).
7. Rajan, K.S., Martell, A.E., Equilibrium studies of uranyl complexes III. Interaction of uranyl ion with citric acid. *Inorg.Chem.,* 4:462-469 (1964).
8. Vaňura, P., Kuča, L., Citrate complexes of uranyl in solutions with high citrate concentrations. *Collect. Czech. Chem. Comm.* 45:41-53 (1980).
9. Gustafson, R., Martell, A., Ultracentrifugation of uranyl citrate chelates.*J. Am. Chem. Soc.*, 85:2571-2574 (1963).
10. Feldman, I., Havill, J., Newman, W., The uranyl-citrate system. II. Polarographic studies of the 1:1 complex. *J. Am. Chem. Soc.*, 73:3593-3595 (1951).
11. Feldman, I., North, C.A., Hunter, H.B., Equilibrium constants for the formation of polynuclear tridentate 1:1 chelates in uranyl-malate, -citrate, and -tartarate systems. *J. Phys.Chem.*, 64:1224-1230 (1960).
12. Adin, A., Klotz, P., Newman, L., Mixed-metal complexes between In(III) and U(VI) with malic, citric and tataric acids. *Inorg.Chem.*, 9:2499-2505 (1970).
13. Markovits, G., Klotz, P., Newman, L., Formation constants for the mixed metal complexes between In(III) and U(VI) with malic, citric and tataric acids. *Inorg.Chem.*,11:2405-2408 (1972).
14. Ohyoshi, E., Oda, J., Ohyoshi, A., Complex formation between the uranyl ion and citric acid. *Bull. Chem. Soc. Jap.*, 48:227-229 (1975).
15. Borkowski, M., Lis, S., Choppin, G.R., Complexation study of NpO_2^+ and UO_2^{2+} ions with several organic ligands in aqueous solutions of high ionic strength. *Radiochimica Acta* 74:117-121 (1996).
16. Feldman, I., Newman, W., The uranyl-citrate system. I. Spectrophotometric studies in acid solution. *J. Am. Chem. Soc.*, 73:2312-2315 (1951).
17. Li, N. C., Lindenbaum, A., White, J. M., Some metal complexes of citric and tricarballylicacids. *J. Inorg. Nucl. Chem.* 12:122-128 (1959).
18. Peppard, D.F., Mason, G. W., Maier, J. L., Driscoll, W.J. Fractional extraction of the lanthanides as their di-alkyl orthophoshates. *J. Inorg. Nucl. Chem.* 4:334 (1954).
19. Erten, H.N., Mohammed, A.K., and Choppin, G.R., Variation of stability constants of thorium and uranium oxalate complexes with ionic strength. *Radiochimica Acta* 66/67:123-128 (1994).
20. Mizera, J., Bond, A.H., Choppin, G.R., Dissociation constants of carboxylic acids at high ionic strengths. Radionuclide Speciation in Real Systems, eds. Reed, D.T., Clark, S., Rao,L., Plenum (1998).
21. Choppin, G.R., Rao L.F., Complexation of pentavalent and hexavalent actinides by fluoride. Raiochimica Acta 37:143-146 (1984).
22. Grenthe,I., Fulger,J., Konings, R.J.M. et al. OECD- Chemical Thermodynamics Vol.1, Chemical Thermodynamics of Uranium. Elsevier Publishers B.V., 683, (1992).
23. Labonne-Wall, N., Choppin, G.R., Lopez, C., Monsallier, J-M., Interaction of uranyl with humic and fulvic acids at high ionic strength, Radionuclide Speciation in Real Systems, eds. Reed, D.T., Clark, S., Rao,L., Plenum (1998).
24. Allen, P., Shuh, D., Bucher, J., Edelstein, N., Reich, T., Denecke, M., Nitsche, H., EXAFS Determinations of uranium structures: the uranyl ion complexed with tartaric, citric, and malic acids. *Inorg.Chem.*, 35:784-787 (1996).

COMPLEXATION AND ION INTERACTIONS IN Am(III)/EDTA/NaCl TERNARY SYSTEM

Jian-Feng Chen[a], Gregory R. Choppin[a], Robert C. Moore[b]

[a]Department of Chemistry
Florida State University
Tallahassee, Florida 32306-3006

[b]Sandia National Laboratory
P.O. Box 5800 M.S. 1341
Albuquerque, NM 87185-1320

ABSTRACT

The complexation of Am(III) by EDTA was investigated over a range of NaCl concentration from 0.30 to 5.0 m by the solvent extraction technique using HDEHP as extractant. The conditional stability constants for the complexation reaction: $Am^{3+} + Y^{4-} = AmY^-$, were determined to be: $\log {}^c\beta_1(I) = 15.1 \pm 0.11$ (0.3 m), 13.96 ± 0.07 (1.0 m), 14.04 ± 0.09 (2.0 m), 13.76 ± 0.02 (3.0 m), 13.89 ± 0.03 (4.0 m), 14.38 ± 0.05 (5.0 m). The data were analyzed by the Pitzer model and the interaction parameters for the Am^{3+}/Cl^-, Na^+/Y^{4-}, Na^+/HY^{3-}, Na^+/H_2Y^{2-}, Na^+/H_3Y^-, Na^+/AmY^- ion pairs as well as the standard chemical potentials of the aqueous species involved were evaluated and are reported.

INTRODUCTION

As a strong complexant, EDTA has been used in separation processes and is present in radioactive wastes from activities of USDOE installations. It has been identified as one of the organic ligands present in the wastes prepared for disposal in the Waste Isolation Pilot Plant (WIPP) repository in New Mexico[1]. Information about actinide-EDTA interaction in high salt media is essential to evaluate the possible effect of EDTA on the migration of actinides in the underground environment of the repository site.

An earlier spectroscopic study of the Am complexation by EDTA[2] proposed the existence of three complexes AmY^-, $Am_2Y_3^{6-}$ and AmY_2^{5-} in different pH regions. The 1:1 mono-nuclear complex was assumed to be the main species in the low pH region (pH 2.0-3.7), while the mono-protonated complex, AmHY, was proposed to occur in more acidic solutions (pH<1)[3,4]. A number of papers have reported conditional stability constants for formation of AmY^- in dilute ionic media by a variety of other methods, such as solvent extraction[5], ion exchange[3,6,7] and electromigration[5,8]. A summary of the stability constants

Actinide Speciation in High Ionic Strength Media, edited by Reed *et al.*
Kluwer Academic / Plenum Publishers, New York, 1999

187

Table 1. Literature data on the stability constants, log $^c\beta_1$, of Am- EDTA complexation at 25 °C.

I, M	Medium	Method	log $^c\beta_1$	Ref.
0.0			19.6	5
0.0			20.8	7
0.0			20.6*	3
0.0			19.8	6
0.06	NH_4ClO_4	iex	17.64	6
0.075	NH_4ClO_4	iex	17.20	6
0.10	NH_4ClO_4	iex	17.38	6
0.10	NH_4ClO_4	sp	18.06	2
0.10	KCl	em	17.0	4
0.10	NH_4Cl	dis	16.91*	5
0.10	NH_4ClO_4	iex	18.16	7
0.10	KNO_3	em	17	8
0.14	NH_4ClO_4	iex	16.59	6
1.0	NH_4Cl	iex	18.03*	3

dis: solvent extraction; iex: ion exchange; sp: spectroscopy; migr: electromigration.
* 20 °C.

of AmY⁻ reported in the literature is given in Table 1, which shows large discrepancies between the values from the various methods and in different ionic media. Moreover, these earlier studies were conducted in potassium and ammonium salt solutions at ionic strengths lower than 1.0 M, but no similar investigation using NaCl media has been reported.

In this paper, a study of the complexation of Am(III) with EDTA by the solvent extraction technique is reported. The stability constants for formation of AmY⁻ in NaCl solutions of different ionic strengths at 25 °C were measured, and the Pitzer formalism employed to analyze the ionic interactions involved in the system.

EXPERIMENTAL

Reagents

EDTA (ethylenediaminetetraacetic acid, denoted as H_4Y hereafter) was prepared as 0.100 M standard solution of Na_2H_2Y (Fisher, ACS reagent grade). Diethylhexylphosphoric acid (HDEHP) was purified by precipitation and crystallization with Cu(II)[9]. A stock solution of 0.010 M HDEHP in heptane was prepared and served as organic phase. All other reagents were also ACS reagent grade.

NaCl stock solutions of 0.30 m - 5.0 m NaCl at pcH 2.40 (pcH = -log [H⁺]) were prepared by dissolving the required amount of NaCl in deionized water, acidifying to pcH 2.40 by addition of 1.0 M HCl with dilution to the final volume with deionized water. EDTA stock solutions of 0.500 mM of total (i.e. H_2Y, HY, Y, etc.) EDTA in 0.30 m to 5.0 m NaCl were prepared by diluting 0.100 M of standard Na_2H_2Y solution with the NaCl stock solutions, and adjusting to pcH 2.40 using concentrated hydrochloric acid. ²⁴¹Am tracer (from Oak Ridge National Lab.), whose radiochemical purity was confirmed by measurement of both alpha and gamma spectra, was prepared in 1.0 mM $HClO_4$ solution and added to the samples in 10 μL aliquot.

Procedures

Following the solvent extraction procedure described in Refs.10 and 11, a total of 5.00 mL of the NaCl stock solution plus the EDTA stock solution at the required volume ratio (5.:0, 4:1, 3:2, 2:3, 1:4, 0:5) were placed in a 20 mL liquid scintillation vial, followed by addition of 5.00 mL of the organic stock solution. The vial was shaken for about 3 hours in a water bath controlled to 25.0 °C, then centrifuged. It had been determined experimentally that 3 hours were sufficient to ensure equilibrium had been reached under our conditions. Aliquots (0.500 mL) of both organic and aqueous phases were taken for γ-activity measurement using an ISOFLEX automatic gamma counter with a window level setting of 10-60 Kev. The remainder of the aqueous phase was used for pH measurement using a Fisher Accumet 950 pH/ion meter with a Corning semi-micro combination glass electrode filled with saturated NaCl inner solution. The pH meter readings were converted to pcH using the calibration equation developed previously[12]: $pcH = spH_m + b$, where $s=1$, and the values of b were taken from Ref. 12.

RESULTS AND DISCUSSION

The distribution of Am(III) between HDEHP in heptane and 0.30 - 5.0 m NaCl aqueous solutions was measured in the presence of 0 to 0.10 mM EDTA at pcH 2.4. In this low pcH region, the solvent extraction reaction can be expressed as[18]:

$$Am^{3+}_{(aq)} + 3H_2L_{2(org)} \overset{K_{ex}}{\Longleftrightarrow} Am(HL_2)_{3(org)} + 3H^+ \tag{1}$$

As can be seen from Figure 1, the extraction of Am(III) in the organic phase decreased with increasing concentration of EDTA, indicating the presence of EDTA complexation. From the previous studies of Am+EDTA[2] and Nd+EDTA[13] (Nd is a chemical analog of Am), AmY^- can be expected to be the predominant species present for our aqueous conditions.

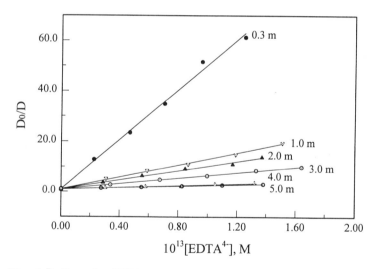

Figure 1. Plot of D_0/D vs. free EDTA concentration, the values at the end of each line are the ionic strength in molality.

For the formation reaction,

$$Am^{3+} + Y^{4-} \Longleftrightarrow AmY^- \tag{2}$$

the conditional stability constant at a particular ionic strength is defined as $^c\beta_1$. The inverse distribution ratio in the presence of EDTA is

$$\frac{1}{D} = \frac{[Am]_{t,aq}}{[Am]_{t,org}} = \frac{1}{D_0}(1 + {}^c\beta_1[Y^{4-}]) \qquad (3)$$

Where, D_0 is the distribution ratio measured in the absence of EDTA. Figure 1 shows a representative plot of D_0/D as a function of the free EDTA concentration ($[Y^{4-}]$) calculated from the conditional protonation constants[14]. The deprotonation of the positively charged species H_6Y^{2+} and H_5Y^+ were not taken into account as they are not present at the pcH used[15]. A linear concentration dependence was found over the entire EDTA concentration range, although the ratio of D_0/D was as high as 60 at 0.30 m ionic strength. This linearity agrees with the formation of only the 1:1 metal to ligand complex in our experiments. The conditional stability constants of AmY^- were calculated from the slopes of the linear plots of D_0/D vs $[Y^{4-}]$ for each ionic strength and the values are summarized in Table 2. Our values at $I = 0.30$ and 0.50 m are about two orders of magnitude lower than those reported in the literature (Table 1) for potassium and ammonium salts at low ionic strength.

Table 2. Stability constants of Am(III)-EDTA complexation in NaCl media at 25 °C.

NaCl		$\log {}^c\beta_1$	$\log {}^{3H}\beta_1$	$\log {}^{2H}\beta_1$
M	m			
0.30	0.30	15.10 ± 0.11	-2.61	-0.18
0.98	1.00	13.96 ± 0.07	-2.98	-0.71
1.92	2.00	14.04 ± 0.09	-3.18	-0.84
2.83	3.00	13.76 ± 0.02	-3.49	-1.26
3.69	4.00	13.89 ± 0.03	-3.85	-1.54
4.52	5.00	14.38 ± 0.05	-4.00	-1.55

Thermodynamic Model for Aqueous Ion Interactions

Pitzer Activity Coefficient Model. The Pitzer formalism has been utilized to analyze the aqueous thermodynamics of the Am/EDTA/NaCl system because: (1) it is applicable for concentrated electrolyte solutions; (2) there is a large database for the ion interaction parameters; (3) it is widely used in many geochemical models for ionic strength corrections. The Pitzer model expresses the excess energy of an aqueous ion solution as a revised Debye-Huckel term plus the terms originating from possible ion interactions, with each described quantitatively by a specific parameter. In this model, the effect of an ion on the excess energy of another ion is proportional to the concentration of the former ion, thus, only the terms for interactions with macro components are necessary to be retained in the calculation of the activity coefficient. This results in a significant reduction of the number of the parameters used for the systems of one background electrolyte and many trace components. In these systems, the Pitzer expression for the activity coefficients of trace components - cation (C) and anion (A) can be simplified, respectively, as:

$$\ln\gamma_C = Z_C^2 f^\gamma(I) + 2\beta_{CCl}^{(1)}\{2g(I) - Z_C^2 g'(I)m\}m + 2(\beta_{CCl}^{(0)} + \phi_{NaC})m$$

$$+(3C_{CCl}^\phi + \frac{3|Z_C|}{2}C_{NaCl}^\phi + \psi_{CNaCl})m^2 \qquad (4)$$

$$\ln\gamma_A = Z_A^2 f^\gamma(I) + 2\beta_{NaA}^{(1)}\{2g(I) - Z_A^2 g'(I)m\}m + 2(\beta_{NaA}^{(0)} + \phi_{ACl})m$$

$$+(3C^{\phi}_{NaA} + \frac{3|Z_A|}{2}C^{\phi}_{NaCl} + \psi_{ANaCl})m^2 \tag{5}$$

where

$$f^{\gamma} = A_{\phi}[\frac{\sqrt{I}}{1+b\sqrt{I}} + \frac{2}{b}\ln(1+b\sqrt{I})] \tag{6}$$

and

$$g(I) = \frac{1-(1+\alpha\sqrt{I})\exp(-\alpha\sqrt{I})}{\alpha^2 I} \tag{7}$$

with A_{ϕ}=-0.392 at 25 °C, b=1.2 Kg$^{1/2}$ mol$^{-1/2}$, and α=2.0 mol$^{-1/2}$. g'(I) is the first derivative of g(I) with respect to I. $\phi = \theta + \theta^E(I)$, and $\theta^E(I)$ is an ionic strength dependent parameter for unsymmetrical mixing. Parameters β and C^{ϕ} are ionic strength independent parameters respectively responsible for binary and ternary ion interactions involved in a single electrolyte solution. Correspondingly, θ and ψ are specific for the ion interactions occurring during mixing of electrolytes.

Calculation of the trace activity coefficient of an ion requires evaluation of the parameters specific for its salt and for its binary mixing with background electrolyte. If the first set of the parameters is available from a separate experiment using a single salt solution, the mixing parameters can be calculated from the present equilibrium constant data. However, this is not a feasible approach for numerous metal/ligand systems because the corresponding salts of many ions cannot be separated from their solutions. For a practical but approximate approach, a convention that has been used neglects the mixing parameters which were shown to have importance only for solutions of high concentration. Another convention uses a neutral species present in aqueous solution as reference state for the ionic species in equilibrium with it. The use of the latter convention is arbitrary but avoids the need to measure the activity coefficient of the neutral species, which is difficult to do in many cases. In our model both of these conventions were followed. All the necessary calculations were performed using the NONLIN program as described in our previous paper[10].

Am(III)/Na$^+$/Cl$^-$ Interaction. The ion pairs involved in the system of Am^{3+}/EDTA/NaCl are:Am^{3+}/Cl$^-$, Na$^+$/H$_3$Y$^-$, Na$^+$/H$_2$Y^{2-}, Na$^+$/HY^{3-}, Na$^+$/Y^{4-}, Na$^+$/AmY$^-$. In order to evaluate the ion interaction parameters for the complex AmY$^-$, it is necessary to include the parameters for the other species. Solvent extraction data measured at different ionic strengths have been used to evaluate the Pitzer parameters for the binary systems of Am^{3+}/ClO$_4^-$ [11] and NpO$_2^+$/Cl$^-$ [16,17]. The basic assumption employed is that the variation of the distribution coefficient with background electrolyte concentration can be ascribed solely to the change of the activity coefficients of the aqueous species involved if the measurements are performed with a uniform organic phase. The solvent extraction data for D_0 from Rosta[18] were used to evaluate the Pitzer parameters for Am^{3+}/Cl$^-$. Since the experiment was performed using Am tracer at fixed concentration of HDEHP and aqueous pcH, the activity coefficients of all the organic components involved can be considered as constants. The extraction data for varying ionic strength can be correlated by the relationship:

$$D = \Lambda \frac{\gamma_{Am^{3+}}}{\gamma^3_{H^+}} \tag{8}$$

Where, Λ is a quantity which includes dependencies on the aqueous pcH, the concentration of HDEHP and the activity coefficients of the organic components. The parameters required for calculation of the proton activity coefficient were taken from Ref. 18. The Pitzer parameters for Am^{3+}/Cl$^-$, as well as the value of Λ were calculated by minimizing the standard deviation, σ, between the calculated and experimental distribution coefficient D. The results are listed in Table 3 and gave an excellent fit up to saturation concentration. In Figure 2, the activity coefficients of Am(III) as a function of molarity of NaCl are compared with those for NaClO$_4$ media deduced in our previous work[10]. The Pitzer parameters for the two systems are in good agreement.

Table 3. The Pitzer ion interaction parameters used in present calculation.

Cation	Anion	$\beta^{(0)}$	$\beta^{(1)}$	C^ϕ	Ref.
H^+	Cl^-	0.1775	0.2945	0.00080	19
Na^+	Cl^-	0.0765	0.2664	0.00127	19
Na^+	ClO_4^-	0.0554	0.2755	-0.00118	19
Na^+	Y^{4-}	1.016	11.60	-0.001	24
Na^+	HY^{3-}	0.546	5.22	-0.048	24
Na^+	H_2Y^{2-}	-0.126	1.74	0.054	24
Na^+	H_3Y^-	-0.235	0.29	0.059	24
Am^{3+}	Cl^-	1.537	4.185	-0.286	pw
Am^{3+}	ClO_4^-	1.592	4.643	-0.242	12
Na^+	AmY^-	0.497	-0.290	-0.098	pw
	$\theta_{H^+Na^+} = 0.036$				19
	$\Psi_{H^+Na^+Cl^-} = -0.004$				19

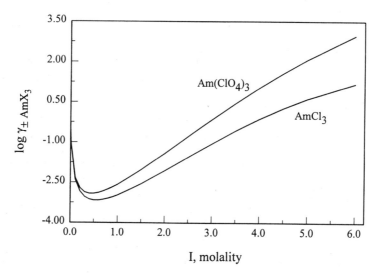

Figure 2. The trace activity coefficients of Am(III) in NaCl and NaClO$_4$ media. The data were calculated from the Pitzer modeling using the parameters listed in Table 3.

A question which remained in this approach is whether the Am^{3+} - Cl^- complexation effect is so weak that it could be neglected. The complexation of Am^{3+} by Cl^- has been studied by a variety of methods[20-21], and values of log $^c\beta_1$ ranging from -0.2 to 0.3 were reported from solvent extraction and cation exchange experiments[21]. However, a much lower value (log $^c\beta_1 \approx -2$) was reported by Barbanel and Mikhailova[20] from measurement of the Am(III) absorption spectra in the presence of 0.0 - 12.6 M or 0 - 10.75 M HCl. This contradiction has also been observed in Np(V)+Cl[16] and Cm(III)+Cl[22] systems and has been discussed in terms of the difference between these methods in the measurement of various factors[16,23]. The spectroscopic method usually measures inner sphere complexation only, while methods such as solvent extraction and ion exchange may take into account an overall effect from many factors such as the activity coefficients, inner sphere complexation, outer sphere complexation, etc. Since it is common practice to incorporate the weak interactions (such as outer sphere complexation) into the activity coefficient, the spectroscopic method should provide a relatively reliable evaluation of the extent of inner sphere metal complexation. If Barbanel's data[20] is applicable, we can estimate that no more than 5 % of Am(III) is present as inner sphere $AmCl^{2+}$ in 5.0 M NaCl solution. Thus, it seems reasonable to associate the variation of the distribution with ionic strength solely with the activity coefficient.

$H_3Y^-/H_2Y^{2-}/HY^{3-}/Y^{4-}/Na^+/Cl^-$ Interactions. For an evaluation of the ion interaction parameters related to these anionic ligands, the most suitable data are the protonation constants measured at different ionic strengths. Many literature data are available for EDTA in a number of ionic media at low ionic strength, however, they vary by at least one logarithm unit, indicating the complexity of the system. The successive protonation constants of Y^{4-} in 0.3 - 5.0 m NaCl solutions were measured recently in this laboratory by Mizera et al.[24], and the Pitzer modeling was made by fixing $^{(0)}\beta_{NaA}$ to an "averaged" value calculated for each type salt due to insufficient data points at low ionic strengths. In comparison with the literature data available, the first protonation constants present by Mizera et al are lower by about two orders of magnitude. Consequently, potentiometric titrations were performed in 0.1 m - 5.0 m NaCl solutions with a 1.0 mM EDTA concentration to prevent significant change in the ionic strength during the titrations. The protonation constants were calculated using the approach we have proposed[12] for the data analysis. In Figure 3, our data are compared with those calculated from the Pitzer paramters (listed in Table 3) and the chemical potentials (listed in Table 4) given by Mizera et al.. The agreement is satisfactory which validates the protonation constants measured.

It must be mentioned that the thermodynamic protonation constant of Y^{4-} derived from the NaCl medium data is log K_1^o =10.35 which is much lower than reported for other ionic media[15]. This raised a concern about possible complexation with Na^+. The Na^+ complexation of EDTA has been investigated previously by potentiometry and spectroscopy and its occurrence used to rationalize the observed difference in the protonation constants[25-27]. A value of 1.82 was reported by Watters et al.[25] for the stability of NaY^{3-} in 0.1 M $(CH_3)_4NCl$ solution, which is in excellent agreement with the value of 1.84 of Daniele et al[27]. To obtain reliable information about NaY^- stability in higher NaCl concentration, we performed ^1H-NMR and potentiometric titration experiments in $(CH_3)_4NCl$. The stability constants calculated from these techniques are in agreement and consistent with the literature values for low ionic strengths. Speciation calculation which includes Na^+ complexation shows that NaY^{3-} can be expected to be predominant over the free ion Y^{4-} at ionic strength higher than 0.1 m. Thus, for a reliable aqueous thermodynamic model, it seems necessary to take the Na^+ complexation into account.

Table 4. Standard chemical potentials of the ion species involved.

Ion	μ^o/RT	Ref.	Ion	$\Delta\mu^o/RT$	Ref.
H^+	0.00	26	H_3Y^{3-}	5.43	23
Am^{3+}	-241.69	26	H_2Y^{2-}	12.46	23
		26	HY^-	27.95	23
			Y^{4-}	51.79	23
			AmY^-	-232.69	pw

$\Delta\mu^o(i) = \mu^o(i) - \mu^o(H_4Y)$.

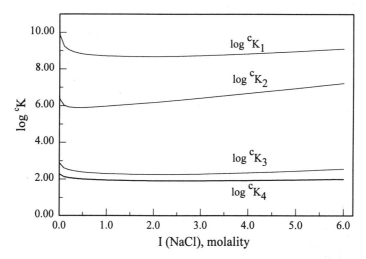

Figure 3. Plot of the conditional protonation constants, log ^cK_n (n=1,4), of EDTA as a function of ionic strength. The solid lines were calculated from the Pitzer modeling using the parameters listed in Table 3.

AmY⁻/Na⁺/Cl⁻ Interactions. The complication caused by the Na^+ complexation led us to test two alternative approaches for the establishment of a reliable aqueous thermodynamic model for the complex AmY⁻. The first approach includes the Na^+ complexation in the activity coefficient factor described by the Pitzer parameters. The conditional stability constants defined in equation 2 were used directly. The parameters, $\beta^{(0)}_{NaAmY}$, $\beta^{(1)}_{NaAmY}$ and C^ϕ_{NaAmY} were calculated using a nonlinear least square root analysis similar to that described in the previous section. Since $\left|\Delta Z^2\right|$ ($\Delta Z^2 = \Sigma \Delta Z_i^2$ (products) - $\Sigma \Delta Z_i^2$ (reactants) where Z_i is the charge of species i) for this reaction is as high as 24, a significant variation of the stability constant with ionic strength can be expected.

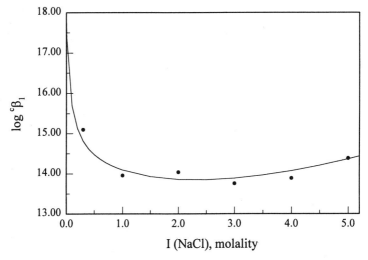

Figure 4. Plot of the conditional stability constants of AmY⁻ as a function of ionic strength. The solid lines were calculated from the Pitzer modeling using the parameters listed in Tables 3 and 4.

The experimental stability constants are given in Figure 4 which shows quite satisfactory agreement for the very dilute region. Thus, it is necessary to measure a sufficient amount of data in that region to accurately calculate both the thermodynamic stability constant of AmY^- and the parameter $\beta^{(1)}_{NaAmY}$. To have a reliable estimate of these two quantities from the present data, a constraint was set for $\beta^{(1)}_{NaAmY}$ ($\beta^{(1)}_{NaAmY}$ =0.29) to assure that the least square fit result would be consistent with those reported in the literature [19] for the 1:1 type of salts. Parameterization of the Pitzer model resulted in $\beta^{(0)}$=-0.235, $\beta^{(1)}$=0.29, $C^\emptyset$=0.059. The thermodynamic stability constant was calculated to be log β^0_1 = 18.59. This value is lower by a log unit than that in the literature (Table 1).

An alternative approach is to avoid the problem of the Na^+ - Y^{4-} complexation. At pcH 2.4, calculation shows that the species H_2Y^{2-} and H_3Y^- are predominant, thus it is reasonable to express the complexation reaction as

$$Am^{3+} + H_2Y \overset{^{2H}\beta_1}{\Leftrightarrow} AmY^- + 2H^+ \qquad (9)$$

or

$$Am^{3+} + H_3Y \overset{^{3H}\beta_1}{\Leftrightarrow} AmY^- + 3H^+ \qquad (10)$$

The corresponding equilibrium constants log $^{2H}\beta_1$ and log $^{3H}\beta_1$ calculated from the conditional stability constants and the protonation constants, are listed in Table 2. An interesting result is that log $^{3H}\beta_1$ is approximately a linear function of ionic strength about 0.3 m and can be calculated by the simple relation:

$$\log {}^{3H}\beta_1 = -(2.60 \pm 0.08) - (0.294 \pm 0.020)m \qquad (\gamma = 0.983) \qquad (11)$$

Pitzer modeling was made using both sets of constants. Initially, three parameters

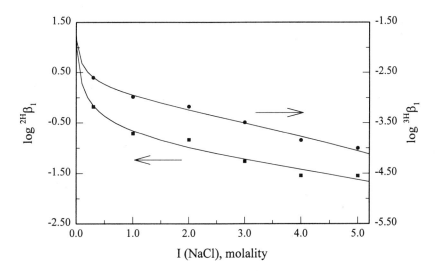

Figure 5. Plots of the equilibrium constants of $Am^{3+} + H_nY^{(4-n)-} = AmY^- + nH^+$ (n=2, 3) as a function of ionic strength. The solid lines were calculated from the Pitzer modeling using the parameters listed in Table 3.

$\beta^{(0)}{}_{NaAmY}$, $\beta^{(1)}{}_{NaAmY}$ and $C^{\phi}{}_{NaAmY}$ were adopted as variables. Discrepancies was observed in the parameters obtained from the two sets of the constants and from the constants defined in eq. 2 . However we found that the activity coefficients of AmY^- calculated from these parameters are identical within calculation errors. Data calculated in the previous approach was used in our final model, and only the thermodynamic equilibrium constant was used as a variable. Excellent fits to the experiment data were obtained (see Figure 5), and the thermodynamic equilibrium constants were calculated to be: $\log {}^{3H}\beta_1^0 = -1.55$ and $\log {}^{2H}\beta_1^0 = 1.49$ respectively. These values, combined with the thermodynamic protonation constants of H_2Y^{3-}, HY^{3-} evaluated in the previous section and the literature recommended value for Y^{4-} [15], were used to calculate the thermodynamic stability constant, $\log \beta_1^0$. We obtained an average value of 19.27, compared to the literature value of 19.8 [6].

We note that the parameter sets in this work can model satisfactorily the effects of EDTA on Am speciation in concentrated chloride solutions. However, the Na^+-$EDTA^{4-}$ parameters may not be applicable to other Na^+ solutions.

CONCLUSION REMARK

The complexation of Am^{3+} by EDTA was investigated over a wide range of NaCl concentration. The conditional stability constants of the complex, AmY^-, were measured, and an aqueous thermodynamic model with ion interaction parameters was proposed to describe accurately the equilibrium data for Am/EDTA/NaCl system up to 5 m. Inclusion of data on the Na^+ - Y^{4-} complexation is necessary for a reliable model in the pH region in which the species Y^{4-} is dominant. More data is needed in the very dilute concentration region if more reliable evaluation of the standard chemical potentials are desired. In this study, the interest was in the conditional constants at high ionic strengths.

ACKNOWLEDGMENTS

This work was performed as part of the Waste Isolation Pilot Plant (WIPP) Actinide Source Term Program, supported at Sandia National Laboratories by the United State Department of Energy under Contract DE-AC04-94AL85000, and at Florida State University under contract AH-5590.

REFERENCES

1. L.H. Brush, in: *Test Plan for Laboratory and Modeling Studies of Repository and Radionuclide ......*
   Chemistry for the Waste Isolation Pilot Plant, Report # SAND90-0266, Sandia National Laboratories, Albuquerque, NM (1990).
2. A. Delle Site and R.D. Baybarz, A spectrophotometric study of the complexing of Am^{3+} with aminopolyacetic Acid, *J. Inorg. Nucl. Chem.* **31**:2201-2233 (1969).
3. A.I. Moskvin, G.V. Khalturin and A.D. Gel'man, Investigation of the complexing of trivalent americium in oxalate and ethylenediaminetetraacetate solutions by the ion exchange method, *Sov. Radiochem.* **1**:67-76 (1960).
4. I.A. Lebedev, A.M. Maksimova, A.V. Stepanov and A.B. Shalinets, Determination of the stability constants of complexes of Am and Cm with EDTA by the method of electromigration, *Radiokhimiya* **9(6)**:707 (1967).
5. J. Stary, Study of complex formation of americium and promethium by the extraction method, *Sov. Radiochem.* **8**:467-470 (1966).
6. A.A. Elesin and A.A. Zaitsev, Ion exchange behavior of trivalent americium, curium and promethium ions in the presence of EDTA, *Sov. Radiochem.* **13**:798 (1971).
7. J. Fuger, Ion exchange behaviour and dissociation constants of americium, curium, and californium complexes with ethylenediaminetetraacetic acid, *J. Inorg. Nucl. Chem.* **5(4)**:332 (1958).

196

8. A.B. Shalinets, Investigation of the complex formation of trivalent actinide and lanthanide elements by the method of electromigration XVI, ethylenediaminetetraacetic acid, *Sov. Radiochem.* **14**:285 (1972).

9. W.J. McDowell, P.T. Perdue and G.N. Case, Purification of di(2-ethylhexyl)phosphoric acid, *J. Inorg. Nucl. Chem.* **38(11)**:2127 (1976).

10. M.S. Caceci and G.R. Choppin, Determination of the first hydrolysis constants of Eu(III) and Am(III), *Radiochim. Acta* **33**:101-104 (1983).

11. G.R. Choppin and J.F. Chen, Complexation of Am(III) by oxalate in NaClO₄ media, *Radiochimica Acta* **74**:105-110 (1996).

12. J.F. Chen, Y.X. Xia and G.R. Choppin, Differential analysis of potentiometric titration data to obtain protonation constants, *Anal. Chem.* 68(22),3973-3978(1996).

13. E. Brucher, R. Kiraly and I. Nagypal, Equilibrium relations of rare earth ethylenediaminetetraacetate complexes in the presence of a ligand excess, *J. Inorg. Nucl. Chem.* **37**:1009-1012 (1975).

14. J.F. Chen, Y.X. Xia and G.R. Choppin, Ionic strength dependence of the deprotonation constants of 1,10-phenanthroline, 8-hydroxyquinoline and some carboxylic acids, Manuscript in preparation.

15. G. Anderegg, *Critical Survey of Stability Constants of EDTA Complexes*, Pergamon Press, Oxford (1977).

16. V. Neck, Th. Fanghanel, G. Rudolph and J.I. Kim, Thermodynamics of neptunium(V) in concentrated salt solutions: chloride complexation and ion interaction (Pitzer) parameters for the NpO₂⁺ ion, *Radiochim. Acta* **69**:39 (1995).

17. C.F. Novak and K.E. Roberts, Thermodynamic modeling of neptunium(V) solubility in concentrated Na-CO₃-HCO₃-Cl-ClO₄-H-OH-H₂O systems, Sandia National Laboratory Report, SAND 94-0802C.

18. L. Rosta, *Ionic Hydration and Interactions*, Ph.D. Dissertation, Florida State University (1987).

19. K.S. Pitzer, *Activity Coefficients in Electrolyte Solutions*, CRC Press, Boca Raton, FL (1991).

20. Yu.A. Barbanel and N.K. Mikhailova, Study of the complex formation of Am(III) with the Cl⁻ ion in aqueous solutions by the method of spectrophotometry, *Sov. Radiochem.* **11**:576 (1969).

21. J. Fuger, I.L. Khodakovsky, E.I. Sergeyeva, V.A. Medvedev and J.D. Navratil, *The Chemical Thermodynamics of Actinide Elements and Compounds: Part 12. The Actinide Aqueous Inorganic Complexes*, Vienna:International Atomic Energy Agency (1992).

22. Th. Fanghanel, V. Neck and J.I. Kim, Thermodynamics of neptunium(V) in concentrated salt solutions: II. ion interaction (Pitzer) parameters for Np(V) hydrolysis species and carbonate complexes, *Radiochim. Acta* **69**:169-176 (1995).

23. G.R. Choppin, Inner vs outer sphere complexation of f-elements, *J. Alloys Compd.* **249**:9-13(1997).

24. J. Mizera, A.H. Bond, G.R. Choppin and R.C. Moore, Dissociation constants of carboxylic acids at high ionic strengths. Radionuclide Speciation in Real System, eds. D.T. Reed and L.F. Rao, Plenum (1997).

25. Wa J. I. Watters and O.E. III Schupp, Acidimetric investigation of complex formation by potassium ion with ethylenediaminetetraacetate, *J. Inorg. Nucl. Chem.* **30**:3359 (1968).

26. J.D. Carr and D.G. Swartzfager, Studies of alkali metal ion complexes of 2,3 -diaminobutane-N,N, N', N'-teraacetic Acid, *J. Amer. Chem. Soc.* **95(11)**:3569 (1973).

27. P.G. Daniele, C. Rigano and S. Sammartano, Ionic strength dependence of formationconstants: alkali metal complexes of ethylenediaminetetraacetate, nitrilotriacetate, diphosphate and tripoly-phosphate in aqueous solution, *Anal. Chem.* **57**:2958-2960 (1985).

INTERACTION OF URANYL WITH HUMIC AND FULVIC ACIDS AT HIGH IONIC STRENGTH

N. Labonne-Wall, G. R. Choppin, C. Lopez and J-M. Monsallier

Department of Chemistry
Florida State University
Tallahassee, Fl 32306

ABSTRACT

The binding of UO_2^{2+} to two different humic acids (one predominantly aliphatic and the other predominantly aromatic) and a fulvic acid has been studied at high ionic strengths (in NaCl medium) by solvent extraction. The results indicate that the aliphatic vs. aromatic nature of humic acids does not influence significantly the binding of UO_2^{2+}. Increases in both the ionic strength and pH cause an increase in the binding constants of the complexes. The complexes with the fulvic acid are weaker than those with the humic acids.

INTRODUCTION

A repository for defense-generated transuranic wastes, the Waste Isolation Pilot Plant (WIPP), under development and evaluation in New Mexico, is located within a natural bedded salt formation approximately 655 meters below the surface. The brines present in the site could be a medium of transport for radionuclides, over a long period of time, from the repository to the environment. Humic substances are ubiquitous organic polyelectrolytes even in brine solutions so they could have a role in actinide transport in WIPP brines. In this study complexation constants for humic substances (humic acids and fulvic acids) with uranyl cation in NaCl medium were studied by solvent extraction, to obtain data for modeling actinide behavior in the WIPP. One fulvic acid and two humic acids were chosen for study. The experiments were conducted at two different pH values (4.8 and 6.2) and two ionic strengths (3 m and 6 m).

EXPERIMENTAL

Reagent

The Lake Bradford Humic Acid (LBHA) was extracted from Lake Bradford (Tallahassee, Florida), at 20 m from the shore and about 3 m depth. It was purified as previously described (Bertha and Choppin 1978, Nash and Choppin 1980). The purified

Actinide Speciation in High Ionic Strength Media, edited by Reed *et al.*
Kluwer Academic / Plenum Publishers, New York, 1999

199

sample was ashless (less than 0.05 %). The Gorleben Humic Acid (GHA) was extracted from Gorleben groundwater (Germany) by Kim and co-workers (1990) (the sample was labeled Gohy-573). Small levels of impurities are present in this GHA sample (less than 1%). The Suwannee River Fulvic Acid (SRFA) was obtained from the IHSS (International Humic Substances Society); the amount of ash was reported equal to 0.68 % (US Geological Survey, 1994).

For use in the titration experiments, carbonate free solutions of NaOH (Fisher Chemical reagent grade) were prepared weekly and standardized with potassium hydrogen phthalate (Fisher reagent grade) and phenolphthalein (Baker reagent grade). Hydrochloric acid solutions (Fisher Chemical concentrated reagent) after preparation were standardized with phenolphthalein and the standardized NaOH solutions.

For the solvent extraction experiments, HEDHP (di(2-ethylhexyl) phosphate) (Sigma) was purified according to the method of Peppard and co-workers (1957) and a methanol solution titrated with a standardized solution of NaOH dissolved in methanol. Acetic acid and MES (4-morpholineethane-sulfonic acid) were used as the pHr buffer (pHr is the value of the pH-meter reading, while pcH is the negative logarithm of the hydrogen ion molarity).

The ^{233}U solution was obtained also from Oak Ridge National Laboratory. In order to purify the sample from fission products, a volume of 3 ml of the uranium solution in 8 M HCl medium was passed through a 1cm diameter $\times$ 50 cm length column loaded with Dowex anionic exchange resin (1×4). The uranyl was eluted with 5 ml of 0.1 M HCl. After evaporation to dryness of the collected uranium solution, 10^{-3} M HCl was added to dissolved the sample to obtain a solution with a final uranium concentration of 2.62×10^{-4} M (determined by liquid scintillation counting). The radiochemical purity of the ^{233}U stock solutions was confirmed by alpha spectrometry, using a silicon surface-barrier detector (ORTEC Instrument), and by gamma spectrometry, using a Ge(Li) detector connected to a Series II Personal Computer Analyzer "The Nucleus Inc.".

Procedure

A carbon-13 nuclear magnetic resonance spectrum of each of the three sample (LBHA, GHA and SRFA) was obtained by the magic angle cross-polarization technique using a Bruker/IBM WP-200SY spectrometer equipped with a solid state probe. The cryogenic magnet had a static field of 4.7 Tesla and its carbon frequency was 50.267 MHz.

Potentiometric titration were performed with an automatic titrator system consisting of a model 950 Accumet Fisher digital pH-meter, a model 665 Metrohm digital auto-burette, a semi-micro combination electrode (Corning), filled with saturated NaCl and a personal computer. Titrations were performed at 25°C in a jacketed vessel under a blanket of N_2. Before each titration, the electrode was calibrated at the same ionic strength as the solution to be titrated. The calibration consisted of a titration at the fixed ionic strength (NaCl), using 0.1 M NaOH, with 0.1 M HCl solutions. A linear equation representing the potential vs. the calculated pcH provided the calibration relationship in pcH for the electrode. The calibrated electrode was used also for the aqueous phase in the solvent extraction experiments.

The proton capacity of the LBHA and SRFA were determined at 0.1 m, 3 m and 6 m NaCl ionic strengths by potentiometric titration. About 0.05 g of humic substance (LBHA or SRFA) was dissolved in 16 ml of I m NaCl with enough of a solution of 0.1 M NaOH added to produce a potential of ca. -280 mV (ca. pcH 11.5). This sample was titrated with a solution of 0.05 M HCl (I m in NaCl, where I is the ionic strength). The automatic titrator made 100 additions of 0.1 ml each of the HCl solution. After each addition, the computer recorded the potential every 120 seconds; when the difference between successive readings became less than 1.2 mV, another addition of 0.1 ml of titrant was made. For each ionic strength, a least two titrations were performed. The titration of the Gorleben humic acid was done at 0.1 M $NaClO_4$ by Kim and co-workers (1990).

For the solvent extraction experiments, the organic phase consisted of 10^{-5} M of HDEHP diluted in toluene which had been pre-equilibrated with 3 m or 6 m NaCl. The aqueous solutions were maintained at constant ionic strengths of 3 m or 6 m with NaCl and at constant pHr with acetate buffer or MES buffer. To obtain pHr = 4.8, 0.01M total acetic acid was used; for pHr = 6.1, 0.01M total MES was used. The aqueous phase contained between 2 mg/L and 10 mg/L of humic substances and 5.24×10^{-7} M of ^{233}U. The preparation of the solutions was accomplished in a dust free environment and in glass vials previously silanized (Caceci and Choppin, 1983) to avoid sorption of the tracers and of the humic material on the walls of the vials. The sealed vials containing 5 ml of each phase were shaken for 24 hours. After centrifugation, the aqueous and organic phases were separated. The aqueous phase was filtered through glass wool for the experiments with humic acids because a thin film of coagulated organic matter formed between the two phases. Duplicate aliquots (1.00 ml) of each phase were taken for counting by liquid scintillation. The pHr was measured in the aqueous phase. Because of the loss of humic acid (LBHA or GHA) in the film between the two phases, the final humic acid concentration was determined in the aqueous phases by spectrophotometry at 238 nm with a CARY 14 using the "OLIS Spectroscopy Operating System". The spectrophotometer was calibrated previously by measuring the absorbance of four freshly prepared solutions of known concentrations of humic substances at the desired ionic strength.

RESULTS AND DISCUSSION

^{13}C - NMR

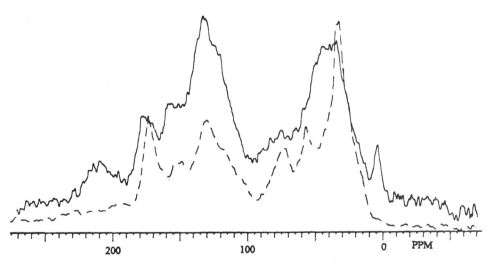

Figure 1. Quantitative carbon-13 nuclear-magnetic-resonance spectrum of LBHA (− −) and GHA (——). Cross polarization magic angle spinning, number of data points collected: 2000, spectral width: ± 25000 Hz, pulse angle: 90°, pulse delay: 3 seconds, number of scans: 1000 (LBHA) or 2300 (GHA), contact time: 1 msecond.

The ^{13}C nuclear-magnetic-resonance spectrum of LBHA and GHA and of SRFA are presented, respectively, in Fig. 1 and Fig. 2. The amounts of the different functional group in each sample, calculated by the peaks surface are listed in Table 1. Two different types of aliphatic carbon are analyzed : aliphatic I represents primarily the carbon bonded to oxygen, such as carbohydrate, alcohol and ether carbons; aliphatic II represents the carbons bonded to other carbons. GHA shows a stronger aromatic character than LBHA, the ratio aromatic/aliphatic is equal to 0.62 in the case of LBHA and is 1.00 for GHA. In the case of SRFA, this ratio is equal to 0.44. The main difference between the fulvic and the humic samples is the amount of carboxylic carbon: 15.90 % in the case of SRFA and ca. 10 % in the case of the humic samples.

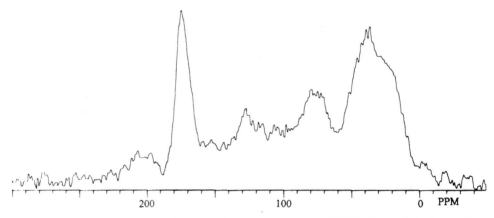

Figure 2. Quantitative carbon-13 nuclear-magnetic-resonance spectrum of SRFA, Cross polarization magic angle spinning, number of data points collected: 2000, spectral width: ± 50325 Hz, pulse angle: 90°, pulse delay: 3 seconds, number of scans: 2000, contact time: 1 msecond.

Table 1. ^{13}C-NMR of LBHA, GHA and SRFA: percent of total carbon of the functional groups.

	Ketone (220 - 180 ppm)	carboxyl (180 - 160 ppm)	aromatic (160 - 90 ppm)	aliphatic II (90 - 60 ppm)	aliphatic I (60 - 0 ppm)
LBHA	5	10	32	13	39
GHA	7	9	42	10	32
SRFA	5	16	24	15	40

Potentiometric Titration

The figures 3a, 3b and 3c show the superposition of titration curves of LBHA, each at different ionic strengths (respectively 0.1 m, 3 m and 6 m (NaCl)) and their first derivative. Perdue (1978) studied a sample of soil humic acid by thermometric titration, obtaining a capacity of 4.1 meq/g of carboxylic group (pKa unknown) and 0.7 meq/g of phenolic hydroxyl group (pKa 10.5). By potentiometric titration, those capacity values were respectively 4.4 meq/g and 2.2 meq/g. Apparently, a fraction of the phenolic hydroxyl group was too weak to be measured by calorimetry. Borggaard (1974) was able to differentiate two carboxylic groups: 1.5 to 3.0 meq/g of a highly acidic carboxylic group (pKa 2.8 - 3.4) and 2.7 to 4.2 meq/g of a moderately acidic carboxylic group (pKa 4.9 - 5.1); his results gave 1.1 to 1.7 meq/g of a weakly acidic phenolic group (pKa 9.4 - 9.7). The capacity (determined as the number of meq OH⁻/g LBHA at the first derivative maxima) and the pKa (determined as the pcH at the half equivalence) values of LBHA at 0.1 m, 3 m and 6 m

(NaCl), presented in Table 2 show good agreement with the values determined by Boggaard, although the phenolic group cannot be observed at low ionic strength. Previous capacity and pKa determinations of LBHA were done at low ionic strength (Choppin and Kullberg, 1978, Bertha and Choppin, 1978). In some of the experiments, two types of functional groups could be observed, carboxylic and phenolic, but only a total capacity of 4.65 meq/g was determined. Torres and Choppin (1984) observed two carboxylic groups (1.47 meq/g and 2.39 meq/g) at the ionic strength 0.1 M NaClO$_4$, but no phenolic group. The analysis of GHA (Kim et al. 1990) reported 4.75 meq/g of total carboxylic sites and 1.86 meq/g of phenol groups.

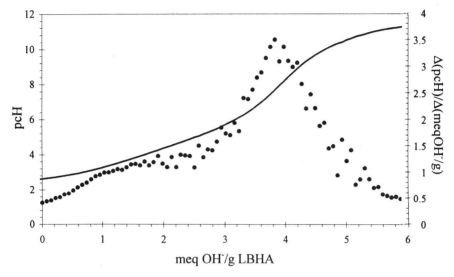

Figure 3a. LBHA titration (——) and first derivative curve (•), at $I = 0.1$ m (NaCl).

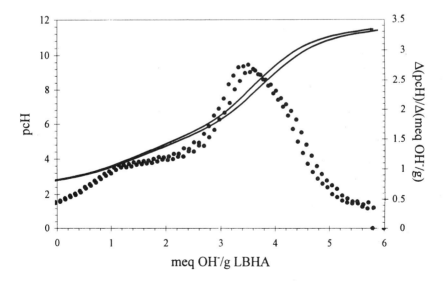

Figure 3b. Duplicate titration (——) and first derivative curve (•), at $I = 3.0$ m (NaCl) of LBHA.

Previous potentiometric titrations of SRFA indicated 4.24 meq/g (± 0.02) of carboxylic groups and 1.41 meq/g (± 0.02) of phenolic groups (Ephraim and co-workers, 1986) and for another sample, 6.1 meq/g of carboxylic groups and 1.2 meq/g of phenolic groups (US Geological Survey, 1994). Figures 4a and 4b present the titration data we performed in this study on SRFA at 3 m and 6 m; the capacity and pKa values are presented in Table 2. These results are in agreement with the literature, although as for LBHA, we can define strong and weak carboxylic groups, which is not reported for SRFA. The total carboxylic capacity is much higher in the case of the fulvic sample than for the humic sample. A similar difference has been reported for LBHA (Bertha and Choppin, 1978).

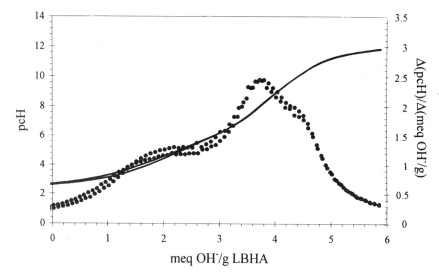

Figure 3c. Duplicate titration (——) and first derivative curve (•), at I = 6.0 m (NaCl) of LBHA.

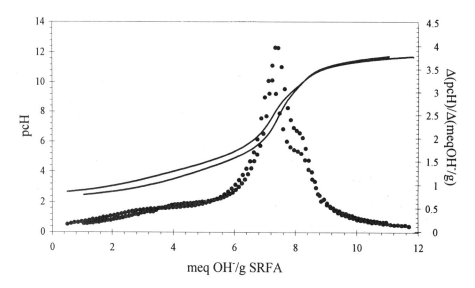

Figure 4a. Duplicate titration (——) and first derivative curve (•), at I = 3 m (NaCl) of SRFA.

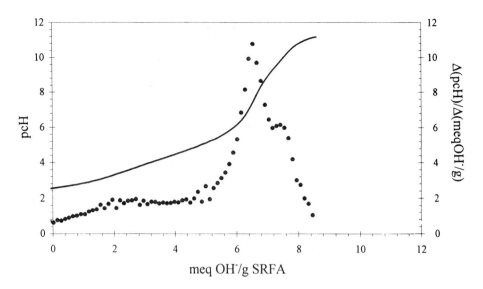

Figure 4b. SRFA titration (——) and first derivative curve (•), at $I = 6$ m (NaCl).

For a better understanding of the complexation behavior of the humic substances, the variation of the degree of ionization of proton exchanging groups as a function of pcH was calculated from the titrations of LBHA and SRFA at the ionic strengths 3 m and 6 m (NaCl). The results are presented on Figure 5.

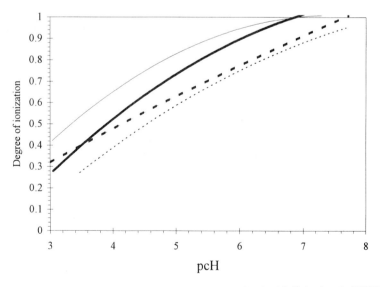

Figure 5. Degree of ionization of LBHA at $I = 3$m, NaCl (- -), at $I = 6$m, NaCl (- -) and of SRFA at $I = 3$m, NaCl (——),at $I = 6$m, NaCl (—) as a fuction of pcH, by direct titration.

Table 2. Proton exchange capacity of the humic substances (meq/g) and pKa associated of LBHA and SRFA at different ionic strengths (NaCl).

I		LBHA			SRFA		
		carboxylic	carboxylic	phenolic	carboxylic	carboxylic	phenolic
0.1 m	meq/g	1.0 ± 0.2	3.0 ± 0.2	-	-	-	-
	pKa	2.8 ± 0.2	5.0 ± 0.1	-	-	-	-
3 m	meq/g	1.12 ± 0.07	2.4 ± 0.1	0.8 ± 0.1	3.5 ± 0.7	3.5 ± 0.3	0.8 ± 0.1
	pKa	3.16 ± 0.06	5.25 ± 0.9	9.0 ± 0.2	2.8 ± 0.5	4.6 ± 0.2	9.0 ± 0.3
6 m	meq/g	1.6 ± 0.2	2.1 ± 0.2	0.8 ± 0.1	2.0 ± 0.1	4.50 ± 0.08	1.1 ± 0.1
	pKa	3.0 ± 0.7	5.5 ± 0.2	9.1 ± 0.3	2.8 ± 0.1	4.6 ± 0.1	9.0 ± 0.2

Solvent Extraction

A modification of the Schubert method (1954) was used to analyze the data from solvent extraction. Assuming reaction with the carboxylate groups, the general equation of the complexation of humic substances with an actinide cation and the equilibrium constant are:

$$An + m\, RCO_2^- \leftrightarrow An\left(RCO_2^-\right)_m \qquad (1)$$

$$\beta_m = \frac{\left[An\left(RCO_2^-\right)_m\right]}{[An] \times \left[RCO_2^-\right]^m} \qquad (2)$$

where $[RCO_2^-]$ is the concentration of deprotonated carboxylate sites, expressed in eq/l.

The distribution coefficient is defined by:

$$D = \left.[An]_{(o)} \middle/ [An]_{(a)}\right. \qquad (3)$$

where $[An]_{(o)}$ is the actinide concentration in the organic phase and $[An]_{(a)}$ is the actinide concentration in the aqueous phase. Defining D_0 as the distribution coefficient in the absence of humic substance; D becomes:

$$D = \left. D_0 \middle/ \left(1 + \left(\sum_{i \geq 1} \beta_i \times \left[RCO_2^-\right]^i \middle/ \alpha_{An}\right)\right)\right. \qquad (4)$$

The value α_{An} takes in account hydrolysis and complexation by carbonate and the buffer ion (B) of the actinide ion:

$$\alpha_{An} = 1 + \sum_i {}^{OH}\beta_i \times \left[OH^-\right]^i + \sum_j {}^{CO_3}\beta_j \times \left[CO_3^{2-}\right]^j + \sum_k {}^{B}\beta_k \times \left[B^-\right]^k \qquad (5)$$

Table 3. Values necessary to calculate $\alpha_{UO_2^{2+}}$ at different ionic strength (NaCl).

ionic strength (NaCl)	3m		6m	
pK_w [1]	13.93 ± 0.02		14.49 ± 0.02	
Acetic acid pKa [1]	4.77 ± 0.003		5.31 ± 0.003	
UO_2^{2+}/Acetic acid : $\log(\beta_1)$ [2]	2.82		3.67	
$HCO_3^-/(CO_2)_a$ pKa_1 [3]	6.02		6.264	
CO_3^{2-}/HCO_3^- pKa_2 [3]	9.46		9.71	
$\log \beta \, (UO_2(OH))^+$ [4]	-5.24		-4.84	
$\log \beta \, (UO_2(OH)_2)^0$ [4]	-12.60		-12.86	
$\log \beta \, (UO_2(CO_3))^0$ [4]	8.15		8.46	
$\log \beta \, (UO_2(CO_3)_2)^{2-}$ [4]	15.19		15.17	
$\log \beta \, (UO_2(CO_3)_3)^{4-}$ [4]	21.54		21.45	
	pHr 4.8	pHr 6.1	pHr 4.8	pHr 6.1
$\alpha_{UO_2^{2+}}$	7.60	26.76	53.67	360.00

with $K = ([H] \times [X]) / [HX]$, $\beta \, (UO_2(OH)_m)^{(2-m)} = [UO_2(OH)_m] \times [H^+]^m / [UO_2]$ and
$\beta \, (UO_2(CO_3)_n)^{(2-2n)} = [UO_2(CO_3)_n] / ([UO_2] \times [CO_3]^n)$.
[1] Calculated from Chen et al. (1996) (a)
[2] Chen et al. (1996) (b)
[3] Thurmond and Millero (1982).
[4] calculated by S.I.T. from OECD (1992)

For the tracer concentration of actinides, the polynuclear species $(An)_{i,\ (i>1)}(OH)_j$ are negligible, and only the mononuclear species $An(OH)_j$ need to be considered. All the values necessary to complete the calculation of α_{An}, at the ionic strengths 3 m and 6 m (NaCl) are presented in Table 3 as well as the values of $\alpha_{UO_2^{2+}}$ at the different pHr and ionic strengths. The concentrations of HCO_3^- and CO_3^{2-} have been calculated assuming equilibrium between the solution and the CO_2 in the air. The value of $[CO_2]_{(a)}$ at 3 m and 6 m (NaCl) is, respectively, 1.3×10^{-7} M and 8.3×10^{-8} M (Yasunishi and Yoshida, 1979). Since very little data have been published on the hydrolysis and the carbonate of U(VI) in NaCl medium at high ionic strength, the constants have been estimated by the Specific Ion interaction Theory (SIT), using the parameters published in OECD - *Chemical thermodynamics* (1992). These values for $UO_2(OH)_n$ and $UO_2(CO_3)$ have a significant uncertainty because of the limitation of the calculation. Consequently, the final values have a larger error than listed and are probably underestimated. At the pcH values of these experiments, the carbonate complexation is insignificant while hydrolysis is dominant as reflected on $\alpha_{UO_2^{2+}}$ in Table 3, so the latter would be the major source of error in our SIT analysis.

The distribution coefficient can be also expressed as:

$$\frac{1}{D} = \sum_{i \geq 1} \phi_i \times \left[RCO_2^- \right] + \phi_0 \tag{6}$$

with

$$\phi_i = \frac{\beta_i}{\alpha_{An}} \times D_0 \tag{7}$$

and

$$\phi_0 = \frac{1}{D_0} \tag{8}$$

In order to obtain a preciser D_0 value, D_0 is calculated from the polynomial regression analysis of $1/D$ vs. $[RCO_2^-]$, instead of determined from one experimental data. The value of $[RCO_2^-]$ is calculated as the product of the final concentration of humic substances in aqueous phase, the capacity and the degree of ionization of the material at the pcH of the experiment. To fit our experimental data, two constants β_1 and β_2 are required, which in complexation for simple ligands implies the presence of the complexes 1:1 and 1:2, metal:ligand donor ratios. However, such an interpretation is unproved thus far in humic acid complexation. The values of β_1 and β_2 were determined from the slope of the linear part of a plot $(D_0/D-1)$ vs. $[RCO_2^-]$ and $(D_0/D-1)/[RCO_2^-]$ vs. $[RCO_2^-]$. The slopes are, respectively, equal to β_1/α_{An} and β_2/α_{An}. Table 4, contains an illustrative set of data for the determination of the UO_2^{2+}-LBHA binding constants at pHr 4.79 ($\alpha = 0.64$), at the ionic strength 3 m (NaCl).

Table 4. Extraction data UO_2^{2+} - LBHA, pHr 4.79 ($\alpha = 0.64$). $I = 3$ m (NaCl); $[HDEHP] = 10^{-5}$ M.

Total humic concentration: [LBHA] (10^{-5} eq/L)			Activity (cpm)	
initial	final	organic	aqueous	
0.000	0.000	2819	150	
0.746	0.160	1320	113	
1.492	0.212	1208	123	
2.238	0.586	1399	145	
2.984	0.936	613	227	
3.730	1.111	621	207	
4.476	1.704	515	298	
5.222	1.842	449	269	
5.968	2.381	411	314	
6.714	3.282	404	593	
7.460	4.389	295	639	

$1/D$ vs. $[RCO_2^-]$

Polynomial regression: $y = 0.174 \times x^2 + 0.268 \times x + 0.054$
correlation: 0.9925
$D_0 = 18.59$

$(D_0/D-1)$ vs. $[RCO_2^-]$

linear regression: $y = 9.27 \times 10^5 \times x - 0.804$
correlation: 0.967
$\log \beta_1 = 6.83$

$(D_0/D-1)/[RCO_2^-]$ vs. $[RCO_2^-]$

linear regression: $y = 3.29 \times 10^{10} \times x + 4.83 \times 10^5$
correlation: 0.932
$\log \beta_2 = 11.38$

The results for the binding constants of UO_2^{2+} with LBHA, GHA and SRFA are presented in Table 5. The results show an increase of the binding constants β_1 and β_2 with pHr and with ionic strength, as reported in the literature for the binding of various actinides with humics (Torres and Choppin 1984, Meunier-Lamy et al. 1986). The increase of binding constant with pHr is explained by the increased ionization of the humic substances, leading to increased [RCO_2^-]. For a specific ionic strength and pH, the binding constants for the humic complexation with UO_2^{2+} are essentially the same for LBHA and GHA. Minai and Choppin (1989) and Kim et al. (1991) have reported that the aromaticity and other compositional differences do not play a major role in binding of actinides by humic acids. The binding constants for the fulvic acids are weaker than those for humic acids, which agrees with earlier reports (Meunier-Lamy et al., 1986, Nash and Choppin, 1980). Unlike humic acid whose analysis requires two binding constants, only one binding constant was obtained for the fulvic acid.

Table 5. Complexation constants of humic substances with UO_2^{2+}, at different pHr and ionic strength.

$I = 3$ m (NaCl)				$I = 6$ m (NaCl)			
pHr	α	$\log \beta_1$	$\log \beta_2$	pHr	α	$\log \beta_1$	$\log \beta_2$
LBHA							
4.9 ± 0.1	0.65 ± 0.01	6.8 ± 0.1	11.3 ± 0.1	4.7 ± 0.1	0.77 ± 0.01	7.5 ± 0.2	12.3 ± 0.4
6.1 ± 0.1	0.83 ± 0.03	7.4 ± 0.2	11.0 ± 0.1	6.5 ± 0.1	0.97 ± 0.00	8.2 ± 0.5	$-$ [1]
GHA							
4.8 ± 0.3	0.65 ± 0.00	6.8 ± 0.1	11.2 ± 0.2	4.8 ± 0.1	0.79 ± 0.01	7.7 ± 0.2	12.4 ± 0.1
6.1 ± 0.5	0.84 ± 0.02	7.7 ± 0.5	$-$ [1]	6.6 ± 0.1	0.98 ± 0.00	8.5 ± 0.1	12.6 ± 0.1
SRFA							
4.8 ± 0.1	0.90 ± 0.02	5.1 ± 0.1	$-$	4.7 ± 0.1	0.91 ± 0.01	5.8 ± 0.1	$-$
6.2 ± 0.2	0.98 ± 0.01	6.6 ± 0.2	$-$	6.5 ± 0.1	0.99 ± 0.01	7.8 ± 0.1	$-$

[1] No binding constant could be determined.

In order to see the influence of the ionic strength on the binding of UO_2^{2+}, at constant pcH (constant degree of ionization of the humic material), additional experiments were performed at various ionic strengths with LBHA. The results in Table 6 show $\log \beta_1$ and $\log \beta_2$ decreasing with I, for the low ionic strengths ($I < 1.0$ m) and increasing with I, for the higher ionic strengths ($I > 1.0$ m). Chen and co-workers (1996, b) have mentionned the same variation of the stability constants with the ionic strength, in the case of other ligands (acetate and lactate). Our observations concerning the variation of the binding constant with pHr and the ionic strength explain the difference between our data and the ones previously published. Shanbag and Choppin (1981) and Meunier-Lamy et al. (1986) studied the binding of humic acids with UO_2^{2+}, at low ionic strength (0.1M) and pHr 4.0 ; both works report two binding constants lower by a factor 10 than those calculated in this paper at high ionic strength (for purified commercial humic acids, $\log \beta_1$ and $\log \beta_2$ are respectively 5.11 and 8.94 and for marine humic acids they are 5.00 and 8.50). Many other authors published binding constant values for the complexes between humic substances and actinides (Moulin et al. 1992, Kim and Czerwinski 1996, Czerwinski et al., 1996), but comparison with our values is difficult as the models applied to the metal-humic binding are not consistent and the experiments are done using different amounts of metal, which is significant in the choice of model for analysis (Choppin and Labonne-Wall, 1997).

Table 6. Complexation constants of LBHA with UO_2^{2+}, at pcH 4.9 ($\pm$ 0.1) and different ionic strength (NaCl).

I (m) (NaCl)	pcH	α	$\log \beta_1$	$\log \beta_2$
0.1	5.00	0.59	6.4 ± 0.2	11.7 ± 0.2
0.5	4.78	0.57	5.89 ± 0.1	10.99 ± 0.1
1.0	4.84	0.58	5.86 ± 0.06	11.35 ± 0.1
1.5	4.91	0.59	5.93 ± 0.1	11.05 ± 0.1
2.0	4.89	0.56	5.90 ± 0.03	11.72 ± 0.1
3.0	4.88	0.56	7.23 ± 0.1	12.37 ± 0.1

ACKNOWLEDGMENTS

This work was supported at Sandia National Laboratory by USDOE under contract DEACO4-94AL85000, and at FSU under subcontract AH5590, under a Sandia-approved quality assurance.

REFERENCES

Bertha, E.L. and Choppin, G.R., Interaction of Humic and Fulvic Acids with Eu(III) and Am(III), *J.Inorg. Chem.* 40:655 (1978).

Borggaard, O.K., Titrimetric determination of acidity and pK values of humic acid, *Acta Chem. Scand.* A28:121 (1974).

Caceci, M.C. and Choppin, G.R., An improved technique to minimize cation adsorption in neutral solutions, *Radiochim. Acta.* 33:113 (1983).

Chen, J.F., Xia, Y.X. and Choppin, G.R., Derivative analysis of potentiometric titration data to obtain protonation constants. *Anal. Chem.*68(22):373 (1996). (a)

Chen, J.F., Xia, Y.X., Pokrovsky, O., Bronikowski, M.G. and Choppin, G.R., Interactions of Am(III), Th(IV), Np(V)O^{2+} and U(VI)O_2^{2+} with Acetate and Lactate in NaCl Solution, *Radiochim. Acta.* Submitted (1996). (b)

Choppin, G.R. and Labonne-Wall, N., Comparison of two models for Metal-Humic Interaction, *J. Radioanal. Nucl. Chem.* 221(1-2):67 (1997).

Choppin, G.R. and Kullberg, L., Protonation thermodynamics of humic acid. *J. Inorg. Nucl. Chem.* 40:651 (1978).

Czerwinski, K.R., Kim, J.I., Rhee, D.S. and Buckau, G., Complexation of trivalent actinide ions (Am^{3+}, Cm^{3+}) with humic acid, *Radiochim. Acta.* 72:179 (1996).

Ephraim J., Alegret S., Mathuthu A., Bicking M., Malcolm R.L. And Marinsky J. A., A united Physicochemical description of the protonation and metal ion complexation equilibria of natural organic acids, *Environ. Sci. Technol.* 20:354 (1986).

Kim, J.I., Buckau, G., Li, G.H., Duschner H., Psarros N., Characterization of Humic and Fulvic Acids from Gorleben Groundwater, *J. Anal. Chem.* 338:245 (1990).

Kim, J.I. and Czerwinski K.R., Complexation of metal ions with humic acid, *Radiochim. Acta.* 73:5 (1996).

Kim, J.I., Rhee, D.S. and Buckau, G., Complexation of Am(III) with humic acids of different origin, *Radiochim. Acta.* 52/53:49 (1991).

Meunier-Lamy, C., Adrian, P., Berthelon, J. and Rouiller, J., Comparison of binding abilities of fulvic and humic acids extracted from recent marine sediments with UO_2^{2+}, *J. Org. Geochem.* 9 (6):285 (1986).

Minai, Y. and Choppin, G.R., Interaction of Americum (III) with Humic Acids and two Synthetic Analogues, *Proceeding of the International Symposium on Advanced Nuclear Energy Research.* (1989)

Moulin, V., Tits, J., Moulin, C., Decambox, P. Mauchien, P. and de Ruty, O., Complexation behaviour of humic substances towards actinides and lanthanides studied by time-resolved laser-induced spectrofluorometry, *Radiochim. Acta.* 58/59:179 (1992).

Nash, K.L., Choppin, G.R., Interaction of Humic and Fulvic Acids with Th(IV), *J. Inorg. and Nucl. Chem.* 42:1045 (1980).

OECD, *Chemical thermodynamics: volume 1*, Elsevier Publishers B.V. (1992).

Peppard, D.F., Mason, G.W., Maier, J.L. and Driscoll, W.J., Fractional extraction of the lanthanides as their di-alkyl orthophosphates, *J. Inorg. and Nucl. Chem.* 4:334 (1957).

Perdue E. M., Solution thermochemistry of humic substances, *Geochimica et Cosmochimica Acta.* 42:1351 (1978).

Schubert, J., The use of ion exchangers for the determination of physical-chemical properties of substances, particularly radiotracers, in solution, *J. Phys. Chem.* 52:340 (1948).

Shanbag, P.M. *Thermodynamics of uranyl humate binding in aqueous solution,* Ph.D. Dissertation, Florida State University (1979).

Shanbag, P.M. and Choppin, G.R., Binding of Uranyl by Humic Acid, *J. Inorg. Nucl. Chem.* 43 (12):3369 (1981).

Thurmond, V. and Millero, F.J., Ionization of Carbonic Acid in Sodium Chloride Solutions at 25°C, *J. Solution Chem.* 11(1):447 (1982).

Torres, R.A. and Choppin, G.R., Europium(III) and Americium(III) Binding constants with Humic Acid, *Radiochim. Acta.* 35:143 (1984).

United States Geological Survey, *Humic substances in the Suwannee river, Georgia,* Water-Supply Paper 2373, Ed. R. C. Averett, J. A. Leenheer, D. M. McKnight and K. A. Thorn (1994)

Yasunishi, A. and Yoshiada, F., Solubility of carbon dioxide in aqueous electrolyte solutions, *J. Chem. and Eng. Data.* 24 (1):11 (1979

III. Actinide Colloidal and Microbiological Interactions

RETARDATION OF COLLOIDAL ACTINIDES THROUGH FILTRATION IN INTRUSION BOREHOLE BACKFILL AT THE WASTE ISOLATION PILOT PLANT (WIPP)

Richard Aguilar,[1] Hans W. Papenguth,[1] and Fred Rigby[2]

[1]Sandia National Laboratories, Nuclear Waste Management Center, P.O. Box 5800, Albuquerque, NM. 87185
[2]Scientific Applications International Corporation (SAIC), Albuquerque, NM. 87106

ABSTRACT

A depth filtration model was used to evaluate filtration of four types of colloids (mineral fragments, humics, microbes, and mature actinide intrinsic colloids) and their agglomerates by borehole backfill material in the event of inadvertent human intrusion to the Waste Isolation Pilot Plant (WIPP). The WIPP is a proposed repository sited in bedded salt for transuranic wastes generated under our nation's defense programs. Under a human intrusion scenario involving two or more boreholes, flow from an underlying brine reservoir could potentially result in the migration of colloids from the repository up a borehole and then outward to the overlying Culebra Dolomite aquifer. However, in the performance assessment of the WIPP it is assumed that any intrusive borehole would be backfilled immediately after the infringement. The borehole filler material is assumed to have the hydraulic properties of either degraded concrete or grout, or silty sand with a maximum permeability of 10^{-11} m^2 (worse case scenario). Mechanisms of filtration were modeled by trajectory analysis with borehole particles regarded as collectors. The dominant filtration mechanisms were diffusion and interception. The collision efficiency (α) of the colloid particles to the collector grains presented the greatest uncertainty for the filtration modeling. Extensive review of the literature indicated that the colloidal particles displaying the lowest collision efficiencies are microbes and particles stabilized by humic substances (0.1 to 0.01 and approximately 0.001, respectively). These collision efficiency values are based upon low ionic strength water; collision efficiencies have been shown to increase, by orders of magnitude in some cases, upon increasing ionic strength. Conservative α values ranging from 10^{-2} to 10^{-3} were used in our model calculations. The model predicts that most of the entrained colloids and colloid agglomerates will be filtered out by the borehole filling within a few meters (or fractions of meters) of brine flow distance. Particles displaying the least efficient filtration (e.g., particles between 1 and 5 μm) would have their concentrations

Actinide Speciation in High Ionic Strength Media, edited by Reed *et al.*
Kluwer Academic / Plenum Publishers, New York, 1999

215

reduced by about an order of magnitude over the travel distance (395 m) between the repository and borehole interface with the Culebra.

Key Words: *nuclear waste repository, transuranic waste, geologic salt waste disposal, resource drilling.*

INTRODUCTION

This work was conducted as part of the Waste Isolation Pilot Plant (WIPP) Colloid Research Program by Sandia National Laboratories. The goals of the program are to: (1) quantify the concentration of colloidal actinides in the proposed underground salt repository, and (2) assess the potential for transport and retardation of colloidal actinides in the Culebra Dolomite Member of the Rustler Formation, an overlying transmissive carbonate rock unit (WIPP Performance Assessment Department, 1992). The only credible mechanisms for escape of radionuclides to the accessible environment have been determined to involve a breach of the repository during exploratory drilling for natural resources within the WIPP site area (WIPP Performance Assessment Department, 1992). The objective of this study was to estimate the degree of retardation of colloidal actinides as they move through the repository waste and backfill material in intrusion boreholes.

Movement of actinide-bearing colloidal particles may be an important mechanism in contaminant transport in groundwaters (Ramsay, 1988; McCarthy and Zachara, 1989; Moulin and Ouzounian, 1992; Smith and Degueldre, 1993). However, suspended colloids and their agglomerates will be mobile only if they are not filtered by porous media. Colloids are important to the performance of the WIPP because of their exceptionally high surface area to unit mass ratio, coupled in some cases with a high actinide sorption capability. Additionally, dissolved actinides themselves may form colloidal-sized particles. The four general categories of colloidal particles which may exist in the WIPP repository environment are: 1) mineral fragments, 2) humic substances, 3) microbes, and 4) actinide intrinsic colloids (Table 1).

Filtration in groundwater can occur by physical straining (i.e., sieving) or by physical-chemical collection by attractive surfaces on the immobile matrix (i.e., chemical retardation). We evaluated the extent to which colloids and their agglomerates might be filtered by repository waste and borehole backfill material, thereby restricting their transport from the WIPP repository to the accessible environment in the event of inadvertent human intrusion (Figure 1). Specific hypotheses evaluated included:

(1) Colloid agglomerates suspended in WIPP brines will be filtered to some extent by waste within the repository. Borehole backfill (assumed to be the deteriorated remains of concrete plugging material) will act as a filter and remove most of the concentration of colloid agglomerates suspended in brine that enters the borehole from the repository. (Brine moving toward the borehole will also be filtered to some extent by the waste within the repository, but this effect is not addressed in this paper).

(2) Given the anticipated low brine flow velocities, only a few meters of waste or borehole backfill material acting as a filter column will effectively remove most of the entrained colloid agglomerates.

(3) Over time, clogging and caking phenomena will lead to filtration of progressively smaller particles until the pore channels in the waste and the borehole backfill are completely plugged, and all brine flow and colloid transport effectively cease.

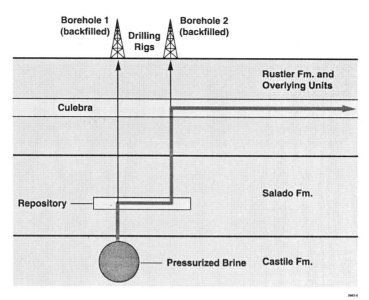

Figure 1. Human intrusion scenario where two boreholes would intersect the WIPP repository and result in brine flow from the underlying Castile Formation through the repository and up through one of the boreholes to the Culebra Dolomite Member. This "two borehole" intrusion scenario would produce the hydrologic gradient needed for waste entrainment to the accessible environment. *[diagrammatic only, not to scale]*

METHODOLOGY

A depth filtration model which combines the effects of particle-particle collision probability and chemical retardation, but excludes sieving, was used to evaluate the potential for filtration of colloids in WIPP brines. Direct calculation of entrainment (e.g., using fluidized bed theory) was not possible because of uncertainties in the hydraulic properties of the repository waste, the brine injection rates and associated turbulence, the hydrodynamic properties of colloids, and energy required for peptizing. Instead, we elected to take an alternative approach in which we assume that colloids and their agglomerates are entrained, and we then investigated the impact of filtration on the movement of these particles after they are transported to an intrusion borehole. The approach we followed is an accepted procedure employed in the waste-water treatment industry (Yao et al., 1971). Our model calculations neglect the effects of sedimentation, filter cake formation, and clogging effects. These mechanisms will tend to further restrict the movement of colloids and their agglomerates over time (Akers and Ward, 1977; Kessler and Hunt, 1994). Additionally, particles with diameters greater than 7-10% the size of the collector grains should be effectively removed from the brine by straining (Beverly, 1993).

Model Calculations

The following equations are all written for Standard International (S.I.) units. The amount of material removed by the filter per distance of fluid flow was expressed by the following relationship:

$$\frac{c}{c_o} = e^{-rx} \qquad \text{(Eq. 1)}$$

Table 1. Colloid particle sizes and anticipated aggregate sizes for the four general categories of colloids expected to be present at the WIPP repository.

Colloid Type	Approximate Colloid Diameter	Approximate Agglomerate Diameter	Specific Gravity
Mineral fragment	1 nm - 1 µm	> 10 µm	2 - 10
Mineral fragment sterically stabilized with organic coating	1 nm - 1 µm	> 10 µm	2 - 10
Microbe	0.5 - 1µm	> 10 µm	1.2
Pu(IV)-polymer, immature	1 - 2 nm	N/A	<10
Pu(IV)-polymer, mature	≥ 2 nm	>10 µm	≅10
Humic substance, ionic	1 - 2 nm	N/A	1.2
Humic substance, precipitated	1 - 10 µm	>1 µm	1.2

where: $\dfrac{c}{c_o}$ = concentration of substance remaining in the fluid,

r = overall trapping coefficient (from Eq. 2) expressed over a given distance (m) of borehole backfill, and

x = distance (m) of fluid movement through filter.

The overall trapping coefficient for particles by the filter was calculated by the following relationship:

$$ r = (N_c)\,(\eta_d + \eta_I)\,(\alpha) \qquad \text{(Eq. 2)} $$

where: r = overall trapping coefficient by the filter,
N_c = number of collector grains encountered (from Eq. 3),
η_d = single collector efficiency for diffusion (from Eq. 4),
η_I = single collector efficiency for interception (from Eq. 5), and
α = collision efficiency or particle sticking factor (from literature).

The likelihood of particles becoming trapped on a collector grain is the product of the total collector efficiency ($\eta_I + \eta_d$) and the collision efficiency or sticking factor (α). The effective number of particle-collector encounters (N_c) along a unit length (for a packed bed of spherical collectors) is related to the size of the collector grains and the porosity of medium:

$$ N_c = \frac{3}{2}(1 - \Phi)\,\frac{L}{d_c} \qquad \text{(Eq. 3)} $$

where: N_c = number of collector grains encountered,
ϕ = porosity,

$$L = \text{unit length (m) of filter medium, and}$$
$$d_c = \text{mean collector grain diameter (m).}$$

Interception and diffusion, the two filtration mechanisms considered important in low velocity flow are defined by formulas for single collector efficiency (Yao et al., 1971). The probability that a particle suspended in a liquid will come in contact with a collector particle through the process of diffusion is described by the following relationship:

$$\eta_d = 0.9 \left(\frac{BT}{\mu V d_p d_c} \right)^{\frac{2}{3}} \qquad \text{(Eq. 4)}$$

where: η_d = single collector efficiency for diffusion,
B = Boltzman's constant,
T = temperature (K),
μ = fluid viscosity,
V = flow velocity (m s^{-1}),
d_p = particle diameter (m), and
d_c = mean collector grain diameter (m).

Flow velocity (V) was calculated from the assumed permeabilities of the borehole backfill using basic hydrological relationships between permeability (k), hydraulic conductivity (K), and Darcy's Law (Freeze and Cherry (1979), pages 26-30).

The probability that a particle suspended in a liquid will come in contact with a collector particle through the process of interception is described by the following relationship:

$$\eta_I = \frac{3}{2} \left(\frac{d_p}{d_c} \right)^2 \qquad \text{(Eq. 5)}$$

where: η_I = single collector efficiency for interception,
d_c = mean collector grain diameter (m), and
d_p = mean particle diameter (m).

The WIPP performance assessment assumes that the boreholes in the human intrusion scenario (Figure 1) will contain backfill material. This backfill is assumed to exhibit the nature of degraded cement, grout, or silty sand, with a permeability ranging from 10^{-14} to (the worst case) 10^{-11} m^2 (Helton et al., 1996). The borehole will thus represent a long column of relatively fine to quite fine grained material forming a filter medium. The principles of the depth filtration model outlined above can be applied with some confidence to this filter medium (given sufficiently conservative assumptions for collector efficiency). The model was run for backfill permeability in the range specified by the WIPP PA and mean grain diameter correlated with these permeability values.

The effects of sedimentation, inertia, and hydrodynamic action on collector efficiency were not considered in the filtration calculations. The results of studies with granular filtration columns (Yao et al., 1971) suggest that these effects are negligible in comparison to diffusion and interception for filtration of very small particles (colloids and their agglomerates) at low flow velocities.

In some other contexts, such as colloid filtration in natural groundwater aquifers, sedimentation has been reported to play a significant role, even for particle sizes and flow rates in the range considered here. However, it is not clear that the characteristics of flow and filtration in a natural aquifer would be applicable to the flow up the filled borehole assumed in the WIPP human intrusion scenario. Furthermore, as shown in Table 1, some of the colloidal particles that may occur in the WIPP brine are assessed to have specific gravities as low as 1.2; this specific gravity is approximately the same as that of the concentrated brine that would exist in the repository and percolate up the borehole in the intrusion scenario depicted in Figure 1. Sedimentation will not occur without a density contrast between the fluid and the particles contained therein. Neglecting sedimentation contributes to development of a conservative assessment of filtration effectiveness for the WIPP Performance Assessment. Given the uncertainty that exists regarding the potential significance of sedimentation for the human intrusion scenario of the WIPP, neglecting the effect seems the most appropriate course in our calculations.

The model was applied to the upper and lower limits in particle size (d_p) for the different colloid types anticipated in WIPP brines (Table 1). Based upon an extensive review of the literature, conservative collision efficiency values (α) for this scenario were considered to range from 10^{-1} to 10^{-3} (Chang & Vigneswaran, 1990; Gross et al., 1995; Harvey & Garabedian, 1991; Jewett et al., 1995). We recognize the uncertainty in our modeling associated with the use of collision efficiency values (α) which are based upon other studies reported in the literature. However, these studies were, for the most part, conducted under low ionic strength conditions and α values have been shown to increase (by orders of magnitude in some cases) upon increasing ionic strength (Chang & Vigneswaran, 1990; Jewett et al., 1995).

RESULTS

Our modeling results suggest that the borehole backfill material will serve as an effective filtration medium. The most effective filtration occurs with lowest borehole backfill permeability (10^{-14} m^2) assumed in the performance assessment of the WIPP (Helton et al., 1996). Within the size range of the types of colloids and their agglomerates anticipated in the WIPP repository (Table 1) the most effective filtration is displayed by the smallest particles; we attribute this to the large contribution of the diffusion process to the overall particle trapping coefficient *(r)*. Filtration of particles larger than 2-5 µm should increase progressively with increasing size because of the increased importance of interception to the depth filtration process.

Filtration of Humic Acid Coated Mineral Fragments

Figures 2a - 2b show the relationship between filtration efficiency (c/c_o) and distance of brine flow (m) for humic acid coated mineral fragments and their agglomerates, using a conservative collision efficiency value of $\alpha = 10^{-3}$. The least efficient filtration prediction occurs when the model is applied using the larger collector grain size ($d_c = 178$ µm) and flow velocity ($V = 5.1 \times 10^{-4}$ m s^{-1}) associated with the higher borehole permeability (Figure 2a). Note, however, that over 10% of the initial concentration of the larger particles ($d_p = 10^{-5}$ m or 10 µm) is still filtered within a 10 m distance of brine flow.

For the collector grain size and brine flow velocity calculated using the lower 10^{-14} m^2 permeability ($d_c = 5.6 \times 10^{-6}$ m and $V = 5.1 \times 10^{-7}$ m s^{-1}, respectively), the larger ($d_p = 10^{-5}$ m) mineral fragments are effectively filtered within a brine flow distance of approximately 0.01

m and the smallest particles ($d_p = 1 \times 10^{-9}$ m) are removed within an even shorter length of borehole backfill (Figure 2b).

Filtration of Microbial Colloids

Figures 3a - 3b show model results of filtration of microbial colloids as a function of brine flow distance through the borehole backfill material. The model predicts that 99% of the initial concentration of 1 μm sized microbes will be filtered within a brine flow distance of approximately 95 m (Figure 3a). For the larger 2 μm sized microbes, 99% of the initial concentration is removed within a brine flow distance of 130 m. With the smaller collector grain size and flow velocity associated with the lower borehole permeability (Figure 3b), the model predicts that concentrations of both the 1 μm and 2 μm size colloids are reduced to <1% of their initial concentrations within 0.003 m and 0.004 m, respectively, of borehole backfill.

Filtration of Humic Substances

Model predictions of the filtration of humic substances by borehole backfill material were similar to those already shown for humic-coated mineral fragments and their agglomerates. For the higher borehole permeability (Figure 4a), the 1 μm colloids are again filtered less efficiently than the smaller particle size (1 nm). Approximately 15% of the initial concentration of the 1 μm sized particles remains after a brine flow distance of 400 m. In contrast, filtration of the 1nm particles is nearly complete within a distance of only 5 m through the borehole. As observed in our other model runs, minimal colloid concentrations remain in the brine after flow distances of only fractions of a meter when the lower brine flow velocity ($V = 5.1 \times 10^{-7}$ m s^{-1}) and smaller collector grain size ($d_c = 5.6 \times 10^{-6}$ m) are used in the model (Figure 4b).

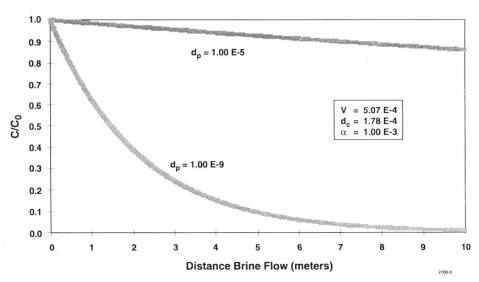

Figure 2a. Filtration of humic-coated mineral fragments and their agglomerates assuming the higher 10^{-11} m^2 permeability for the borehole filling. The smaller 1 nm particles ($d_p = 10^{-9}$ m) have a much higher filtration efficiency than the larger 10 μm ($d_p = 10^{-5}$ m) diameter particles.

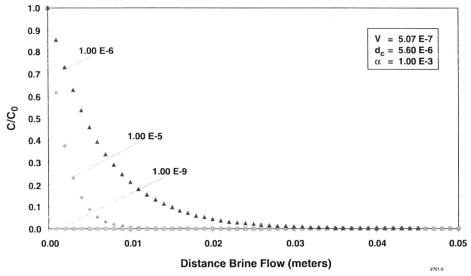

Figure 2b. Filtering of humic-coated mineral fragments and their agglomerates, (d_p = 10 μm, 1 μm, & 1 nm) by the borehole filling, assuming the lower permeability of 10^{-14} m^2.

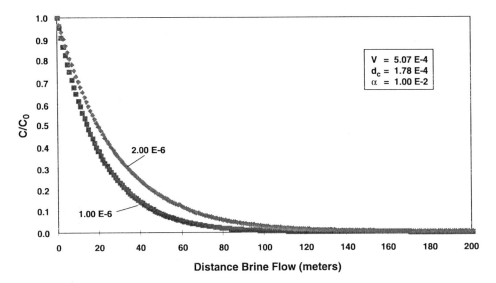

Figure 3a. Filtering of microbes (d_p = 1 μm & 2 μm) by the borehole filling, assuming the higher permeability of 10^{-11} m^2.

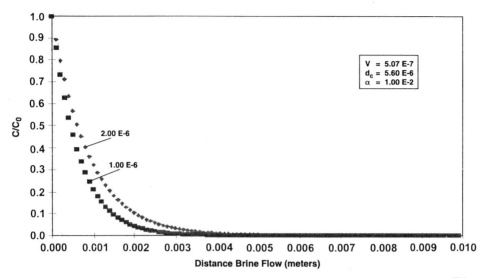

Figure 3b. Filtering of microbes (d_p = 1 μm & 2 μm) by the borehole filling, assuming the lower permeability of 10^{-14} m^2.

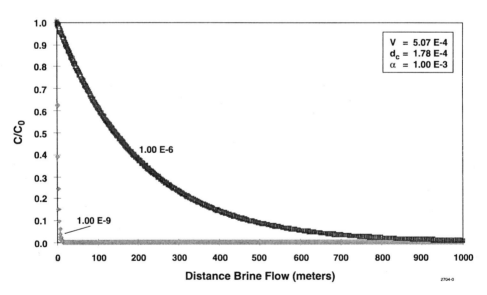

Figure 4a. Filtering of humic substances (d_p = 1 nm & 1 μm) by the borehole backfill, assuming the higher permeability of 10^{-11} m^2.

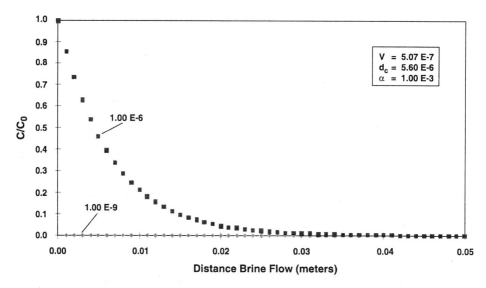

Figure 4b. Filtering of humic substances ($d_p = 1$ nm & 1 μm) by the borehole filling, assuming the lower permeability of 10^{-14} m^2.

CONCLUSIONS

Depth filtration of brine and entrained colloids is likely to play a significant role in the retardation of actinide transport from the WIPP repository under the potential conditions proposed for human intrusion scenarios. We recognize the uncertainty in our modeling associated with the use of collision efficiency values (α) based upon other studies reported in the literature. These studies were, for the most part, conducted under low ionic strength conditions and α values have been shown to increase (by orders of magnitude in some cases) upon increasing ionic strength. We therefore consider the use of α values ranging from 10^{-1} to 10^{-3} appropriate for conservative modeling of filtration of high ionic strength WIPP brines and associated colloids by borehole backfill material at the WIPP in the event of future inadvertent human intrusion. Our modeling of depth filtration through borehole backfill material leads us to the following conclusions:

(1) the process of diffusion insures that the smaller colloid particles (1 nm to 0.1 μm dia.) would frequently come in contact with pore walls and filtration of these particles would likely be greater than 99% within a brine flow distance of 5-10 m,

(2) given the extensive vertical distance (approximately 400 m) from the top of the repository to the base of the overlying Culebra aquifer, even colloid agglomerates with particle sizes larger than 1 μm would have their concentrations reduced by about an order of magnitude by the time the brines reach the borehole interface with the overlying Culebra aquifer, and

(3) filtration efficiency for particles larger than 2-5 μm should increase progressively with increasing size because of the increased importance of interception to the depth filtration process.

ACKNOWLEDGMENTS

This work was supported by the U.S. Department of Energy under Contract DE-AC04-94AL85000. Sandia is a multiprogram laboratory operated by Sandia Corporation, a Lockheed Martin Company, for the United States Department of Energy.

LITERATURE CITED

Akers, R.J. and A. S. Ward. 1977. "Liquid Filtration Theory and Filtration Pretreatment," *Filtration: Principles and Practices. Part I.* Ed. C. Orr. New York, NY: Marcel Dekker, Inc. pp. 170-250

Beverly, R.P. 1993. "Granular Filter Media," *Fluid/Particle Separation Journal.* Vol. 6, no. 1, 27-44.

Chang, J.S., and S. Vigneswaran. 1990. "Ionic Strength in Deep Bed Filtration," *Water Research.* Vol. 24, no. 11, 1425-1430.

Freeze, R.A., and J.A. Cherry. 1979. *Groundwater.* Englewood Cliffs, NJ: Prentice-Hall, Inc.

Gross, M.J., O. Albinger, D.G. Jewett, B.E. Logan, R.C. Bales, and R.G. Arnold. 1995. "Measurement of Bacterial Collision Efficiencies in Porous Media," *Water Research.* Vol. 29, no. 4, 1151-1158.

Harvey, R.W., and S.P. Garabedian. 1991. "Use of Colloid Filtration Theory in Modeling Movement of Bacteria Through a Contaminated Sandy Aquifer," *Environmental Science and Technology.* Vol. 25, no. 1, 178-185.

Helton, J. C., D. R. Anderson, B. L. Baker, J. E. Bean, J.W. Berglund, W. Beyeler, K. Economy, J. W. Garner, S. C. Hora, H. J. Iuzzolino, P. Knupp, M. G. Marietta, J. Rath, R. P. Rechard, P. J. Roache, D. K. Rudeen, K. Salari, J. D. Schreiber, P. N. Swift, M. S. Tierney, and P. Vaughn. 1996. "Uncertainty and Sensitivity Analysis Results Obtained in the 1992 Performance Assessment for the Waste Isolation Pilot Plant," *Reliability Engineering and System Safety.* Vol. 51, 53-100.

Jewett, D.G., T.A. Hilbert, B.E. Logan, R.G. Arnold, and R.C. Bales. 1995. "Bacterial Transport in Laboratory Columns and Filters: Influence of Ionic Strength and pH on Collision Efficiency," *Water Research.* Vol. 29, no. 7, 1673-1680.

Kessler, J.H., and J.R. Hunt. 1994. "Dissolved and Colloidal Contaminant Transport in a Partially Clogged Fracture," *Water Resources Research.* Vol. 30, no. 4, 1195-1206.

McCarthy, J.F., and J.M. Zachara. 1989. "Subsurface Transport of Contaminants," *Environmental Science & Technology.* Vol. 23, no. 5, 496-502.

Moulin, V., and G. Ouzounian. 1992. "Role of Colloids and Humic Substances in the Transport of Radioelements through the Geosphere," *Applied Geochemistry.* Supplemental Issue No. 1, 179-186.

Ramsay, J.D.F. 1988. "The Role of Colloids in the Release of Radionuclides from Nuclear Waste," *Radiochimica Acta.* Vol. 44-45, pt. 1, 165-170.

Smith, P.A., and C. Degueldre. 1993. "Colloid-Facilitated Transport of Radionuclides through Fractured Media," *Journal of Contaminant Hydrology.* Vol. 13, 143-166.

WIPP Performance Assessment Department. 1992. *Preliminary Performance Assessment for the Waste Isolation Pilot Plant, December 1992. Volume 1: Third Comparison with 40 CFR 191, Subpart B.* SAND92-0700/1. Albuquerque, NM: Sandia National Laboratories.

Yao, K., M.T. Habibian, and C.R. O'Melia. 1971. "Water and Waste Water Filtration: Concepts and Applications," *Environmental Science and Technology,* Vol. 5, no. 11, 1105-1112.

CONTRIBUTION OF MINERAL-FRAGMENT TYPE PSEUDO-COLLOIDS TO THE MOBILE ACTINIDE SOURCE TERM OF THE WASTE ISOLATION PILOT PLANT (WIPP)

John W. Kelly, Richard Aguilar, and Hans W. Papenguth

Sandia National Laboratories, Albuquerque, NM. 87185

INTRODUCTION

The Waste Isolation Pilot Plant (WIPP) is being developed in southeastern New Mexico by the U.S. Department of Energy (DOE) as a repository for transuranic waste produced by defense programs. Regulations promulgated by the U.S. Environmental Protection Agency (EPA) place limits on the cumulative release of radionuclides to the accessible environment over 10,000 years and require performance assessments to demonstrate WIPP compliance with these regulatory standards.

The proposed WIPP repository has been sited approximately 650 m below the desert floor within the thick evaporite layers (approximately 500 m of halite with occasional layers of anhydrite) of the Permian-aged Salado Formation. A potential mechanism for escape of radionuclides to the accessible environment involves a breach of the repository during exploratory drilling for natural resources. In a human intrusion scenario (Figure 1), the pressurized brines present in the Castile sediments underlying the repository are envisioned to be transported through the repository waste and then up through borehole backfill material to the overlying transmissive carbonate rock unit, the Culebra Dolomite Member of the Rustler Formation. Additional radionuclide release to the environment could result from the movement of contaminated drilling muds and fluids directly to the surface by pressurized fluids in the repository. Performance assessment calculations have shown that the actinides Th, U, Np, Pu, and Am are the most significant radionuclides in terms of potential total release to the accessible environment.

The focus of this study was to assess the potential concentration of actinides that might be transported by mobile mineral fragment-type pseudo-colloids likely to be present in the repository environment. Recent compilations of the types of colloidal particles present in surface waters and shallow and deep groundwaters have shown that mineral fragment colloids are ubiquitous and may often dominate the total number of particles in some cases. At the WIPP, mineral colloids, which may act as substrates for actinide sorption, may be present in natural groundwaters, or in the waste itself. Additional sources may result from waste canister corrosion, waste degradation, or reprecipitation of natural constituents and waste products.

Actinide Speciation in High Ionic Strength Media, edited by Reed *et al.*
Kluwer Academic / Plenum Publishers, New York, 1999

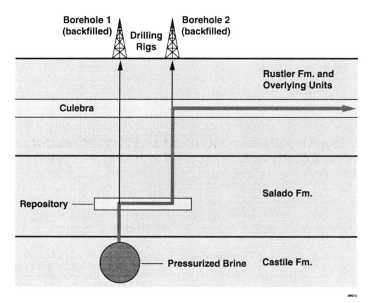

Figure 1. Human intrusion scenario considered in the WIPP Performance Assessment (Helton et al., 1996). In this "two-borehole scenario," brine flow from the underlying Castile Formation would migrate through the repository and up one (or both) of the boreholes to the relatively transmissive Culebra Dolomite Member. The established hydraulic gradient could potentially transport colloidal actinides suspended in the brine to the accessible environment. *[diagrammatic only, not to scale]*

Destabilization of hydrophobic-type surfaces and collapse of a mineral colloid's electrolyte double layer with even moderate ionic strength water has long been recognized. The ionic strength at which rapid destabilization and agglomeration of mineral colloids occurs is typically quite distinct, and is referred to as the *critical coagulation concentration*, or *c.c.c.* Empirically determined *c.c.c.* values for symmetrical electrolytes are summarized in Table 1.

Considering that all groundwaters in the WIPP environment are brines (Figure 2), we employed a bounding approach in the laboratory, rather than conduct costly direct groundwater sampling and analysis, to assess the stability of mineral colloids in the brines likely to be involved in the transport of actinides at the WIPP. This approach was preferred for two reasons: 1) subsurface sampling provides information on present-day conditions, not conditions potentially present after waste is placed within the proposed repository, and 2) results of such a sampling program can be inconclusive because of sampling complications. In the case of the WIPP, the use of scientifically-based projections of what might be anticipated in the WIPP repository environment after closure was deemed the most sound approach.

Table 1. Summary of empirically determined critical coagulation concentrations (after van Olphen, 1991, p. 24).

Counter-ion Valence	*c.c.c.* (mM)	Example Counter-ion	*c.c.c.* (mg/L)
+1	2.5 to 150	Na^+	600 to 3,500
+2	0.5 to 2	Ca^{2+}	20 to 80
+3	0.01 to 0.1	Al^{3+}	0.3 to 3

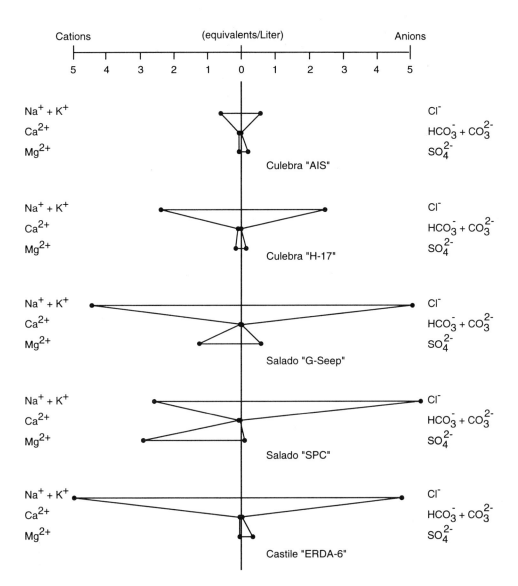

Figure 2. Stiff diagrams illustrating major constituents and concentrations of WIPP groundwaters. The most dilute brine, Culebra "AIS," has an ionic strength of ~0.8 m, similar to sea water. The most concentrated brine, Salado "SPC," has an ionic strength of 8 m, approximately 10 times greater than sea water.

Two types of experiments were conducted: coagulation series experiments in which the *c.c.c.* was quantified, and residual colloid population experiments in which the number of colloids in a destabilized system was measured.

The coagulation series experiments were conducted with a broad selection of mineral colloids. The residual concentration experiments, in which colloid population and size were measured, were conducted as an indirect approach to estimate associated actinide concentration. Assuming a spherical particle, the amount of actinide sorbed to a particle is simply:

$$[An]_p = \frac{\pi \Phi^2 N_s}{N_A}$$
(Eq. 1)

where:

$[An]_p$	=	adsorbed actinide (moles/particle)
Φ	=	colloid diameter (nm)
N_s	=	adsorption site density (sites/nm^2)
N_A	=	Avogadro constant

Assuming reasonable values for adsorption site densities (e.g., Kent et al., 1988), the number of mineral colloids required to mobilize a significant concentration of actinides is

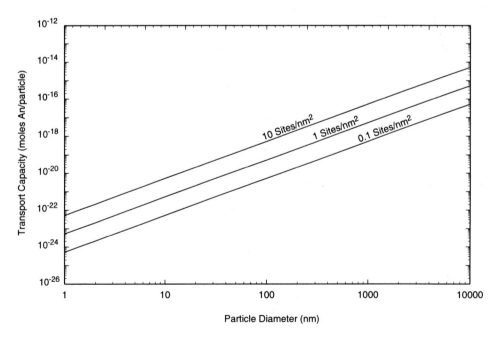

Figure 3. Estimates of sorbed actinide concentrations on single spherical particles based upon site binding densities. Site binding densities for strongly sorptive materials may be as high as 2 to 20 sites per nm^2 but are generally lower for most materials (Kent et al., 1988).

rather large (Figure 3). However, mobilization of contaminants is still dominated by these smaller particles simply because of their greater population density in natural waters as compared to larger particles. With residual population measurements, Equation 1 can be used to estimate transport capacity in terms of moles of actinide per liter of dispersion.

EXPERIMENTAL PROCEDURES

Colloidal dispersions were prepared by milling (ground to ~0.5 μm diameter particles with a McCrone Micronizing Mill) or by chemical precipitation. Dispersions were then ultrasonicated and injected into sequential dilutions of synthetic WIPP brine under acidic (pH 3 to 4), neutral (pH 6 to 8), or basic (pH 9 to 12) conditions. Coagulation was quantified by measuring turbidity and particle-size distribution in the upper portion of the test vessel (standard 10 mL test tube flasks) as a function of time with a Coulter N4MD submicron particle-size analyzer. Measurements were made at 1-min, 20-min, 1-hr, and 24-hr after the start of the experiment; some experiments were conducted for several weeks.

For select experiments, residual concentrations of colloidal particles were quantified with a particle spectrometer (Particle Measurement Systems Model HSLIS S100 particle spectrometer).

At the end of each experiment, electrophoretic mobilities were measured (Coulter DELSA 440), and pHs were confirmed.

Colloids included in the experiment were potential corrosion products (hematite, goethite, magnetite, limonite, and siderite), naturally occurring minerals (dolomite, gypsum, anhydrite, calcite, magnesite, strontianite, quartz, illite, and pyrite), waste constituents (paper towels, diatomaceous earth, kaolinite, and vermiculite), precipitation products (brucite), and possible backfill materials (bentonite and montmorillonite).

RESULTS

Coagulation Series Experiments

As expected, experiments in sequential dilutions of the synthetic WIPP brines showed that colloidal dispersions were destabilized at dilutions of 100× to 10,000×, depending upon the mineral type, brine chemical composition, and solution pH. Coagulation and gravitational settling was usually evident within 20 minutes after introducing the dispersion into the brine, and distinct effects of these processes were observed after 24 hours (Figure 4). The pH conditions of the experiment noticeably affected the *c.c.c.* due to variations in the points of zero pH charge of the different materials.

Residual Concentration Experiments

Longer duration experiments were conducted to measure residual concentrations in the synthetic WIPP brines. These experiments showed that most coagulation and settling occurred within the first few hours, with subsequent setting slowly leveling off after about one week (Figure 5).

The rate of agglomeration in solutions is proportional to the particle concentration, and so the observed rate of decrease in residual concentration becoming progressively slower over time in our experiments was not surprising. Using Equation 1, the estimated associated actinide concentrations associated with the residual concentrations of the three mineral colloids are estimated to be about 10^{-9} moles per liter of dispersion (see Table 2).

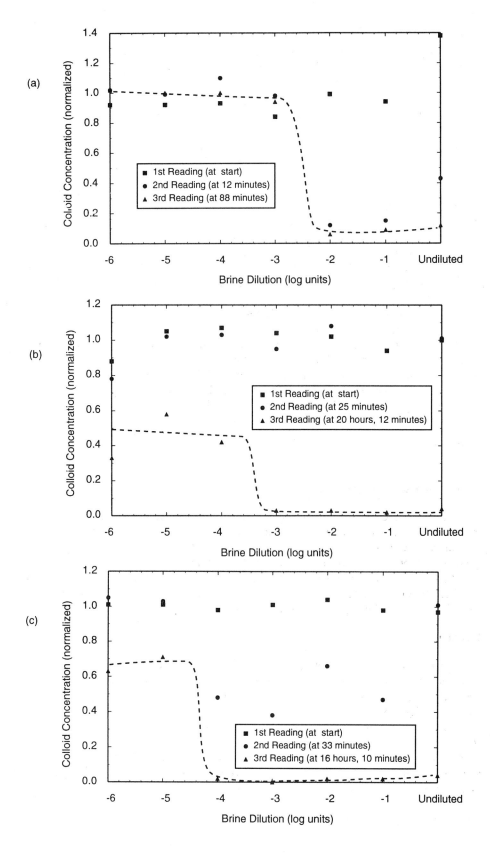

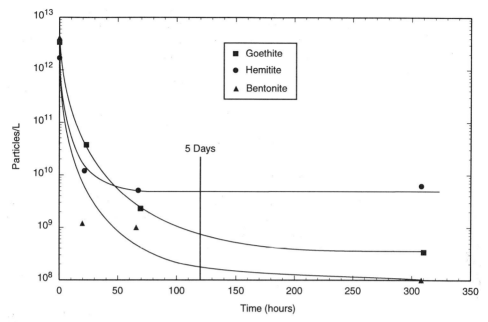

Figure 5. Residual concentrations of three mineral colloids in 0.1 M NaCl synthetic brine over a two-week period. Mean particle sizes: goethite = 0.7 μm; hematite = 0.7 μm; Aldrich bentonite = 0.6 μm. The majority of particle settling for all three minerals occurred during the first week.

Table 2. Estimated actinide concentration associated with three mineral fragment pseudo-colloids (refer to Figure 5), assuming 1.0 actinide sorption site per nm^2 of colloid surface area.

Mineral	Final Population (particles/L)	Estimated Actinide Concentration (moles/L dispersion)
hematite	6.2×10^9	1.6×10^{-8}
goethite	3.4×10^8	9.5×10^{-10}
bentonite	9.8×10^7	1.6×10^{-10}

DISCUSSION

Behavior of Mineral Colloids in Groundwaters at Other Sites

Groundwater sampling and characterization work at the Grimsel Test Site in Switzerland (Degueldre et al., 1989) has resulted in an extremely high quality data set

←—————————————————————————————————————

Figure 4. Typical coagulation series experiments: a) hematite (Fe_2O_3; 0.2 μm particle diameter) under acidic pH conditions, b) limonite ($Fe \cdot OH \cdot nH_2O$, 0.4 μm) under acidic pH conditions, and c) montmorillonite (0.8 μm) under neutral pH conditions. All experiments shown were conducted in sequential 10x dilutions of Salado "SPC" brine (see Figure 2). The *c.c.c.* occurs at dilutions of 100× to 10,000×, corresponding to ionic strengths of about 80 to 0.8 mM.

describing naturally occurring colloidal particles in the deep subsurface. Most of the colloids identified in Grimsel groundwater consisted of silicate mineral fragments (McCarthy and Degueldre, 1992, p. 293). Colloid concentration at Grimsel was described by Degueldre et al. (1989) using the following power function:

$$[\text{coll}] = 10^{15.82(\pm 0.36)} \Phi^{-3.17(\pm 0.16)} \qquad \text{(Eq. 2)}$$

where:

$[\text{coll}]$ = colloid concentration (particles/L)

Φ = colloid diameter (nm)

Simple manipulation of Equation 2 provides a means to estimate the potential concentration of sorbed contaminants associated with the colloid population. The size distribution equation (derivative of Equation 2) is:

$$\frac{d[\text{coll}]}{d\Phi} = -10^{16.32} \Phi^{-4.17} \qquad \text{(Eq. 3)}$$

can be combined with the actinide sorption expression (Equation 1) to estimate the actinide concentration associated with the colloid suspensions as a function of mean particle size:

$$\frac{d[\text{An}]}{d\Phi} = -10^{16.32} \Phi^{-4.17} \left(\frac{\pi \Phi^2 N_S}{N_A} \right) \qquad \text{(Eq. 4)}$$

where:

$[\text{An}]$ = total actinide concentration (moles/L)

Φ = colloid diameter (nm)

N_S = adsorption site density (sites/nm²)

N_A = Avogadro constant

The total colloidal actinide concentration over the 10 to 10,000 nm size range can be estimated by integrating Equation 4:

$$[\text{An}] = \frac{10^{16.32} \pi N_S}{(1.17) N_A} \left[\Phi^{-1.17} \right]_{\Phi=10}^{\Phi=10000} \qquad \text{(Eq. 5)}$$

Solving Equation 5 for the total concentration of divalent actinides in Grimsel groundwater (assuming a sorption density of 1 site/nm²) yields the rather low integrated actinide concentration estimate of 6×10^{-9} moles actinide per liter of dispersion. WIPP brines are between 600 and 4,000 times higher in ionic strength than Grimsel groundwater. Consequently, relating the outcome of the Grimsel calculations to the brines present in the WIPP environment, mineral fragment colloids in WIPP brines would carry only a fraction of the estimated 6×10^{-9} moles of actinide per liter of dispersion. Of the various colloids in Grimsel groundwater reported by Degueldre et al. (1989), only natural organic matter and bacteria might be stable in WIPP brines.

In general, inspection of data produced by groundwater chemical analysis may provide a quick screening approach for the behavior of mineral pseudo-colloids. Monovalent cations will have little effect on mineral colloid abundance because of the high

Table 3. Compilation of Ca^{2+} and Mg^{2+} in groundwaters and surface waters in various geologic terrain (from Tables in Holland, 1978). Values are mg/L or mg/kg (approximately equal). GW = groundwater; RW = river water; OW = ocean water.

Category	Terrain	$[Ca^{2+}]$	$[Mg^{2+}]$	Description
GW	carbonate	47	24	spring 1, central Penn.
GW	carbonate	91	35	well 59, central Penn.
GW	carbonate	55	15	well 66, central Penn.
GW	carbonate	56	29	well 332, central Penn.
GW	carbonate	41	26	well SC-11, central Penn.
GW	carbonate	75	35	well 303, central Penn.
GW	granitic	1.7	0.6	Norway
GW	granitic	5.8	2.4	Vosges
GW	granitic	1	0.4	Alrance Spring F
GW	granitic	0.7	0.3	Alrance Spring A
GW	granitic	4.6	1.3	Central Massif
GW	granitic	4.4	2.6	Brittany
GW	granitic	8.1	4	Corsica
GW	granitic	40	na	Sahara
GW	granitic	8.3	3.7	Senegal
GW	granitic	8	2.5	Chad
GW	granitic	1	0.1	Ivory Coast, dry season
GW	granitic	0.4	0.12	Malagasy
GW	granitic	6.5	2.6	West Warwick, R.I.
GW	granitic	13	4.3	McCormick Co., S.C.
GW	granitic	27	6.2	Ellicott City, Maryland
GW	quartz monzonite	34	7.3	W. of Clayton, Idaho
GW	granitic	87	14	Spokane, Washington
GW	gabbro	5.1	2.3	Waterloo, Maryland
GW	gabbro	32	16	Harrisburg, N. Carolina
GW	basalt	13	9	Camas, Washington
GW	basalt	24	15	Farmington, Oregon
GW	basalt	29	19	Moses Lake, Washington
GW	basalt	48	14	Shoshone, Idaho
GW	basalt	62	28	Hyderabad, India
GW	ultramafic	9.6	35	Pretoria district, South Africa
GW	peridotite	2.5	7.7	Webster, N.Carolina
GW	serpentine	9.5	51	Lake Roland, Maryland
GW	serpentine	2.1	76	Nottingham, Penn.
OW	open ocean	412	1294	35% salinity
RW	N. American	21	5	mean composition
RW	S. American	7.2	1.5	mean composition
RW	European	31.1	5.6	mean composition
RW	Asian	18.4	5.6	mean composition
RW	African	12.5	3.8	mean composition
RW	Australian	3.9	2.7	mean composition
RW	World	15	4.1	mean composition

concentrations of these ions required for mineral colloid destabilization. Trivalent cations, e.g., Al^{3+} and Fe^{3+}, are likely to be hydrolyzed and, thus, will not play a major role in destabilization. Divalent cations (e.g., Ca^{2+} and Mg^{2+}) are likely to be the most important ions controlling concentrations of mineral colloids suspended in natural groundwaters. Of particular importance is Ca^{2+}, because this ion is generally present in large concentrations and has been shown to have a notable effect on coagulation (see Degueldre et al., 1989). A cursory review of groundwater compositions from a variety of geologic environments shows it is not uncommon that Ca^{2+} and Mg^{2+} are present at concentrations ranging from 20 to 80 mg/L; this divalent cation range is readily within the conditions required for destabilization of mineral colloids (compare Tables 1 and 3). At the WIPP site, as in many other sites, natural organic matter (NOM), NOM-stabilized mineral colloids, and bacteria are likely to be the most important colloidal transport agents in groundwater.

CONCLUSIONS

Substantially high concentrations of colloidal particles are required to mobilize significant concentrations of actinides in groundwater. Because of the rather low electrolyte concentration required for destabilization of mineral pseudo-colloids, particularly in the case of divalent cations, ion concentrations in the groundwaters at many sites may minimize the role of mineral colloids in contaminant transport. From this standpoint, the most important colloidal particles are more likely to be bacteria and humics (and perhaps natural organic matter-coated mineral fragments).

In a human intrusion scenario for the WIPP, actinide release to the accessible environment could result from the direct movement of contaminated drilling muds and fluids to the surface by pressurized fluids in the repository. Over the longer term, actinides in repository waste are assumed to be mobilized to the overlying unit (Culebra Dolomite Member of the Rustler Formation) by a hydrologic gradient driven by pressurized brines in the Castile Formation underlying the repository. Given the high ionic strength brines present at the WIPP site, mineral fragment colloids are potentially capable of mobilizing only a very small concentration of actinides. The presence of mineral colloids may actually improve the performance of the WIPP by immobilizing associated actinides within the repository.

ACKNOWLEDGMENTS

This work was supported by the DOE under Contract DE-ACO4-94-AL85000. Sandia is a multi-program laboratory in Albuquerque, NM operated for the DOE by Sandia Corporation, a Lockheed Martin Company.

LITERATURE CITED

Degueldre, C., G. Longworth, V. Moulin, P. Vilks, C. Ross, G. Bidoglio, A. Cremers, J. Kim, J. Pieri, J. Ramsay, B. Salbu, and U. Vuorinen. 1989. *Grimsel Colloid Exercise, An International Intercomparison Exercise on the Sampling and Characterisation of Groundwater Colloids*. PSI-Bericht Nr. 39; EUR-12660-EN; NAGRA-NTB-90-01. Würenlingen und Villigen: Paul Scherrer Institut; Luxembourg: Commission of the European Communities; Wettingen, Switzerland: NAGRA, National Cooperative for the Disposal of Radioactive Waste. (PSI version available from the National Technical Information Service (NTIS), Springfield, VA as DE90614233/XAB.)

Helton, J.C., D.R. Anderson, B.L. Baker, J.E. Bean, J.W. Berglund, W. Beyeler, K. Economy, J.W. Garner, S.C. Hora, H.J. Iuzzolino, P. Knupp, M.G. Marietta, J. Rath, R.P. Rechard, P.J. Roache, D.K. Rudeen, K. Salari, J.D. Schreiber, P.N. Swift, M.S. Tierney, and P. Vaughn. 1996. "Uncertainty and Sensitivity Analysis Results Obtained in the 1992 Performance Assessment for the Waste Isolation Pilot Plant," *Reliability Engineering and System Safety*. Vol. 51, no. 1, 53-100.

Holland, H.D. 1978. *The Chemistry of the Atmosphere and Oceans.* New York, NY: John Wiley & Sons.

Kent, D.B., V.S. Tripathi, N.B. Ball, J.O. Leckie, and M.D. Siegel. 1988. *Surface-Complexation Modeling of Radionuclide Adsorption in Subsurface Environments.* SAND86-7175; NUREG/CR-4807. Albuquerque, NM: Sandia National Laboratories.

McCarthy, J.F., and C. Degueldre. 1992. "Sampling and Characterization of Colloids and Particles in Groundwater for Studying Their Role in Contaminant Transport," *Environmental Particles.* Eds. J. Buffle and H.P. van Leeuwen. Environmental and Physical Chemistry Series. Boca Raton, FL: Lewis Publishers, Inc. Vol. 2, 247-315.

van Olphen, H. 1991. *An Introduction to Clay Colloid Chemistry: For Clay Technologists, Geologists, and Soil Scientists.* 2nd ed. Malabar, FL: Krieger Pub. Co. (Reprint of 1977 - 2nd edition originally published by John Wiley & Sons.)

LABORATORY EVALUATION OF COLLOIDAL ACTINIDE TRANSPORT AT THE WASTE ISOLATION PILOT PLANT (WIPP): 1. CRUSHED-DOLOMITE COLUMN FLOW EXPERIMENTS

W. G. Yelton, Y. K. Behl, J. W. Kelly, M. Dunn, J. B. Gillow,
A. J. Francis, and H. W. Papenguth

Colloid-facilitated transport of Pu, Am, U, Th, and Np has been recognized as a potentially important phenomenon affecting the performance of the Waste Isolation Pilot Plant (WIPP) facility being developed for safe disposal of transuranic radioactive waste. In a human intrusion scenario, actinide-bearing colloidal particles may be released from the repository and be transported by brines (~0.8 to 3 molal ionic strength) through the Culebra, a thin fractured microcrystalline (mean grain size 2 µm) dolomite aquifer overlying the repository. Transport experiments were conducted using sieved, uniformly packed crushed Culebra rock or nonporous dolomite cleavage rhombohedra. Experiments with mineral fragments and fixed and live WIPP-relevant bacteria cultures showed significant levels of retardation due to physical filtration effects. Humic substances were not attenuated by the Culebra dolomite. Comparison of elution curves of latex microspheres in columns prepared with microcrystalline rock and nonporous rock showed minimal effect of Culebra micropores on colloid transport. These data form part of the basis (see also Lucero et al., this volume) to parameterize numerical codes being used to evaluate the performance of the WIPP.

INTRODUCTION

Under the authorization of Public Law 96-164 (1979), the U.S. Department of Energy (DOE) has been developing the Waste Isolation Pilot Plant (WIPP) facility, located approximately 42 km east of Carlsbad, New Mexico for the safe disposal of nuclear wastes produced by the defense nuclear-weapons program. The U.S. Environmental Protection Agency has established the regulatory standards (U.S. EPA, 1993) for cumulative radioactive release to the accessible environment over 10,000 years to demonstrate the WIPP facility compliance.

The only credible potential mechanism for escape of radionuclides to the accessible environment involves human intrusion scenarios, i.e., repository breach during drilling for natural resources. In human intrusion scenarios, the brines present in the repository together with introduced drilling muds are transported up the intrusion boreholes under the driving force of pressurized fluids in the disposal area. Some of the radioisotopes could reach the surface, while others could move laterally toward the WIPP Site boundary, principally through the Culebra Dolomite Member of the Rustler Formation (Culebra). Performance assessment calculations (WIPP PA Division, 1991a,b) showed that releases of radioactive isotopes of the actinide elements Th, U, Np, Pu, and Am would contribute the most to the total release of radioactivity, with Pu and U being the largest contributors. These calculations did not include colloid-facilitated transport.

Actinide Speciation in High Ionic Strength Media, edited by Reed *et al.*
Kluwer Academic / Plenum Publishers, New York, 1999

Transport of actinides by colloidal particles has been recognized as a phenomenon of critical importance to the performance of nuclear waste repositories (Jacquier, 1991; Avogadro and de Marsily, 1984). Colloidal particles, which are generally defined as particles with sizes between 1 nm and 1 μm in a liquid dispersant, will be generated in the repository environment as a result of microbial degradation of cellulosics, corrosion of steel waste containers and waste constituents, by the hydrodynamic entrainment of colloidal-sized mineral fragments, and several other mechanisms. Those colloidal particles may sorb dissolved actinides or the dissolved actinides themselves may form colloidal-sized particles. Additional colloidal particles may be present in natural Culebra groundwater and could form additional actinide-bearing colloidal particles.

The objective of the WIPP colloid research program is to provide the WIPP PA Department with sufficient information to quantify the concentration of colloidal actinides that reaches the accessible environment (Papenguth and Behl, 1996). One of the tasks required to meet this objective includes the quantification of colloid-facilitated transport of actinides in the overlying Culebra, in the event of a repository breach. The laboratory experimental program identified two types of transport experiments. Crushed-rock column flow experiments, the subject of this paper, were used to provide quantification of retardation factors and filtration coefficients for WIPP colloids, in the absence of actinides. Intact core column flow experiments were used to confirm results of crushed-rock column flow experiments for humic colloids and live microbes, in the presence of some of the important actinides (see Lucero et al. (this volume)).

CRUSHED-ROCK COLUMN FLOW EXPERIMENTS DESIGN

Columns used for crushed-rock column experiments were 1 cm in diameter and generally 5 to 44 cm in length. Columns were hand packed with crushed and treated Culebra rock collected at Culebra Bluffs near Carlsbad, New Mexico, and from the horizontal cores collected from the Air Intake Shaft for use in the intact core column flow experiments (Lucero et al., 1995). In addition, some non-porous mineralogically pure crystalline dolomite cleavage rhombohedra (referred to as Butte) were obtained commercially (Ward's Natural Science Establishment, Inc., Rochester, NY, part number 47E2715). Crushed-rock column packing material was prepared in three size fractions using pairs of sieves: 125-250 μm, 250-500 μm, and 500-1000 μm. The material was cleaned, crushed, rinsed thoroughly, acid washed with dilute HCl, rinsed thoroughly, and dried at 40°C.

Culebra brine simulants were prepared to closely match brines from the Culebra at the H-17 well (H-17 brines) or from seeps in the Air Intake Shaft (AIS brines). These two brines are primarily NaCl in composition with ionic strengths of approximately 3 and 0.8 molal, respectively. The pCO_2 was generally kept at atmospheric ($10^{-3.5}$ atm), but was scrubbed in some experiments. For experiments designed to isolate certain phenomena, relatively dilute NaCl solutions were used, with concentrations of 1 or 10mM. Flow rates were 0.1 to 0.5 mL/min. Colloidal particles (described below) consisting of a variety of humic substances, preserved and live microbes, mineral fragments, and latex microspheres (non-functional) were injected as spikes or as step injections. The concentration of colloids and tracers in the effluent were analyzed using inductively coupled plasma-atomic emission spectroscopy (ICP-AES), scanning fluorometry, fixed wavelength fluorometry, epifluorescence microscopy, and laser particle spectrometry. For some experiments, effluent samples were collected in generally 2-mL aliquots with the aid of an automated fraction collector. The particle spectrometry (Particle Measurement Systems model HSLIS S100) was used on-line to count number populations in real time.

Porosity of the packed columns was calulated from gravimetric and volumetric measurements using a grain density of dolomite of 2.82 g/cm^3 (Kelley and Saulnier, 1990). These results were confirmed with transport experiments using a nonsorbing tracer, fluorescein. These tests indicated that effective and actual porosities (typically about 45%) were equivalent, and so it was not necessary to perform a conservative tracer test with each crushed-rock column.

A series of tests were conducted to test the experiment configuration and to evaluate the effect of potential artifacts. In those tests column lengths, flow rates, injection mode, packing diameter, and colloid diameter were varied. The tests demonstrated that the test design was robust, in that varying test configuration did not have an appreciable effect on results. In some of the filtration experiments, the direction of flow was reversed to confirm that filtration was irreversible.

INTERPRETATION

In the column transport experiments, the main phenomena affecting peak elution time and peak breadth are the following: advection; dispersion and diffusion; reversible sorption; and filtration including interception and sieving, sedimentation, and irreversible sorption. To describe transport of colloidal bound actinides through the crushed-rock columns, the advection-dispersion-filtration equation with adsorption is (Harvey and Garabedian, 1991):

$$\frac{dC}{dt} = D\frac{d^2C}{dx^2} - \frac{v}{R}\left(\frac{dC}{dx} + \gamma C\right) \tag{1}$$

It reduces to the advection-dispersion equation with adsorption for $\gamma = 0$:

$$\frac{dC}{dt} = D\frac{d^2C}{dx^2} - \frac{v}{R}\frac{dC}{dx} \tag{2}$$

where C is the contaminant concentration; D is the dispersion coefficient; v is the average linear flow velocity; R is the retardation coefficient for reversible sorption; and γ is the filtration coefficient.

The solution of Eq. (1) can be fitted to the experimental breakthrough curves (BTC) to estimate parameters of interest. However, the experimental BTCs were often noisy and were not suitable for automated curve fitting procedures. Therefore, to interpret the crushed-rock column flow experiments conducted with colloidal particles, the following steps were taken. First the retardation factor was estimated from the peak position of the elution curve. A conservative value for the dispersion was estimated on the basis of suggestions provided in Parker and van Genuchten (1984), tests with conservative tracers, and sensitivity analyses. For the tests described herein, values for the dispersion coefficient were 200, 300, and 500 cm^2/d for the columns which ranged in length from 5 to 44 cm. The larger dispersion values were used for the longer columns. The theoretical peak height without filtration was then calculated using CXTFIT (Parker and van Genuchten, 1984) with the retardation and dispersion values determined above. The difference in the calculated peak height without filtration and the observed peak height were attributed to filtration effects. The value of the filtration term was then determined as follows (Harvey and Garabedian, 1991):

$$\text{observed peak concentration} = \text{theoretical peak concentration} * \exp(-\gamma L) \qquad (4)$$

where γ is the filtration coefficient and L is the column length.

RESULTS AND ANALYSIS

Crushed-Rock Column Flow Experiments with Latex Microspheres

To design experiments for WIPP-relevant colloids, a set of twelve experiments were performed using monodisperse non-functionalized latex microspheres with diameters of 0.48 and 1.05 μm (Table 1). Non-functionalized microspheres were used to minimize chemical interaction with the column packing material and isolate filtration effects. Two different sizes of latex microspheres were used to study the effects of colloid size on physical retardation and filtration. To test the possibility that the intercrystalline porosity of the Culebra enhances physical retardation, parallel tests were conducted with Culebra rock and non-porous dolomite (containing no intercrystalline porosity). Culebra rock was crushed to sizes which preserved a significant fraction of intercrystalline pore throats (crushed particle sizes ranged from 125 to 1000 μm, compared to microcrystalline grain sizes of 2 μm, and mean pore throat diameters of 0.63 μm). An ionic strength of 0.01 M was selected for the eluant to minimize coagulation and settling.

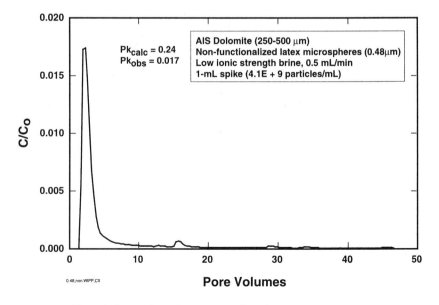

Figure 1. Latex microsphere transport through a crushed-rock column.

Table 1. Latex Microsphere transport Experiments

Column - Test Number	Packing Material	Packing Diameter (um)	Latex Particle Diameter (um)	Retardation Factor	Mass Balance (%)
I-a	Butte	125-250	1.053	No discernible peak	0.072
I-b	Butte	125-250	0.48	1.9	0.007
IV-a	Butte	250-500	1.053	1.5	0.64
IV-b	Butte	250-500	0.48	1.5	1.18
III-a	Butte	500-1000	1.053	3.3	3.44
III-b	Butte	500-1000	0.48	1.7	6.09
V-a	Culebra	125-250	1.053	1.6	0.11
V-b	Culebra	125-250	0.48	4.2	0.11
II-a	Culebra	250-500	1.053	1.6	1.37
II-b	Culebra	250-500	0.48	2.2	10.9
VI-a	Culebra	500-1000	1.053	1.3	8.95
VI-b	Culebra	500-1000	0.48	1.6	8.5

In general, elution curves for the latex microsphere experiments appeared qualitatively like typical elution curves for a dissolved solute spikes injected onto a column with a retardation factor of about 2.2, based on the peak position being at 2.2 pore volumes of the effluent (Figure 1). Note, however, that the peak concentration in the effluent at 2.2 pore volumes is only about 2% of the injected concentration. Typically, in dissolved solute transport, a much higher retardation factor is observed with such a large reduction in the peak concentration, suggesting that most of colloid particles are removed by physical filtration or irreversible adsorption.

The small elution peaks occurring intermittently during the remainder of the test and in the elution curves for other tests show that the colloidal particles trapped in pores created by packing of spherical particles are released back into the flowing water in spikes of varying sizes. However, the total number of particles in the effluent remains small. For example, in one typical experiment (II-a, Table 1), after 46.5 pore volumes were eluted, the total number of particles released is still only 1.4% of the injected particle concentration.

Physical filtration effects are reflected in the number population of latex microspheres observed in the effluent. The mass balances (integrated counts under the first peak and over the entire elution curve) indicate that about 11% or less of the injected colloidal particles were released from the columns (Table 1). In the case of columns packed with the smallest dolomite particle size fraction, essentially all the latex microspheres were physically trapped within the columns. The total number of trapped latex microspheres decreases with increasing size of the crushed-rock particle size fraction. That inverse relationship probably occurs because the average pore throat size is directly proportional to the size of packing material; the smaller pore throats trap particles more effectively than the larger pores.

Entrapment of latex microspheres appeared to be sensitive to the size of the microspheres as well. Comparison of each pair of experiments involving small and large microspheres indicates that a higher percentage of small microspheres are released in the effluent. That could be partly due to the fact the injected concentration of small microspheres in these experiments was about an order of magnitude higher than that of large colloids. However, the trend of enhanced releases of small microspheres is expected even at equivalent number populations of large and small particles, because large particles require larger spaces to move around due to their larger cross-sectional area. Once trapped, it would be harder for large microspheres to break free than the small microspheres.

The small latex microspheres were expected to show enhanced retardation in the Culebra dolomite relative to the non-porous Butte dolomite, but they did not (Table 1). This expectation was based on the fact that the Culebra dolomite has intercrystalline porosity with a mean pore throat diameter of 0.63 μm (Kelley and Saulnier, 1990), but the Butte dolomite is nonporous. Therefore, the microspheres should experience physical retardation (i.e., diffusion in the pores) as well as chemical retardation in the Culebra dolomite, but only chemical retardation in the Butte dolomite.

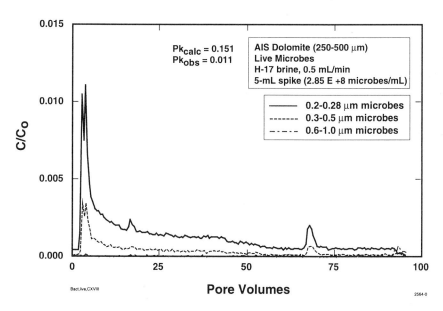

Figure 2. Microbe transport through a crushed-rock column.

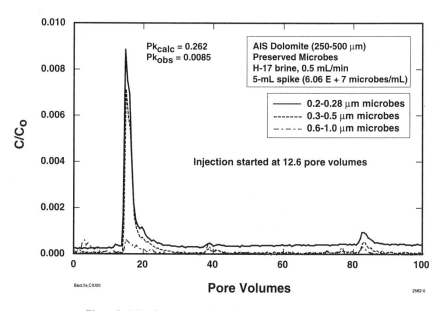

Figure 3. Microbe transport through a crushed-rock column.

Crushed-Rock Column Flow Experiments with Microbes

Crushed-rock column flow experiments were conducted with live and preserved microbe cultures developed from samples collected from several sources, including G-Seep, WIPP muck-pile salt, and saline lakes near the WIPP Site (Francis and Gillow, 1994). Naturally occurring halophilic or halotolerant microbes collected at the WIPP and WIPP vicinity are rod-shaped (aspect ratio of about five) and range in length from about 0.5 to 2 µm. Within a particular class of microbes, size will vary depending on the stage of their reproduction and also nutrient conditions. The outer surface of microbes consists of complex organic molecules with exposed surface functional groups. The surface chemistry of microbes is an important characteristic which affects how they interact with the host rock during transport, as well as how they bioaccumulate dissolved actinides. Measurements of electrophoretic mobility of microbes in WIPP brines, however, showed that the surface charges of microbes is near neutral, suggesting limited chemical interaction with rock surfaces.

Most of the experiments were conducted with preserved microbes harvested during the stationary phase of the growth curve, and fixed with formalin. Some experiments were also conducted with live microbes of the WIPP-1A culture in nutrient-free environment. Experiments in which nutrients were added along with the live microbes resulted in nearly complete stoppage of flow through the column and therefore were not used for developing parameters. Tests were conducted with 10^7 to 10^8 cells/mL spike injections; these concentrations are similar to concentrations of microbes observed in natural environments at the WIPP. Step injection tests conducted with 10^6 to 10^7 cells/mL resulted in column plugging and were not used for determination of parameter values.

Generally, H-17 brine simulant (primarily NaCl with an ionic strength of 3 molal) was used as the eluant. Experiments were conducted with 250-500 µm sized packing to acquire data for determination of parameter values.

The concentration of microbes eluting from the column was usually measured in real time, with a particle spectrometer connected to the effluent line. Aliquots were collected periodically for direct counting using epifluorescence microscopy as an independent check that the particle spectrometer was counting accurately. The particle spectrometer records number concentrations in several size intervals. A recurring observation in the microbe tests was that larger microbes (0.6 to 1.0 µm; defined by counting-channel-width of the spectrometer) were more effectively filtered than smaller microbes (0.2 to 0.28 µm; defined by counting-channel width) (Figures 2 and 3). Under nutrient-poor conditions, such as might occur in the Culebra down gradient from the intrusion borehole, microbes become reduced in size. The elution curve representing transport of the smaller microbes was used to develop parameter values. That approach is likely to result in some degree of conservatism.

The results of the tests were interpreted using the approach described earlier. Representative elution curves from tests conducted with live microbes and preserved microbes are presented in Figures 2 and 3, respectively. A compilation of the values used and results of the numerical

Table. 2 Selected Crushed-Rock Column Experiments

Colloid	size fraction (um)	Column Length (cm)	Packing porosity	Pump Flow Rate (mL/min)	Spike Volume (mL)	Brine	R	Flow Velocity (cm/min)	Est Dispersion (cm^2/d)	γ (1/cm)
Preserved microbes	0.2 to 0.28	5.1	0.44	0.5	0.5	H-17	2.3	1.46	200	0.67
Live microbes	0.2 to 0.28	5.2	0.46	0.5	0.5	H-17	3.8	1.40	200	0.51
Goethite	0.2 to 0.28	5.2	0.45	0.5	1.0	1 mM*	4.9	1.42	200	0.10
Quartz	0.2 to 0.28	5.5	0.47	0.5	1.0	1 mM*	2.5	1.34	200	0.26
Latex microspheres	0.48	12.0	0.50	0.5	1.0	10 mM*	2.2	1.28	300	0.22
Humic acid	entire range	44.1	0.43	0.111	5.0	H-17	1.0	0.33	500	0.02
Fulvic acid	entire range	44.1	0.43	0.113	5.0	H-17	1.0	0.33	500	0.01

* Pure NaCl

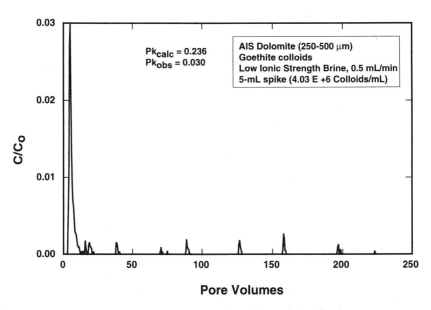

Figure 4. Mineral fragment transport through a crushed-rock column.

interpretation is presented in Table 2. Retardation factors for microbes themselves (actinides were not used in these tests) was about 4; this low value is consistent with our expectations. Filtration coefficients were 0.67 and 0.51 cm^{-1} for preserved and live microbes, respectively (Table 2).

Results of the crushed-rock column flow tests on microbes were evaluated with a flow experiment on intact core under simulated subsurface conditions and were consistent with results of the intact core column flow tests (see Lucero et al., this volume).

Crushed-Rock Column Flow Experiments with Mineral Fragments

Dispersions of mineral fragment colloidal particles were made by mechanically milling quartz (SiO$_2$) or through chemical precipitation of goethite (FeO·OH). A 1mM NaCl solution was used to minimize the agglomeration of particles due to Brownian motion or induced by hydrodynamic motion. After allowing dispersions to agglomerate and settle, an aliquot of the supernatant was extracted. The aliquot was filtered with a 1.2 μm filter to remove suspended agglomerates and injected as a spike onto a crushed-rock column. The residual concentration of colloidal particles in the supernatant was on the order of 10^6 particles/mL.

The elution curve profiles for mineral colloids (Figure 4) were similar to the latex microsphere experiments and the microbe experiments (Figures 1, 2, and 3, respectively). Elution curves were analyzed using the approach described earlier. As with microbes, the filtration coefficients obtained show effective attenuation of the mineral fragment colloids by the crushed rock (Table 2). A retardation factor (for the colloid; actinides were not used in the test) on the order of 4 to 6 was observed in the tests (Table 2).

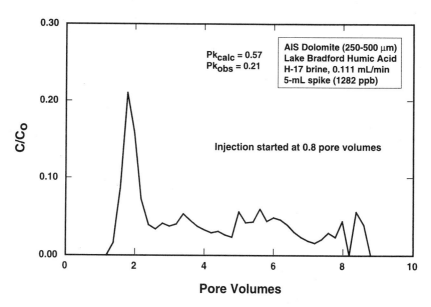

Figure 5. Humic acid transport through a crushed-rock column.

Crushed-Rock Column Flow Experiments with Humic Substances

Humic substances represent a variety of high-molecular weight organic compounds. They form by a variety of processes in terrigenous and marine environments, resulting in compounds with different geochemical behaviors. It is believed that humic substances will form after emplacement of the waste at the WIPP, but the nature of the humics cannot be predicted. To span the likely range of behaviors, three types were used in the colloid source term and transport work: aromatic humic acid, aliphatic humic acid, and fulvic acid. Samples of those types were collected by Florida State University, purchased commercially, obtained from colleagues, or purchased from the International Humic Substances Society (Golden, Colorado).

Typical results are illustrated in Figure 5. Elution curve profiles from the final experiments were analyzed using the approach described earlier. Very little filtration of the humic substances was observed (Table 2). This result was anticipated because humic substances, in their ionic form, are dissolved macromolecules and are essentially not susceptible to physical (straining) filtration as are microbes and mineral fragment. The experiments showed that the humic substances eluted at the same number of pore volumes as fluorescein dye, the nonsorbing tracer, indicating that enhanced transport of humic substances did not occur. A hypothesis that was rejected based on these tests was the possibility that the >Mg and >Ca surface functional groups of the dolomite would have reacted with the humics, resulting in sorption.

CONCLUSIONS

Transport experiments using sieved, uniformly packed crushed Culebra rock demonstrated that mineral fragments and fixed and live WIPP-relevant bacteria cultures were significantly retarded due to physical filtration effects, but humic substances were not attenuated by the Culebra dolomite. Comparison of elution curves of latex microspheres in columns prepared with microcrsytalline rock and nonporous rock showed minimal effect of Culebra micropores on colloid transport.

REFERENCES

Avogadro, A., and G. de Marsily. 1984. "The Role of Colloids in Nuclear Waste Disposal," *Scientific Basis for Nuclear Waste Management VII, Materials Research Society Symposia Proceedings, Boston, MA, November 14-17, 1983*. Ed. G.L. McVay. New York, NY: North-Holland. Vol. 26, 495-505.

Francis, A.J., and J.B. Gillow. 1994. *Effects of Microbial Processes on Gas Generation Under Expected Waste Isolation Pilot Plant Repository Conditions. Progress Report Through 1992.* SAND93-7036. Albuquerque, NM: Sandia National Laboratories.

Harvey, R.W., and S.P. Garabedian, 1991. "Use of Colloid Filtration Theory in Modeling Movement of Bacteria through a Contaminated Sandy Aquifer," *Environmental Sciences and Technology,* Vol. 25, p. 178-185.

Jacquier, P. 1991. "Geochemical Modelling: What Phenomena are Missing?" *Radiochimica Acta* Vol. 52/53, pt. 2, 495-499.

Kelley, V.A., and G.J. Saulnier, Jr. 1990. *Core Analyses for Selected Samples from the Culebra Dolomite at the Waste Isolation Pilot Plant Site.* SAND90-7011. Albuquerque, NM: Sandia National Laboratories.

Lucero, D.A., F. Gelbard, Y.K. Behl, and J.A. Romero. 1995. "Test Plan for Laboratory Column Experiments for Radionuclide Adsorption Studies of the Culebra Dolomite Member of the Rustler

Formation at the WIPP Site". SNL Test Plan TP 95-03. Albuquerque, NM: Sandia National Laboratories.

Papenguth, H.W., and Y.K. Behl. 1996. "Test Plan for Evaluation of Colloid-Facilitated Actinide Transport at the Waste Isolation Pilot Plant." SNL Test Plan TP 96-01. Albuquerque, NM: Sandia National Laboratories.

Parker, J.C., and M.T. van Genuchten. 1984. *Determining Transport Parameters from Laboratory and Field Tracer Experiments,* Virginia Agricultural Experimental Station Bulletin. VAESB 84-3. Blacksburg, VA: Virginia Agricultural Experimental Station. 1-96.

Public Law 96-164. 1979. "Department of Energy National Security and Military Applications of Nuclear Energy Authorization Act of 1980."

U.S. EPA (Environmental Protection Agency), 1993. "40 CFR Part 191: Environmental Radiation Protection Standards for the Management and Disposal of Spent Nuclear Fuel, High-Level and Transuranic Radioactive Wastes; Final Rule," *Federal Register.* Vol. 58, no. 242, 66398-66416.

WIPP PA (Performance Assessment) Division. 1991a. *Preliminary Comparison with 40 CFR 191, Subpart B for the Waste Isolation Pilot Plant, December 1991—Volume 1: Methodology and Results.* SAND91-0893/1. Albuquerque, NM: Sandia National Laboratories.

WIPP PA (Performance Assessment) Division. 1991b. *Preliminary Comparison with 40 CFR 191, Subpart B for the Waste Isolation Pilot Plant, December 1991—Volume 2: Probability and Consequence Modeling.* SAND91-0893/2. Albuquerque, NM: Sandia National Laboratories.

ACKNOWLEDGEMENTS

This work was supported by the United States Department of Energy under Contract DE-ACO4-94-AL85000. Sandia is a multiprogram laboratory operated by Sandia Corporation, a Lockheed Martin Company, for the United States Department of Energy.

LABORATORY EVALUATION OF COLLOID TRANSPORT UNDER SIMULATED SUBSURFACE CONDITIONS AT THE WASTE ISOLATION PILOT PLANT (WIPP): 2. LARGE-SCALE-INTACT-CORE COLUMN FLOW EXPERIMENTS.

D. A. Lucero[1], Y. K. Behl[2], G. O. Brown[3], K. G. Budge[1], M. Dunn[4], A. J. Francis[4], J. B. Gillow[4], and H. W. Papenguth[1]

[1] Sandia National Laboratories, Albuquerque, NM 87185
[2] SciRes, Albuquerque, NM 87122
[3] Oklahoma State University, Stillwater, OK 74078
[4] Brookhaven National Laboratory, Upton, NY 11973

INTRODUCTION

The Waste Isolation Pilot Plant (WIPP), located approximately 42 km southeast of Carlsbad, New Mexico, is being developed by the U.S. Department of Energy (DOE) as a disposal facility for transuranic waste produced by defense nuclear weapons programs. Regulations promulgated by the U.S. Environmental Protection Agency (1985) place limits on the cumulative radioactive release to the accessible environment over 10,000 years and require performance assessments to demonstrate WIPP compliance with regulatory standards. The only credible mechanisms for escape of radionuclides to the environment involve a breach of the repository during drilling for natural resources. The brines present in the repository, together with drilling mud and cuttings, may be transported through intrusion boreholes to a overlying transmissive carbonate rock unit, the Culebra Dolomite Member of the Rustler Formation, or directly to the surface, by pressurized fluids in the repository. Performance assessment calculations have shown that thorium, uranium, neptunium, plutonium, and americium are the most significant radionuclides in terms of total release of activity.

Previous work has been performed on soluble actinide sorption and retardation in the Culebra (Lucero et al., 1994). In those tests, uranium has undergone moderate retardation, $(2 < R < 20)$ while americium and plutonium have never eluted from the column at measurable activities. Sterile conditions were not maintained in the core or injection solution and measurable bacteria numbers are always present in the column effluent. At least some of those bacteria are believed to be native to the cores themselves. Large concentrations of humic compounds are not present in the cores or brine.

The objective of this work was to quantify possible colloid-facilitated transport of actinides in large, intact Culebra cores under pressure and flow conditions similar to the natural environment. Two types of colloids were tested, a mixture of humic and fulvic acid and live microbes. Since bacteria are naturally present in the core effluent, the microbes

Actinide Speciation in High Ionic Strength Media, edited by Reed *et al.*
Kluwer Academic / Plenum Publishers, New York, 1999

tests are a measure of the effects of a microbe enrichment. These results can be compared to crushed-rock retardation studies performed by Yelton et al., (1996).

PROCEDURES

Two separate tests were performed, one using humics and the second using a enriched bacteria population (Papenguth and Behl, 1996). Intact cores were collected by horizontal drilling at a depth of approximately 220 meters inside the Air Intake Shaft at the WIPP. The microbe test was performed with Core E, a subsample of core VPX27-7A; while the humics test used core D, a subsample of core VPX25-8A. Both cores were 14.5 cm in diameter and 10 cm long, and oriented in the direction of the current groundwater flow. Several types of porosity which may affect hydraulic and transport properties were present in each, including microcrystalline, porous carbonate interbeds, vugs, fractures, and brecciated zones. The cores, which are shown in Figure 1, were similar, with the exception that Core D showed more vugs and carbonate interbeds. The dolomite matrix initially formed with interbeds of silty dolomite and large vugs which were later connected by subvertical fractures. Bacteria are generally larger than the microcrystalline and carbonate interbed porosity. However, the vugs, fractures and brecciated zones were visible to the naked eye and could possibly allow bacteria transport. The humic compounds were small enough to fit through all pore features.

A schematic of the test apparatus used is shown in Figure 2. Its major components were column core holders, syringe pumps, brine injection accumulator and effluent collectors. The aluminum column barrel had teflon coated, brass end fittings, with the core held in a flexible sleeve between the two distributor plates. Overburden loads were simulated by applying pressure in the annulus between the column barrel and the sleeve. That pressure was also applied to the top distributor plate which was free to slide up and down, while the bottom plate was fixed. Thus, the core was subjected to the reservoir pressure in all three directions while allowing the effluent to exit at atmospheric pressure. The applied triaxial confining pressure of 725 psi was estimated assuming a rock density of 1 psi/ft above the borehole (Mercer and Snyder, 1990).

Brine was pumped from an accumulator through the core at the rate of 0.1 ml/min, which produced a Darcy flux of 10^{-5} cm/s, the upper range of Culebra flow. Double, high pressure, liquid chromatography syringe pumps drive deionized water into one end of an accumulator, which forces brine out the other. The accumulator, a simple cylinder and piston sealed at both ends, prevents contamination and fouling of the pumps, while the double pump configuration allows automatic filling of the syringe pumps and insures continuous operation. Effluent was collected with a fraction collector set to provide approximately 5 ml per sample.

The humic acid mixture used was equal parts of aliphatic humic (isolated from sediments collected from Lake Bradford, Florida), aromatic humic and fulvic acids (both isolated from the Suwannee River and purchased from the International Humic Substances Society, Golden, Colorado) added to the synthetic brine at 0.1 mg/l concentration. Bacteria used to enrich the injection were WIPP-1A; live, halophilic, rod-shaped cells. They were added to the synthetic brine at 10^6 cells/ml. That concentration was roughly equal to the native population of the synthetic brine. While no specific steps were taken to inoculate the synthetic brine, it was equilibrated with crushed Culebra samples and occasionally left open to the lab air. As a result, bacteria similar to WIPP-1A can be found in the injection fluid. Both colloids were introduced as step injections to allow greater time for colloid-actinide-rock interaction and to enhance the possibility of demonstrating some effect of the colloids. A conservative tracer, ^{22}Na, was also added to the step injections to quantify column porosity and dispersion. Actinides were introduced by two methods. Uranium was

Core D - VPX25-8A

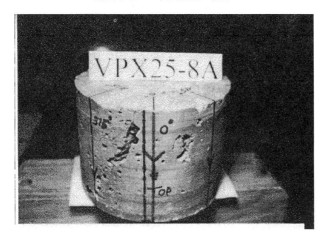

Core E - VPX27-7A

Figure 1. Intact core samples used.

introduced with the colloid step injection to maximize its contact colloid time, while americium and plutonium were injected as spikes directly on the rock to eliminate sorption on the experimental apparatus. Test conditions for both injections are listed in Table 1.

The use of live microbes required additional procedures to ensure cells population did not change during testing and analysis. The accumulator that held the brine-microbe mix was insulated and chilled to $10^{\circ}C$ to minimize cell growth or mortality. Due to the necessity of maintaining a uniform flow, samples could not be taken from the accumulator once the test began and prevented direct monitoring of the microbes being injected. Therefore, a sample of bacteria enriched brine was placed in a mock injection vessel and maintained under similar conditions. The mock vessel was sampled at several times during the injection and the samples subjected to cell assays along with the effluent.

Most column effluent was analyzed by gamma ray spectrometry for ^{22}Na, ^{232}U, and ^{241}Am and liquid scintillation for ^{232}U and ^{241}Pu at Sandia National Labs following standard methods (Lucero et al., 1994). The radioassay tests are insensitive to colloids or actinide chemical state, and their results indicate only total sample activity. Selected effluent samples from the enriched bacteria test were fixed with formalin solution, chilled and shipped overnight to Brookhaven National Laboratory where they underwent cell

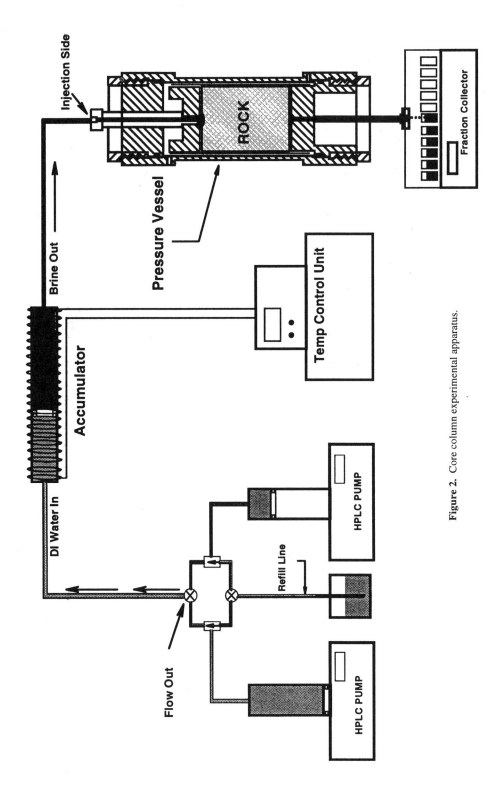

Figure 2. Core column experimental apparatus.

Table 1. Test conditions and results.

Test	Humics	Bacteria
Colloids	0.1 gm/l	10^6 cells/ml
Core and Brine	VPX25	VPX27
Brine Composition		
Boron (gm/l)	0.026	0.027
Bromine (gm/l)	0.054	0.024
Calcium (gm/l)	0.88	0.86
Inorganic Carbon (gm/l)	0.011	0.17
Chloride (gm/l)	19	19
Magnesium (gm/l)	0.45	0.45
Potassium (gm/l)	0.32	0.32
Sodium (gm/l)	13	13
Sulfate (gm/l)	6.7	6.9
pH	7.7	8.1
Step Injection		
Volume (ml)	4,000	3,900
^{232}U (uCi/ml)	0.020	0.034
^{22}Na (uCi/ml)	0.004	0.004
Initial Actinide Spike		
^{241}Am (uCi)	4.0	13
^{241}Pu (uCi)	16	---
Fitted transport parameters		
Porosity	0.12	0.15
Dispersivity (cm2/s)	0.00079	0.00076
Uranium retardation	12.8	2.7

assays by direct counting using epifluorescence microscopy. Bacteria length and width was measured for 15 to 53 individuals in each sample and the population volumes computed.

RESULTS

Figure 3 presents cell concentration and volume as functions of time for both the column effluent and the mock injection brine. The overall number of cells in the mock injection vessel increased after the start of the test to 7×10^6 cells/ml at 600ml and then decreased to 2×10^6. This would indicate an initial growth followed by a die-off to equilibrium levels. However, cell size in the mock injection vessel was relatively constant throughout the experiment.

Effluent cell counts were an order of magnitude lower than injection and ranged from 1.7×10^5 to 7.7×10^5. Cell numbers exhibited an erratic behavior with the largest peak at 700 ml. Early effluent, from 0 to 1,100 mls, contained large cells which may have resulted from some enriched cell breakthrough. Bacterial cell volume was greatest at 400 ml, but after 1,500 ml the effluent bacteria were indistinguishable from the normal effluent production. There were many eukaryotic organisms in the effluent, that appear to be generated in the column itself. These flagellated microbes may have eaten a small portion of the injected bacteria.

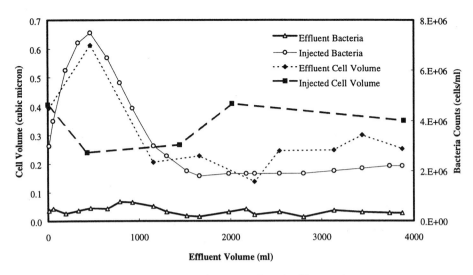

Figure 3. Live microbe cell numbers, and volume in effluent and mock injection.

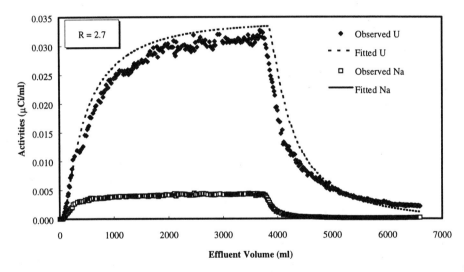

Figure 4. ^{22}Na and ^{232}U effluent activity for live microbe tests.

Figures 4 and 5 show the breakthrough curves for the humics and bacteria enriched tests. Each graph shows the measured effluent curves for ^{22}Na and ^{232}U and retardation fittings for the single porosity model of Parker and van Genuchten (1984). Fitted parameters are listed in Table 1. For the enriched bacteria test, the fitted uranium retardation was 2.7 compared to 2 for the same core without enrichment. For the humic test the fitted uranium retardation was 12.8 compared to 9.4 and 21 for normal conditions. For both tests the variation is within the normal range observed, indicating the colloids did not impact the uranium transport. Americium did not elute at detectable activities in either test, and plutonium did not elute in the humic test. Thus, the colloids did not effect the transport of those actinides at measurable levels.

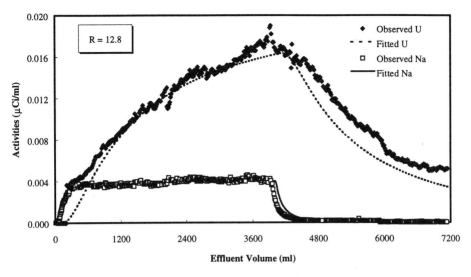

Figure 5. 22Na and 232U effluent activity for humics test.

CONCLUSIONS

The presence of an enriched microbe population had no significant impact on the retardation of uranium and americium in the intact Culebra cores. Even through the cores contained some relatively large pores, microbe cell counts were reduced an order of magnitude over the 10 cm column. This reduction is consistent with the filtration observed by Yelton et al., (1996). Likewise, the presence of humic compounds had no significant impact on the retardation of uranium, plutonium and americium in the Culebra cores.

ACKNOWLEDGMENTS

Charles Heath, Fred Salas and Lori Montano provided technical and analytical support for this project. We appreciate the guidance and support received form Butch Stroud and Dick Lark of DOE, Carlsbad Area Office. This work was supported by the United States Department of Energy, under Contract DE-AC04-94AL85000. Sandia is a multiprogram laboratory operated by Sandia Corporation, a Lockheed Martin Company, for the United States Department of Energy.

REFERENCES

Lucero, D. A., F. Gelbard, Y. K. Behl, and J. A. Romero. 1995. *Test Plan for Laboratory Column Experiments for Radionuclide Adsorption Studies of the Culebra Dolomite Member of the Rustler Formation at the WIPP Site.* WIPP Test Plan, TP 95-03. WPO#22640-October 12, 1994. Sandia National Laboratories, Albuquerque, NM.

Papenguth, H. W., and Y. K. Behl. 1996. *Test Plan for Evaluation of Colloid-Facilitated Actinide Transport at the Waste Isolation Pilot Plant. WIPP Test Plan, TP 96-01.* WPO#31337. Sandia National Laboratories, Albuquerque, NM.

Parker, J. C., and M. T. van Genuchten. 1984. *Determining Transport Parameters from Laboratory and Field Tracer Experiments. Virginia Agricultural Experiment Station Bulletin VAESB 84-3.* Virginia Agricultural Experiment Station, Blacksburg, VA.

US Environmental Protection Agency. 1985. "Environmental Standards for the Management and Disposal of Spent Nuclear Fuel, High-Level, and Transuranic Radioactive Waste; Final Rule, 40 CFR Part 191," *Federal Register* 50: 38066- 38089.

Yelton, W. G., Y. K. Behl, J. W. Kelley, M. Dunn, J. B. Gillow, A. J. Francis, H. W. Papenguth. 1996. Laboratory evaluation of colloidal actinide transport at the Waste Isolation Pilot Plant (WIPP): 1. Crushed-dolomite column flow experiments, in *Experimental and Modeling Studies of Radionuclide Speciation in Real Systems*, D. T. Reed, S. Clark and L. Ruo (eds), Plenum Publishing Corp., NY.

TOXICITY OF ACTINIDES TO BACTERIAL STRAINS ISOLATED FROM THE WASTE ISOLATION PILOT PLANT (WIPP) ENVIRONMENT

Betty A. Strietelmeier[1], Jeffrey B. Gillow[2], Cleveland J. Dodge[2], Maria E. Pansoy-Hjelvik[1], Suzanne M. Kitten[1], Patricia A. Leonard[1], Inés R. Triay[1], A.J. Francis[2], and Hans W. Papenguth[3]

[1]Los Alamos National Laboratory
[2]Brookhaven National Laboratory
[3]Sandia National Laboratories

INTRODUCTION

The possibility of toxic effects from several actinide elements to bacteria isolated from the Waste Isolation Pilot Plant (WIPP) site has been investigated. This study is part of an extensive ongoing research program that endeavors to validate the suitability and safety of the WIPP site as a transuranic (TRU) radioactive waste repository. The motivation for the toxicity studies was to determine the eventual fate of the actinides after their contact with microorganisms relevant to the WIPP site. The toxicity studies investigated possible adverse effects to the growth of the bacteria due to actinide toxicity. Actinide interactions with the bacteria will impact actinide transport or retardation by the bacterial species.

This work was performed in a collaborative effort between Brookhaven National Laboratory (BNL) and Los Alamos National Laboratory (LANL). The two laboratories have unique technical capabilities required for this study. Researchers at BNL have studied microbial activity at the WIPP site with respect to gas generation (Francis et al., 1994; Francis et al., 1990) as well as species characterization. They have isolated from the WIPP environs the pure and mixed bacterial cultures used in these studies. At LANL, researchers have the expertise and special facilities to work with plutonium and americium at levels possibly toxic to these bacteria. In addition, other work at LANL has focussed on determining the viability of bacterial populations inoculated into 54 different test containers which hold actual TRU waste similar to waste that could be stored at the WIPP site (Villarreal and Phillips, 1993; Pansoy-Hjelvik et al., 1997).

The research presented here is the first in which the toxic effects to halophilic, facultative anaerobes (denitrifiers) due to Th, U, Np, Pu and Am have been investigated. These actinides are present in existing radioactive waste destined for long-term storage at the WIPP site. The bacteria relevant to these studies were isolated from muck pile salt, hypersaline lake brine and sediment slurry taken from areas surrounding and underground the WIPP site. In the unlikely event of a groundwater intrusion into the WIPP repository in the future, the bacteria could perhaps come into contact with the actinides in the waste. From this standpoint, it is important to determine if the levels of actinides expected to be present in the repository could have an inhibitory or lethal effect on the bacteria. If the levels of actinides are not lethal to the bacteria, the radiological and chemical toxicity can still result in an inhibitory effect to the bacterial growth or morphological changes to the cells.

Any effects to bacterial growth or morphology due to an actinide toxicity must be investigated because actinide retardation or transport by bacteria depends on the size and cell surface characteristics of the bacteria. Large cells, greater than ~0.3 μm in one dimension,

Actinide Speciation in High Ionic Strength Media, edited by Reed *et al.*
Kluwer Academic / Plenum Publishers, New York, 1999

261

will have the potential to be filtered out or removed from the groundwater by the rock matrix due to the size and structure of matrix pores. In fractured rock, however, the transport of larger bacterial cells could be more prevalent, since there will be paths of larger diameter, allowing for transport of larger bacteria at greater flow rates. The bacterial cells are likely to act in a manner similar to colloids under these conditions, and may transport much faster than dissolved species.

With respect to actinide transport by the bacteria, it is also important to determine the extent of any actinide-cell association such as sorption, surface particulate formation, cell surface complexation or internal bioaccumulation. Bacteria are known to actively bioaccumulate actinides intracellularly and can provide passive extracellular surface sorption sites (Francis, 1990). In work performed concurrently with the toxicity studies, ultra-microfiltration techniques were utilized to determine the amount of actinide associated with different size fractions of the cultures, down to 0.03 μm. The results from the ultramicrofiltration work are presented separately in this publication (Gillow et al., 1997).

Toxic effects to the two cultures of bacteria due to the actinides were studied by measuring changes to turbidity and total cell counts, as well as pH, as a function of incubation time in growth media with various initial concentrations of actinide. Any toxic effect could be observed, specifically, as fluctuations in the turbidity and total cell counts relative to controls without actinide. In the experiments where the actinide remained in solution, relatively little toxic effect was observed. However, in experiments where the actinide was present both in solution and solid phases, dramatic effects to cell growth were observed. These effects are speculated to be a result of an ancillary versus a direct toxicity.

EXPERIMENTAL

Bacterial Cultures

WIPP1A. WIPP1A is a gram-positive rod with dimensions of approximately 1-2 μm in length by 0.5-0.75 μm in width. It is grown and maintained in WIPP1A growth medium. It is a facultative anaerobe, uses nitrate as an electron acceptor under anaerobic conditions, and reaches static (maximum cell number) growth by 72 hours. Within 32 hours after the bacteria has been put into fresh growth medium, some of the bacteria gradually settles out of the medium fluid colum.

The WIPP1A culture was isolated from an inoculum developed for the WIPP Gas Generation Program (Francis, 1990). The inoculum consisted of muck pile salt, hypersaline lake brines, and sediment slurry taken from the WIPP environs.

Mixed Culture - BAB. The mixed culture is a stable combination of 3-5 bacterial cell types. The cells vary in shape: long thin rods, triangular species, oval-shaped cells and very small cocci are present and easily distinguishable by epifluorescence microscopy. Together they are grown and maintained in BAB growth medium. The consortium of bacteria are all facultative anaerobes, and as with WIPP1A, use nitrate as an electron acceptor under anaerobic conditions. The growth behavior of the mixed culture is different from WIPP1A, however; it reaches static growth by 15 days and remains suspended in the growth medium fluid column over long periods of time.

The BAB culture was isolated from an inoculum developed by BNL for the Actinide Source Term Waste Test Program (STTP) (Villarreal and Phillips, 1993). The inoculum came from the same areas from which the WIPP1A inoculum was obtained.

Preparation of Radionuclides (An)

Stock solutions of 2mM ^{239}Pu, ^{243}Am and 20 mM ^{237}Np were electrochemically purified and provided as the uncomplexed species. Stock solutions of An:EDTA complexes were prepared by mixing equal volumes of equimolar solutions of the actinide stock and aqueous EDTA. After mixing, the pH of the An:EDTA solutions were adjusted to approximately pH = 6.0 in order to ensure complex formation. Appropriate volumes of the An:EDTA stock solutions were added to the growth media in order to obtain final An concentrations in the two different growth media as follows: [Pu] = 5 x 10^{-5} M, 1 x 10^{-5} M, 1 x 10^{-6} M and 1 x 10^{-7} M; [Am] = 5 x 10^{-6} M, 5 x 10^{-7} M and 5 x 10^{-8} M; [Np] = 5 x 10^{-4}

M, 5 x 10^{-5} M and 5 x 10^{-6} M.

The EDTA maintained the solubility of Pu and Am species during the time-course of the experiment. However, in the growth medium containing Brine A and Np:EDTA, a fine white precipitate formed after 24 hours. Diffuse reflectance spectroscopy analysis indicated that the precipitate was most likely a magnesium salt of Np:EDTA, formed due to the high Mg content in Brine A.

The preparation of solutions containing 1:1 complexes of Th:EDTA, Th:NO$_3$, or U:NO$_3$ involved mixing purified Th or U stock solutions with equal volumes of equimolar solutions of EDTA and nitrate. Targeted initial concentrations of the Th and U complexes were 4 x 10^{-4} M, 1 x 10^{-3} M, 2 x 10^{-3} M, and 4 x 10^{-4} M.

Control experiments involving the An:complex/growth media with no added inoculum enabled us to resolve whether decreases in actinide solubility were a result of An association with bacteria or a result of the formation of an insoluble An species. More detail regarding the control experiments will be discussed below.

WIPP1A Growth Medium - WIPP Brine and Nutrients

The pure culture, WIPP1A, was grown under nitrate-reducing conditions in a gas-purged 20% (w/v) WIPP halite sterile medium containing as nutrients (g/L): sodium succinate, 5.0; KNO$_3$, 1.0; K$_2$HPO$_4$, 0.25; yeast extract, 0.5; with an initial, unadjusted pH of approximately 6.3. Radionuclides were added to the medium at the appropriate concentrations, while maintaining the pH between 6.1 and 6.5. The sterile medium was aliquoted in an anaerobic glovebox and inoculated with a 24-hour culture prior to sealing, followed by incubation at 30°C for the required time interval. Uninoculated controls were included in each experiment. Samples were removed at appropriate intervals from both control and test containers after transfer from the incubator into the anaerobic glovebox. Oxygen was excluded from the cultures at every stage in the procedure. Samples were aliquoted into separate containers for the various analytical procedures described.

BAB Growth Medium - Brine A and Nutrients

The mixed culture, BAB, was also grown under nitrate-reducing conditions, with the same nutrient concentrations as with the pure culture. However, the brine, Brine A, used contained (g/L): NaCl, 100.1; Na$_2$SO$_4$, 6.2; Na$_2$B$_4$O$_7$·10H$_2$O, 1.95; NaHCO$_3$, 0.96; NaBr, 0.52; KCl, 57.2; MgCl$_2$·6H$_2$O, 292.1; CaCl$_2$, 1.66; with an initial, unadjusted pH of approximately 6.5. Procedures similar to those used with the pure culture were used throughout these experiments. Brine A simulates brine that is commonly encountered at the WIPP site.

Turbidimetric and pH Measurements

A 1 ml aliquot of the unfiltered culture was used to measure the turbidity at 600 nm using a dual-beam Spectrophotometer. The turbidity was measured against deionized water using disposable polystyrene cuvettes. Cultures were well mixed prior to taking absorbance readings. The spectrophotometer is self-calibrating.

A pH meter with a sealed, gel-filled micro-electrode was used for all pH measurements. The pH meter was calibrated before each use using a 3-point calibration at pH 4, 7 and 10, with re-analysis of the NIST-traceable calibrators following each batch of samples. Samples were filtered using a 0.2 μm syringe filter prior to obtaining pH measurements.

Bacterial Enumeration by Epifluorescence Microscopy

Total bacterial counts were performed using an epifluorescence microscope with an excitation wavelength of 340 nm and a fluorescence wavelength of 488 nm. A minimum of 1000 cells were counted on each slide, using multiple fields, when necessary. Slides were prepared in duplicate for each control and sample. Control slides were checked for contamination but were not quantitated. The specific fluorochrome used to render the bacteria fluorescent was the deoxyribonucleic acid (DNA)-specific stain 4',6-diamidino-2-phenylindole dihydrochloride (DAPI).

An aliquot of each sample was preserved with 10% formalin at the time of sampling. In order to provide accurate measurements of cell numbers in the time-dependent experiments, cell preservation was required immediately after the sample was taken to prevent further growth of cells. Since the cells were added to deionized water for staining by DAPI, preservation was also required to prevent cell lysis due to drastic changes in ionic strength. The preserved sample was stored at 4°C until slides could be prepared.

The slides were prepared by first obtaining a small volume of the preserved sample. Approximately 10-40 μl of a 1 mg/ml DAPI solution and 2 ml of Nanopure water were added to the sample solution. After the sample was placed in the dark for 5-7 minutes, the sample solution was vacuum-filtered onto a 0.20 μm poresize black Poretics membrane. Filtration was performed quantitatively and in a manner such that the sample material was filtered uniformly across the membrane. The membrane was mounted on a microslide and a coverslip placed over the membrane was sealed into place with silicon sealant.

Radionuclide Analysis

Analysis of solubilized Np, Pu and Am radionuclides was performed by measuring alpha radiolytic activity in the solutions via liquid scintillation counting (LSC) using a Packard LS Analyzer. The alpha activity of Pu and Am was quantitated directly while alpha activity of neptunium was determined by using an alpha/beta discrimination mode in order to eliminate the contribution to total activity from beta decay of the protactinium daughter of neptunium.

The LSC measurements were performed by first separating any 1) precipitates from the controls or 2) precipitates and bacteria from the inoculated samples by 0.20 μm filtration. LSC analysis was then performed on the filtered solution.

Uranium and Th in solution were quantitated by ICP/AES measurement following 0.20 μm filtration.

Test Container Sampling

The culture bottles were well-shaken before aliquots were taken for analysis. This was to assure that any inhomogeneities occurring as a result of the static growth conditions were not manifested during analysis. Measured quantities were thus an averaged estimation of concentrations in an homogeneous medium.

RESULTS AND DISCUSSION

Two distinct growth medium conditions were observed in these studies, and these conditions differed primarily in the solubility of the actinides in the test solutions. The two bacterial cultures used in these studies contain facultative anaerobes (denitrifiers), and under static growth conditions the bacteria will be distributed throughout the growth medium. Actinide elements will either be associated with cell surfaces (through sorption, precipitation, etc.), remain in the solution phase, or be precipitated. The presence of precipitated actinide (primarily in the Np, U and Th cases), in the form of a fine white dispersion, made interpretation of the effects seen in these studies difficult. The results obtained with soluble vs. insoluble actinide are discussed separately, with more emphasis placed on those experiments involving solubilized actinide (Pu and Am).

A large percentage of the Pu and Am initially added remained in solution throughout the duration of the experiment. The solution phase actinide concentrations were quantified by LSC analysis of the uninoculated control growth media, and confirmed by the lack of detectable turbidity. Growth curves for WIPP1A and BAB microbial cultures containing either Pu and Am are presented in Figure 1. An examination of these growth curves indicates that there is little apparent toxic effect from either actinide. However, two noticeable effects were detected: 1) a decrease in maximum cell numbers in the Pu/ WIPP1A and Pu/BAB experiments, Figures 1a and 1c,   and   2)   an   extension   of   the   lag

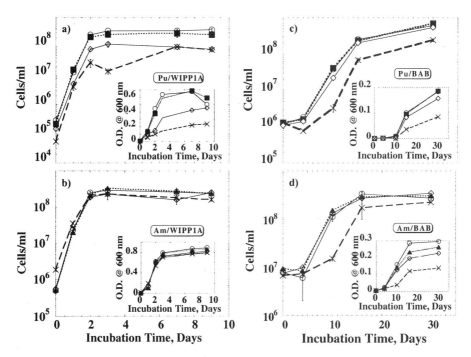

Figure 1. Growth curves of WIPP1A and BAB cultures exposed to Pu and Am at various concentrations. The cell counts presented are the mean of analysis of at least three sample slides, 5-30 fields counted per slide. Error bars shown are for one Standard Deviation. At inoculation, concentration (M) of [PuEDTA] = 0 (open circles); 1×10^{-7} (closed squares); 1×10^{-6} (open diamonds); 1×10^{-5} (X); [AmEDTA] = 0 (open circles); 5×10^{-8} (closed triangles); 5×10^{-7} (open diamonds); 5×10^{-6} (X). The inset graphs are the turbidimetric analysis results which are obtained by measuring the optical density (O.D.) of the samples at 600 nm. Turbidimetric results are consistent with the cell counts data. The results show that the bacterial cultures suffer little overall toxicity due to the presence of the actinides.

phase portion of the Am/BAB growth curve, Figure 1d. The inset graphs in Figure 1 present the results obtained using turbidimetric analysis. This data is consistent with the trends seen in the microscopy (total cell counts) measurements. The effects seen in the growth curves derived from the turbidimetric analysis, however, are more dramatic than those seen in the cell counts data, due to the uncertainty of light scattering effects in turbidity measurements. Aggregation of cells and cell shape variations will affect these measurements, but these types of effects can be excluded when direct counting is done. Many instances of cell aggregation, as well as cell morphology changes (Pansoy-Hjelvik et. al., 1997), were observed in these experiments, again consistent with the above interpretation. The microscopy and turbidity measurements are complimentary techniques, and the agreement between these measurements additionally ensures that differences in cell counts at different time points in the experiment are not a result of variations in cell counts at inoculation.

Table I lists the theoretical initial concentrations and associated α-activities of Pu and Am used for both WIPP1A and BAB experiments. For the same order-of-magnitude concentration of both actinides, Am exhibits significantly more α-activity, which is expected.

Presented in Figure 2 are the concentrations of Pu and Am for both the inoculated and control experiments. The concentrations of soluble Pu and Am were derived from LSC measurements after the sample was 0.20 µm filtered. The results from the control experiments indicate that most of the Pu and Am remain in solution during the time-course

Table I. Radiological Activity and Initial [An] in Growth Media

[Pu] 1:1 EDTA Complex	DPM/ml	[Am] 1:1 EDTA Complex	DPM/ml
0	0	0	0
1.0×10^{-7}	3.3×10^3	5.0×10^{-8}	5.4×10^3
1.0×10^{-6}	3.3×10^4	5.0×10^{-7}	5.4×10^4
1.0×10^{-5}	3.3×10^5	5.0×10^{-6}	5.4×10^5

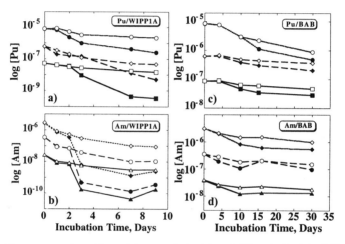

Figure 2. The concentration of Pu and Am was derived from LSC analysis of the 0.20 μm filtered sample. At inoculation [Pu] = 1×10^{-7} (squares), 1×10^{-6} (diamonds), and 1×10^{-5} (circles); [Am] = 5×10^{-8} (triangles); 5×10^{-7} (circles), and 5×10^{-6} (diamonds). The closed symbols represent data for [An] in the inoculated samples; open symbols represent data for [An] in the uninoculated samples. The difference between [An] for the inoculated and uninoculated control experiments roughly approximate actinide associated with the bacteria. Both Pu and Am appear to associate with the bacteria.

of the experiment. The difference between the actinide concentrations for the inoculated and control experiments approximate the amount of actinide associated with the bacteria, although the preferred method for determination of the dependence of cell size and actinide association is ultra-microfiltration of the various bacterial size fractions. As mentioned previously, the results of ultra-microfiltration experiments performed to more accurately determine the size fraction, down to 0.03 μm, associated with the actinide are presented separately in this publication (Gillow et al., 1997).

In a comparison of Figures 1 and 2, Am appears to associate more with the bacteria (thus exposing a cell to higher levels of ionizing energy), yet the effects to bacterial growth are negligible. Similarly, although Pu associated with the bacteria, the effect on cell growth was also minimal, as only a slight decrease in the maximum cell numbers was observed. These results suggest that WIPP1A and BAB microorganisms were not greatly affected by the radiological activity of Pu or Am at the soluble actinide concentrations used in these studies.

In the experiments involving Np, U and Th, the actinide species were in both precipitated and/or solubilized form. The addition of insoluble actinide was important for several reasons. First, it was necessary to maintain the concentration of actinide in solution at or near the saturation limit so that the bacteria experienced a maximum toxic effect, if any, from the actinide. Second, it is generally expected that soluble actinides have the greatest toxic effect on a bacterium because they can be taken up into the cell interior; however, in the case of actinides, the radiological effect from a precipitated material may be as significant as that of a soluble species, thus effects from both solubilized and precipitated actinide must be considered. Third, it was important to investigate whether the bacteria could solubilize insoluble actinide by some process.

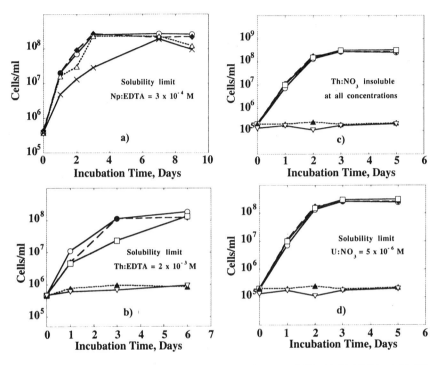

Figure 3. Growth curves of WIPP1A in the presence of a) Np:EDTA, b) Th:EDTA, c) Th:NO$_3$, and d) U:NO$_3$. Initial [Np:EDTA] = 0 (open circle), 5 x 10^{-6} (closed diamond), 5 x 10^{-5} (open triangle), 5 x 10^{-4} (X); [Th:EDTA], [Th:NO3], and [U:NO3] = 0 (open circle), 4 x 10^{-4} (closed diamond), 1 x 10^{-3} (open square), 2 x 10^{-3} (closed triangle), 4 x 10^{-3} (open triangle). Bacterial growth is affected at higher initial concentrations of actinide.

There was minimal effect on the BAB growth at all starting concentrations of Th, U, or Np. There was a minor decrease in cell numbers at the high concentrations of Np:EDTA at 5 x 10^{-4} M and Th:EDTA at 5 x 10^{-4} M. At these concentrations, both complexes were near or above their solubility limit. No effects to BAB cell growth were observed when the majority of the Th and Np remained in solution, yet a decrease in cell number was observed as soon as a precipitate started to appear. This was an indication that the decrease in maximum cell numbers achieved for the experiments with Th and Np was perhaps tied to the precipitate.

For the WIPP1A experiments with Th, U, or Np, much different results were obtained. Figure 3 shows WIPP1A growth curves as a function of varying concentrations of a) Np:EDTA; b) Th:EDTA; c) Th:NO$_3$ and; d) U:NO$_3$. Turbidimetric data for these systems are available but complicated to interpret due to the presence of the precipitate and are not shown. For NpEDTA at 5 x 10^{-4} M, Figure 3a, an extended lag phase is observed in the growth curve, followed by a slight decrease in maximum cell numbers. The NpEDTA concentration of 5 x 10^{-4} M is above the solubility limit for this actinide complex, so the species was in both soluble and insoluble forms at this concentration. For ThEDTA, presented in Figure 3b, the WIPP1A maximum cell numbers steadily decreased as the concentration of ThEDTA increased past its solubility limit. These two results indicate that as the concentration of the actinide complex approaches its respective solubility limit, the WIPP1A maximum cell numbers decrease.

For the WIPP1A experiments with the actinide supplied as An:NO$_3$, a fundamentally different result was obtained. The Th:NO$_3$ was insoluble in all experiments and U:NO$_3$ was soluble only up to 5 x 10^{-6} M. In both experiments, WIPP1A growth was not affected at the lower initial concentrations of both actinides; yet, at the higher concentrations, total toxicity to WIPP1A (no cell growth) occurred. While there are several potential explanations for these observed results, it is probable that the change in WIPP1A cell growth was due to loss of essential nutrients from the growth media due to co-precipitation of iron or calcium or complex formation between phosphate and precipitated actinide. Thus, while no cell growth is certainly a toxic effect, the effect is more likely ancillary and dependent on a possible co-precipitation process versus being a result of a direct actinide toxicity to the cells. Further experiments are required in order to demonstrate this conclusively.

SUMMARY

The toxic effects of Th, U, Np, Pu and Am to the growth of two WIPP-relevant bacterial cultures in high-ionic strength brines has been investigated. The bacterial cultures were isolated from the WIPP site, and the brines simulate the waters in the WIPP environs.

The growth of the WIPP1A and BAB cultures have been shown to be affected negligibly by radiological effects due to Pu and Am (and Np) even though the two actinides do associate with the bacteria.

The WIPP1A growth is dramatically affected by the presence of Th:NO$_3$ and U:NO$_3$. It is speculated that the observed toxic effect in this case is related to the co-precipitation of nutrients with insoluble actinide. The co-precipitation process renders the nutrients unavailable to the bacteria.

ACKNOWLEDGEMENTS

This work was conducted under the auspices of the U.S. Department of Energy. The authors gratefully acknowledge a critical review of this work by L. Hersman.

REFERENCES

Francis, A.J., and Gillow, J.B., 1994, *Effects of Microbial Processes on Gas General Under Expected WIPP Repository Conditions*, Sandia National Laboratory Report No. SAND93-7036.

Francis, A.J., 1990, Microbial transformations of toxic metals and radionuclides in mixed wastes, *Experentia*, 46:840-851.

Gillow, J.B., Strietelmeier, B.A., Dodge, C.J., Mantione, M., Dunn, M., Francis, A.J., Pansoy-Hjelvik, M.E., Kitten, S.M., Triay, I.R., and Papenguth, H.W., 1997, *Radionuclide Speciation in Real Systems*, Reed, Clark, and Rao, eds., (Am. Chem. Soc. Proc., Orlando, FL, 1996), in this issue.

Pansoy-Hjelvik, M.E., Strietelmeier, B.A., Paffett, M.T., Kitten, S.M., Leonard, P.A., Dunn, M., Gillow, J.B., Dodge, C.J., Villarreal, R., Triay, I.R., and Francis, A.J., (1997). Enumeration of microbial populations in radioactive environments, in *Scientific Basis for Nuclear Waste Management XX*, Triay, I.R. and Gray, W.J., eds. (MRS Proc., Boston, MA, Fall, 1996).

Villarreal, R. and Phillips, M., (1993). *Test Plan for Actinide Source Term Waste Test Program (STTP)*, Los Alamos National Laboratory Document No. CLS1-STP-SOP5-012/0, May.

INDEX